A Practical Approach to Computational Fluid Dynamics Using OpenFOAM®

Giovanni Caramia · Elia Distaso

A Practical Approach to Computational Fluid Dynamics Using OpenFOAM®

Springer

Giovanni Caramia
Department of Mechanics
Mathematics and Management
Politecnico di Bari
Bari, Italy

Elia Distaso
Department of Mechanics
Mathematics and Management
Politecnico di Bari
Bari, Italy

ISBN 978-3-031-88956-1 ISBN 978-3-031-88957-8 (eBook)
https://doi.org/10.1007/978-3-031-88957-8

Translation from the Italian language edition: "La simulazione numerica delle macchine a fluido. Le basi per l'uso di OpenFOAM®" by Giovanni Caramia and Elia Distaso, © Authors 2024. Published by Città Studi Edizioni. All Rights Reserved.

The original submitted manuscript has been translated into English. The translation was done using artificial intelligence. A subsequent revision was performed by the author(s) to further refine the work and to ensure that the translation is appropriate concerning content and scientific correctness. It may, however, read stylistically different from a conventional translation.

Extended, English-language Edition

This Springer imprint is published by the registered company Springer Nature Switzerland AG
The registered company address is: Gewerbestrasse 11, 6330 Cham, Switzerland

Preface

In the hope that it will prove to be a valuable resource for both teaching and learning, this volume is designed for undergraduate and graduate students in Engineering who are exploring computational fluid dynamics for the first time as a tool for analysing fluid machines and devices.

The main objective is to facilitate a gradual and effective learning process, significantly reducing the time needed to acquire the minimum level of knowledge required to produce acceptable results for most general interest case studies. For detailed insights, readers are referred to the texts cited in the bibliography—and the bibliographic references contained therein—whose reading and understanding will certainly be more straightforward once familiarity with the material presented here is gained.

To ensure the topics covered are both interesting and comprehensible, the authors have, in certain cases, prioritised clarity of exposition over mathematical rigour, continuously referring to the physical meaning of the mathematical formulas considered. Researchers and engineers will also find in this book a collection of basic concepts essential for the correct configuration of any fluid dynamics simulation software.

To enable as many readers as possible to put into practice the content presented in this volume, the authors have chosen to focus on the free and open-source software OpenFOAM®, which is supported by a large and vibrant community of developers and users. Note, however, the not insignificant difficulty in accessing and using the related manuals; this volume represents the result of painstaking research and the collection of information scattered across numerous sources, including books, academic articles, multimedia notes published online, specialized forums, and more.

The initial part of this text (Chap. 1) aims to provide the necessary tools for those wishing to perform fluid dynamic analysis using finite volume methods for complex systems such as fluid machines. This is followed by an introduction to the main governing equations of fluid dynamics (Chap. 2), which represent the core of the physical description for the practical problems under study. An extensive yet pragmatic description of the finite volume approach is then provided, starting with the description of the main discretization methods (Chap. 3), moving on to the numerical methodologies used to solve the systems of equations generated in these processes (Chap. 4), and concluding with the pressure-velocity coupling problems (Chap. 5). Having acquired the necessary knowledge to undertake practical case analysis, the computational code OpenFOAM® is introduced (Chap. 6), with its structure and potential described through practical examples of setup. The book concludes with a discussion on issues related to setting boundary conditions in various cases (Chap. 7) and the turbulence closure problem (Chap. 8).

Bari, Italy

Giovanni Caramia
Elia Distaso

Competing Interests The authors have no competing interests to declare that are relevant to the content of this manuscript.

Contents

Chapter 1
Preliminary Concepts

Embarking on the study of finite volume methods and their application to the conservation equations governing fluid dynamics requires mastering certain fundamental concepts. Therefore, this chapter provides an essential overview covering *differential operators* (notation, definition, and their physical significance), elements of *fluid mechanics*, generalities and physical applications of *differential equations*, elements of *gas dynamics* useful for studying acoustic waves, and finally, basic concepts related to *numerical computation*. The approach is pragmatic and does not claim to be exhaustive. The aim is to provide the reader with the necessary tools to be effectively utilised in subsequent chapters. From another perspective, this chapter serves as an initial reference for delving deeper into these concepts through the reading and study of more specific texts.

1.1 Differential Operators

1.1.1 Gradient

The gradient is a differential operator denoted by the word $grad$ or the symbol ∇, called "nabla". The gradient of a scalar function f of three variables (x_1, x_2, x_3) is a vector that, at each point in space, allows the calculation of the directional derivative of f in the direction of a generic unit vector $\mathbf{v}$ through the dot product between $\mathbf{v}$ and the gradient of the function at that point. In the case of an orthonormal Cartesian reference system, the gradient of $f(x_1, x_2, x_3)$ is the vector whose components are the first partial derivatives evaluated at the point:

$$grad\ f = \nabla f = \frac{\partial f}{\partial x_1}\mathbf{i} + \frac{\partial f}{\partial x_2}\mathbf{j} + \frac{\partial f}{\partial x_3}\mathbf{k}$$

G. Caramia and E. Distaso, *A Practical Approach to Computational Fluid Dynamics Using OpenFOAM®*, https://doi.org/10.1007/978-3-031-88957-8_1

where **i**, **j**, **k** represent the unit vectors along the coordinate axes. Using vector notation, ∇ can be defined as the vector whose components are the partial derivative operators along the coordinate axes. The gradient will then be the product between the vector ∇ and the scalar function f. In formulas:

$$\nabla f = \begin{pmatrix} \frac{\partial}{\partial x_1} \\ \frac{\partial}{\partial x_2} \\ \frac{\partial}{\partial x_3} \end{pmatrix} f = \begin{pmatrix} \frac{\partial f}{\partial x_1} \\ \frac{\partial f}{\partial x_2} \\ \frac{\partial f}{\partial x_3} \end{pmatrix}.$$

The gradient of the scalar function f is a vector whose direction is that of the maximum variation of the function itself, and its magnitude provides the measure of such variation. For example, considering $x_1 = x$ and $x_2 = y$, below is the calculation of the gradient of the scalar function $f(x, y) = x^2 + y^2 + b$, whose plot is shown in Fig. 1.1, along with the vector field formed by its gradient.

$$\nabla f = \begin{pmatrix} \frac{\partial f}{\partial x} \\ \frac{\partial f}{\partial y} \end{pmatrix} = \nabla f = \begin{pmatrix} 2x \\ 2y \end{pmatrix}.$$

Continuing to observe Fig. 1.1, it can be noticed that the vector field corresponding to the gradient of f develops in the xy-plane. Moreover, as one moves away from the origin, there is an increase in both the function f and its gradient values.

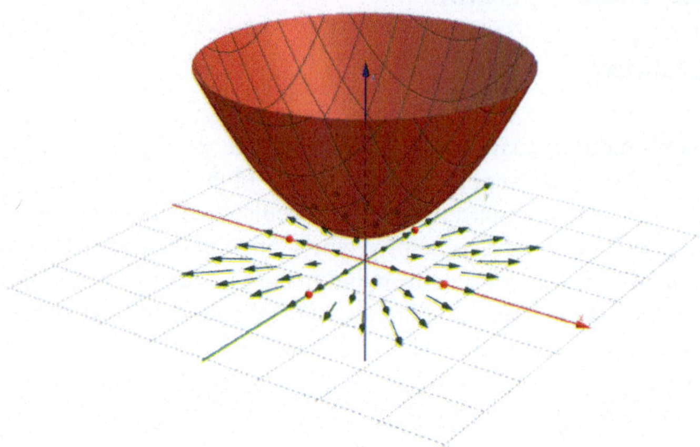

Fig. 1.1 Representation of the scalar function $f(x, y) = x^2 + y^2 + b$ together with its gradient

The gradient of a vector function $\mathbf{f} = (f_x, f_y, f_z)$ is defined as:

$$\nabla \mathbf{f} = \begin{pmatrix} \dfrac{\partial f_x}{\partial x} & \dfrac{\partial f_y}{\partial x} & \dfrac{\partial f_z}{\partial x} \\ \\ \dfrac{\partial f_x}{\partial y} & \dfrac{\partial f_y}{\partial y} & \dfrac{\partial f_z}{\partial y} \\ \\ \dfrac{\partial f_x}{\partial z} & \dfrac{\partial f_y}{\partial z} & \dfrac{\partial f_z}{\partial z} \end{pmatrix}.$$

The transpose matrix of the gradient of a vector function is:

$$\mathbf{J} = (\nabla \mathbf{f})^T = \begin{pmatrix} \dfrac{\partial f_x}{\partial x} & \dfrac{\partial f_x}{\partial y} & \dfrac{\partial f_x}{\partial z} \\ \\ \dfrac{\partial f_y}{\partial x} & \dfrac{\partial f_y}{\partial y} & \dfrac{\partial f_y}{\partial z} \\ \\ \dfrac{\partial f_z}{\partial x} & \dfrac{\partial f_z}{\partial z} & \dfrac{\partial f_z}{\partial z} \end{pmatrix}$$

The transpose matrix of the gradient of a vector function is called the *Jacobian matrix*, while the determinant of the Jacobian matrix is referred to as the *Jacobian*.

1.1.2 Divergence

The divergence is a differential operator indicated by the symbol div or by the symbol $\nabla\cdot$. The divergence of a three-dimensional vector $\mathbf{b} = (b_1, b_2, b_3)$ is

$$div\, \mathbf{b} = \nabla \cdot \mathbf{b} = \sum_{i=1}^{3} \frac{\partial}{\partial x_i} b_i = \frac{\partial b_1}{\partial x_1} + \frac{\partial b_2}{\partial x_2} + \frac{\partial b_3}{\partial x_3} \tag{1.1}$$

where b_i are the components of the vector $\mathbf{b}$ and x_i are the coordinates of the chosen reference system. From a mathematical perspective, the application of the divergence differential operator to a vector can be thought of as the dot product between the operator $\nabla = \left(\frac{\partial}{\partial x_1}, \frac{\partial}{\partial x_2}, \frac{\partial}{\partial x_3}\right)$ and the vector $\mathbf{b}$.

To better understand the physical meaning of this differential operator, one can consider the velocity field of a fluid in motion in its vector representation, as indicated in Fig. 1.2, where at a limited number of points, the representative vector of the fluid velocity is drawn. In the same figure, the curve delimiting a generic area inside which it is desired to understand whether the fluid is accumulating or not is also depicted.

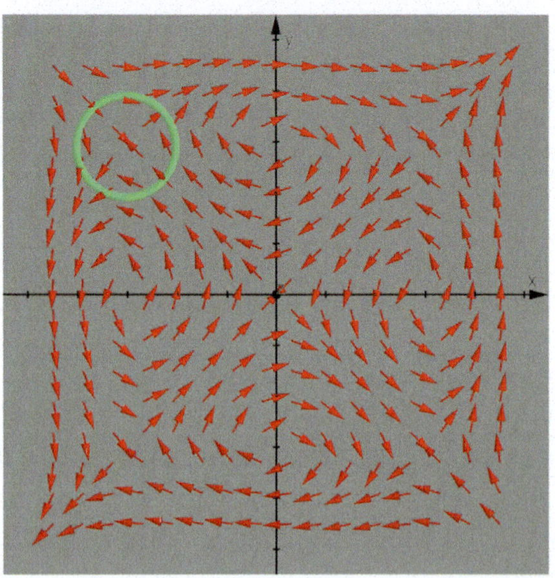

Fig. 1.2 Vector representation of a fluid motion field

For simplicity, this area can be considered as a circle centred at the origin of the axes, as shown in Figs. 1.3 and 1.4, where a flow field is visualised. In Fig. 1.3, the fluid tends to move particles away from the origin, while in Fig. 1.4, it tends to move particles towards the origin. In Fig. 1.3, it is said that the function representative of the vector field has positive divergence at the origin, and the origin is called a *source*. In Fig. 1.4, it is said that the function representative of the vector field has negative divergence at the origin, and the origin is called a *sink*. Considering a generic area in space where the function is defined, it will be said that for this area, the divergence is:

- Positive: if the balance of particles crossing its boundary is in favour the particles exiting the area (dispersion of particles).
- Negative: if the balance of particles crossing its boundary is in favour of the particles entering the area (accumulation of particles).
- Zero: if the number of particles entering equals the number exiting (particle conservation).

Therefore, divergence can be interpreted as an indicator of the extent to which particles tend to converge or spread (diverge) from a generic area in space.

As an example, let's calculate the divergence of the vector function $\mathbf{V} = (xy, y^2 - x^2)$, whose representation is shown in the Fig. 1.5. Applying the definition (1.1), it is

$$div\,\mathbf{V} = \nabla \cdot \mathbf{V} = \sum_{i=1}^{2} \frac{\partial}{\partial x_i} V_i = \frac{\partial v_1}{\partial x_1} + \frac{\partial v_2}{\partial x_2} = \frac{\partial xy}{\partial x} + \frac{\partial (y^2 - x^2)}{\partial y} = y + 2y = 3y.$$

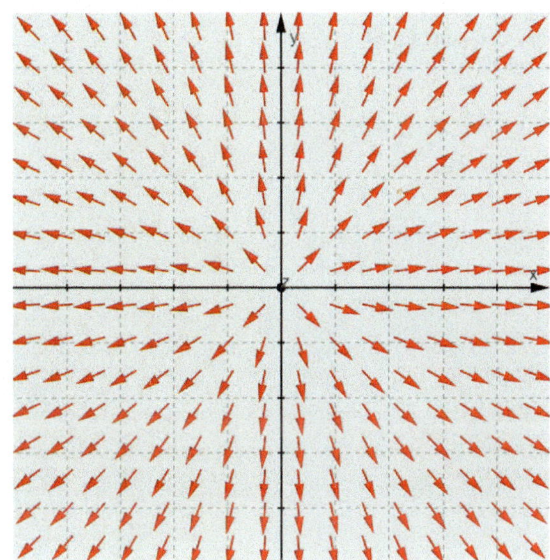

Fig. 1.3 Positive divergence

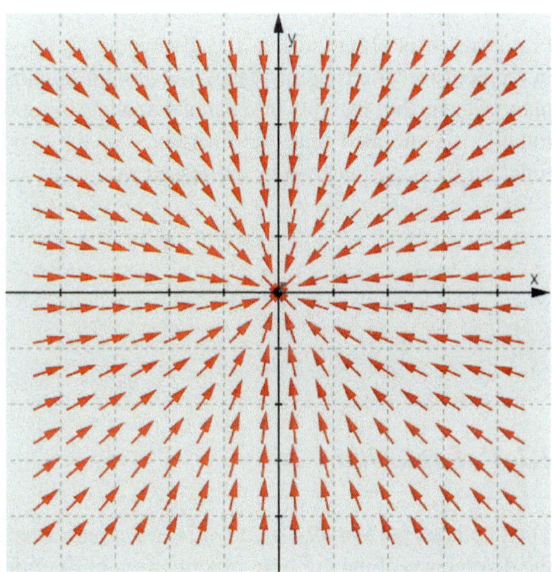

Fig. 1.4 Negative divergence

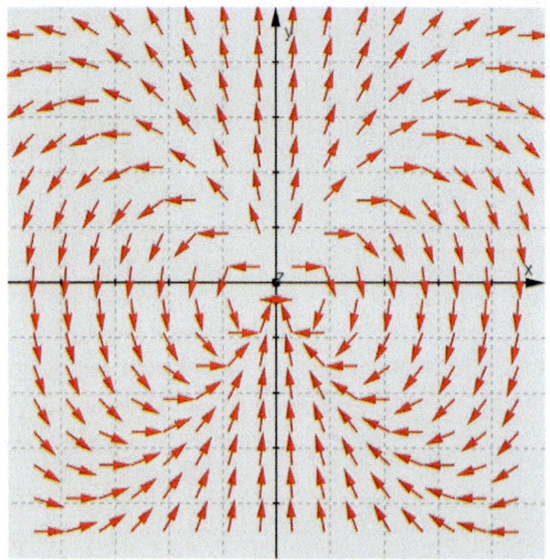

Fig. 1.5 Vector representation of the function **V**

From this, it is observed that the divergence is zero on the x-axis, always positive in the positive y-axis semi-plane, and always negative in the negative y-axis semi-plane. For further insights into the physical meaning of divergence, refer to Sect. 2.3. Finally, it is noticed that the divergence of the product of a vector **b** by a scalar function $\phi\,(x_1, x_2, x_3)$ can be expressed as:

$$\nabla \cdot (\mathbf{b}\phi) = \mathbf{b} \cdot \nabla\phi + \phi\nabla \cdot \mathbf{b}$$

where the symbol $\nabla\phi$ denotes the gradient vector of ϕ, defined as: $\nabla\phi \equiv \left(\frac{\partial\phi}{\partial x_1}, \frac{\partial\phi}{\partial x_2}, \frac{\partial\phi}{\partial x_3}\right)$.

1.1.3 Laplacian

The Laplace operator, or *Laplacian*, is a second-order differential operator defined as the divergence of the gradient of a function in Euclidean space. The most significant way to denote the Laplacian is using the vector differential operator ∇^2. The Laplace operator applied to a function $f(\mathbf{x})$ in Euclidean space is the divergence of the gradient of f:

$$\nabla^2 f(\mathbf{x}) = \nabla \cdot \nabla f(\mathbf{x}),$$

where $\mathbf{x} = (x_1, x_2, \ldots, x_n)$ represents the set of coordinates. The Laplace operator in Cartesian coordinates, in an n-dimensional space, is given by:

$$\nabla^2 = \frac{\partial^2}{\partial x_1^2} + \cdots + \frac{\partial^2}{\partial x_n^2} = \sum_{i=1}^{n} \frac{\partial^2}{\partial x_i^2}.$$

Given a vector function $\mathbf{F}$ defined in a three-dimensional Euclidean space, the Laplacian is defined as the vector whose components are the scalar Laplacians of each of the component functions of $\mathbf{F}$:

$$\overline{\nabla}^2 \mathbf{F} = \left\{\nabla^2 F_x, \nabla^2 F_y, \nabla^2 F_z\right\}.$$

To better understand the meaning of this operator, consider the function $f(x, y) = \sin(x)\cos(y) + 2$, whose graph is represented in Fig. 1.6 along with the vector field of its gradient. In Fig. 1.7, only the gradient ∇f is represented. Calculating the Laplacian of $f(x, y)$ involves computing the divergence of the vector field of the gradient of $f(x, y)$. Referring to the discussion above regarding divergence, the regions with positive divergence (where vectors diverge) and the regions with negative divergence (where vectors converge) are evident in Fig. 1.7.

Observing Fig. 1.6, it is noted that the regions with positive divergence of the gradient correspond to a minimum of the function $f(x, y)$, while the regions with negative divergence of the gradient correspond to a maximum of $f(x, y)$. The Laplacian can therefore be thought of as the equivalent of the second derivative in the case of a single-variable function in determining whether a point with zero first derivative is a maximum or a minimum. In other words, the Laplacian provides the sign of concavity and the "measure" of curvature.

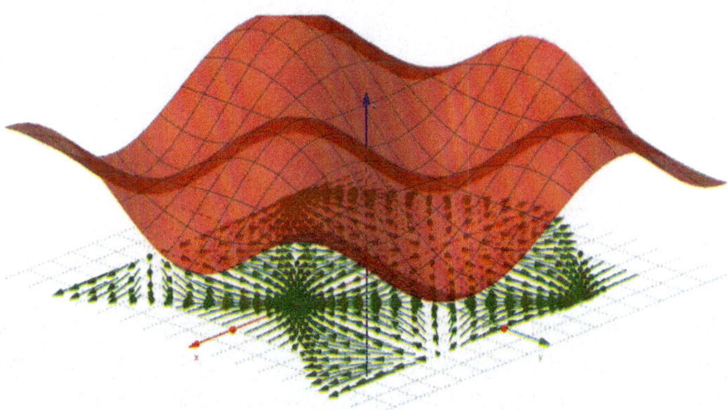

Fig. 1.6 Representation of the function $f(x, y) = \sin(x)\cos(y) + 2$ and the corresponding gradient function

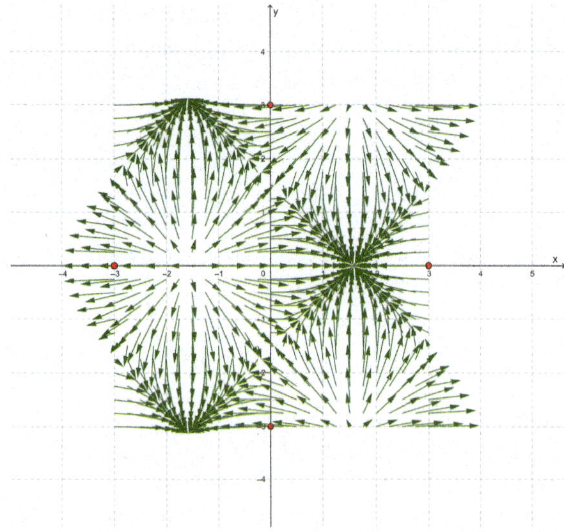

Fig. 1.7 Representation of the gradient of the function showed in Fig. 1.6

1.1.4 Curl

The curl is a differential operator indicated by the symbols *curl*, *rot*, or $\nabla\times$. From a mathematical perspective, the application of the curl differential operator to a generic vector $\mathbf{b}$ can be thought of as the cross product between the differential operator $\nabla = \left(\frac{\partial}{\partial x_1}, \frac{\partial}{\partial x_2}, \frac{\partial}{\partial x_3}\right)$ and the vector $\mathbf{b}$:

$$\nabla \times \mathbf{b} = \begin{vmatrix} \mathbf{i} & \mathbf{j} & \mathbf{k} \\ \dfrac{\partial}{\partial x_1} & \dfrac{\partial}{\partial x_2} & \dfrac{\partial}{\partial x_3} \\ b_1 & b_2 & b_3 \end{vmatrix}$$

that is

$$\nabla \times \mathbf{b} = \left(\frac{\partial b_3}{\partial x_2} - \frac{\partial b_2}{\partial x_3}\right)\mathbf{i} + \left(\frac{\partial b_1}{\partial x_3} - \frac{\partial b_3}{\partial x_1}\right)\mathbf{j} + \left(\frac{\partial b_2}{\partial x_1} - \frac{\partial b_1}{\partial x_2}\right)\mathbf{k}$$

where $\mathbf{i}$, $\mathbf{j}$, $\mathbf{k}$ represent the unit vectors along the coordinate axes, and b_i represent the component functions of the vector $\mathbf{b}$. To better understand the physical meaning of this differential operator, consider the velocity field of a fluid in motion in its vector representation, as indicated in Fig. 1.8, where at a limited number of points, the representative vector of the fluid velocity is drawn. In the same figure, circles are drawn. Near the circle positioned on the positive x-semiaxis, the vector field rotates counterclockwise; in this case, it is said that the curl is positive. Near the circle positioned on the positive y-semiaxis, the vector field rotates clockwise: in

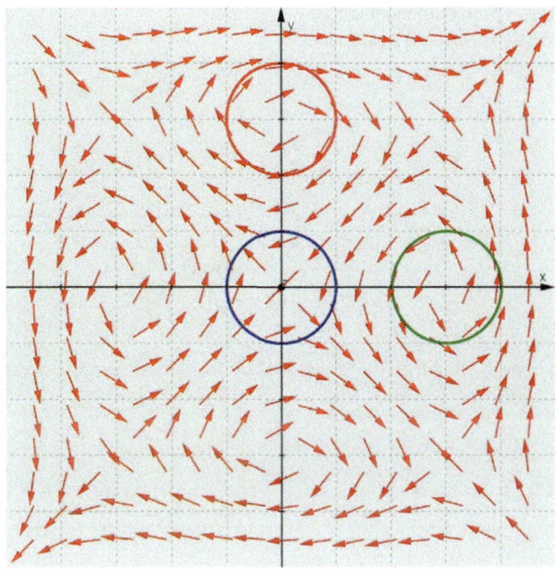

Fig. 1.8 Vectorial representation of a fluid motion field

this case, it is said that the curl is negative. The circle positioned at the origin of the axes is near four vectors oriented two by two clockwise and counterclockwise: in this case, it is said that the curl is zero. The curl can therefore be interpreted as an indicator of the extent to which particles tend to rotate around a point, clockwise or counterclockwise, in a generic area of space. The derivation of the curl formula from its physical meaning is simpler in the two-dimensional case. Let's consider a vector function **b** defined as:

$$\mathbf{b}(x, y) = \begin{bmatrix} P(x, y) \\ Q(x, y) \end{bmatrix}$$

where $P(x, y)$ is the function describing the behaviour of the component along the x-direction of the function **b**, $Q(x, y)$ is the function describing the behaviour of the component along the y-direction of the function **b**. Considering what has been said about Fig. 1.8, to obtain a positive value of the curl and considering the area of interest centred at the origin, the representative vectors of the function **b** should be oriented as in Fig. 1.9. In particular, the two vectors positioned on the x-axis—on the circumference centred at the origin of Fig. 1.8—would not have any component along the same axis, and therefore the vector function **b** would reduce to only the value of the scalar function Q; in this case, the derivative of Q with respect to x would be positive because it would change from a negative value at negative abscissas to a positive value at positive abscissas. Correspondingly, the derivative of P with respect to y would be negative. Expressing what is represented in Fig. 1.9 into mathematical formulas, the curl can be defined as:

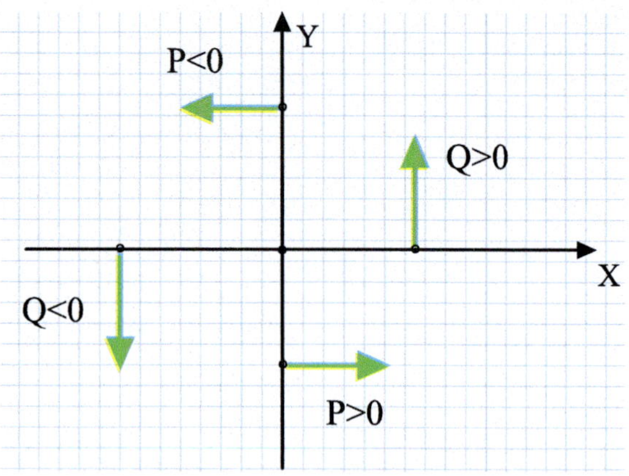

Fig. 1.9 Components of vector function **b** near the origin

$$curl\ \mathbf{b} = \frac{\partial Q}{\partial x} - \frac{\partial P}{\partial y}$$

where the negative sign is inserted to ensure that the curl is positive (i.e., counter-clockwise rotations) when the derivative of P with respect to y is negative. Assuming the vector field shown in Fig. 1.8 to be described by the vector function

$$\mathbf{b}(x, y) = \begin{bmatrix} P(x, y) \\ Q(x, y) \end{bmatrix} = \begin{bmatrix} y^3 - 9y \\ x^3 - 9x \end{bmatrix},$$

the curl is

$$curl\ \mathbf{b} = \frac{\partial Q}{\partial x} - \frac{\partial P}{\partial y} = 3x^2 - 9 - \left(3y^2 - 9\right) = 3x^2 - 3y^2. \tag{1.2}$$

Willing to calculate the value of the curl at the point with coordinates ($x = 3$, $y = 0$), we obtain the value 27, which is consistent—in terms of sign—with the fact that, referring to Fig. 1.8, the curl at the circumference centred at the point ($x = 3$, $y = 0$) has a positive sign. The same considerations apply to the value of the curl at the points with coordinates ($x = 0$, $y = 3$) and ($x = 0$, $y = 0$). The vector field represented in Fig. 1.8 can be depicted in a three-dimensional reference system, as shown in Fig. 1.10. Using the right-hand rule, we can associate a direction and orientation to the calculated curl value using Eq. 1.2, as illustrated in the same Fig. 1.10. It is now possible to redefine the function **b** as

$$\mathbf{b}(x, y, z) = \begin{bmatrix} P(x, y, z) \\ Q(x, y, z) \\ R(x, y, z) \end{bmatrix} = \begin{bmatrix} y^3 - 9y \\ x^3 - 9x \\ 0 \end{bmatrix}.$$

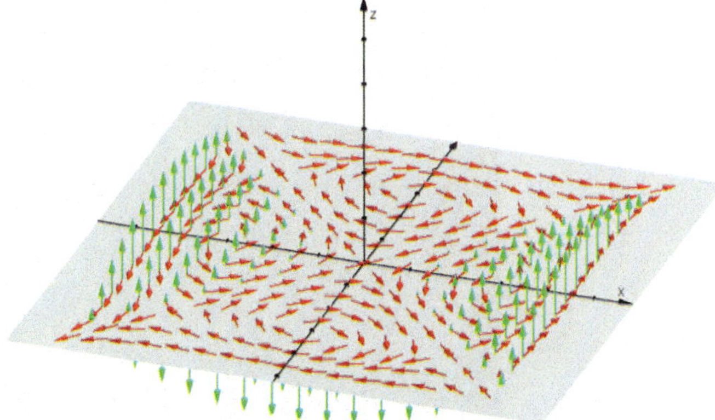

Fig. 1.10 Representation of the vector function **b** (vectors parallel to the xy-plane) together with its corresponding curl (vectors parallel to the xz-plane)

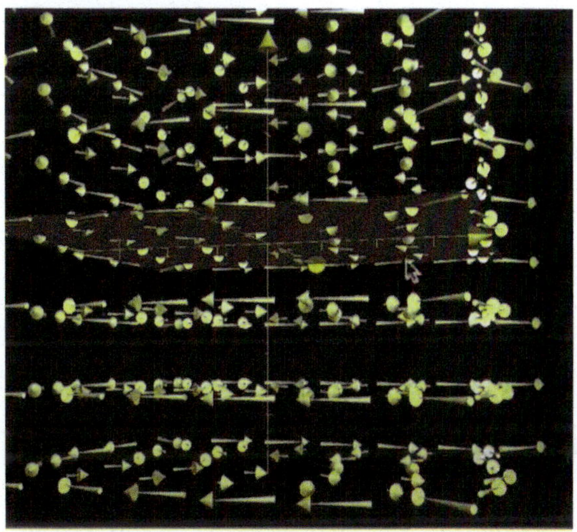

Fig. 1.11 Representation of the vector function **b** redefined

For each value of the z-coordinate, this function will be equal to itself, as shown in Fig. 1.11. The curl of this function will be the three-dimensional vector function:

$$curl\ \mathbf{b}(x, y, z) = \begin{bmatrix} 0 \\ 0 \\ 3x^2 - 3y^2 \end{bmatrix} \tag{1.3}$$

whose representation consists of all parallel vectors, as shown in the Fig. 1.12. In conclusion, referring to the analogy that associates the motion of a fluid with the vector field under consideration, the curl provides a measure of the intensity with

Fig. 1.12 Representation of the curl of the function **b** redefined

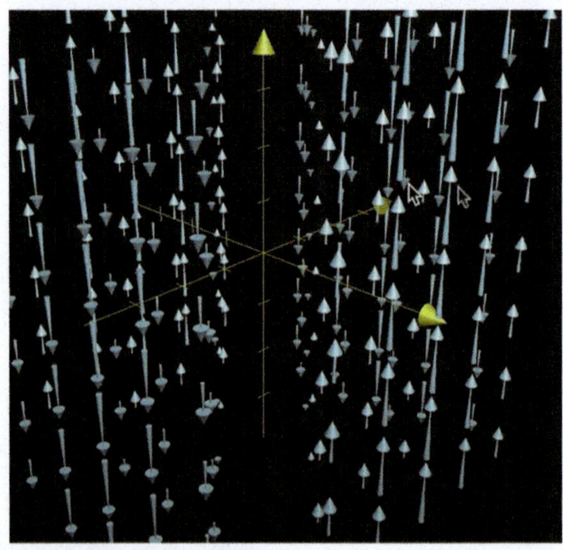

which the fluid rotates around the generic point considered, while also indicating the direction along which this rotation occurs through the use of the right-hand rule.

1.1.5 *Vector and Tensor Notation*

Here, we assume to be in an orthogonal Cartesian coordinate system (O, x_1, x_2, x_3) or equivalently (O, x, y, z), where the axes are identified by the unit vectors **i**, **j**, **k**.

In both notations, a generic scalar quantity (e.g., pressure, temperature, turbulent kinetic energy, etc.) will always be indicated by a symbol without subscripts (for example, ϕ). A vector quantity (e.g., velocity) will be indicated using:

- a bold symbol in vector notation; for example, **a**, being $\mathbf{a} = a_x\mathbf{i} + a_y\mathbf{j} + a_z\mathbf{k}$;
- a symbol with a single subscript in tensor notation; for example, $a_i,\ i = 1, 2, 3$, being $a_i = a_x\mathbf{i} + a_y\mathbf{j} + a_z\mathbf{k}$.

As seen above, the gradient of a scalar function ϕ is a vector expressible:

- in vector notation as $\nabla\phi$, with $\nabla\phi = \frac{\partial\phi}{\partial x}\mathbf{i} + \frac{\partial\phi}{\partial y}\mathbf{j} + \frac{\partial\phi}{\partial z}\mathbf{k}$;
- in tensor notation $\frac{\partial\phi}{\partial x_i}$, with $\frac{\partial\phi}{\partial x_i} = \frac{\partial\phi}{\partial x}\mathbf{i} + \frac{\partial\phi}{\partial y}\mathbf{j} + \frac{\partial\phi}{\partial z}\mathbf{k}$.

The dot product of two vectors **a** and **b**—giving a scalar as result—can be expressed:

- in vector notation as $\mathbf{a} \cdot \mathbf{b}$, with $\mathbf{a} \cdot \mathbf{b} = a_1b_1 + a_2b_2 + a_3b_3$;
- in tensor notation as a_jb_j with $a_jb_j = a_1b_1 + a_2b_2 + a_3b_3$.

Note that, for tensor notation, the *Einstein notation* or *Einstein summation convention* has been used. It is a convention for summing over repeated indices: each index that appears more than once in a term is summed over all possible values it can take. In this specific case, the notation $a_j b_j$ indicates the following summation

$$\sum_{j=1}^{3} a_j b_j .$$

Here, the subscript j has been used instead of the subscript i to emphasise the need for a subscript whose symbol can be chosen arbitrarily. Consistently, $a_i b_i c_j$ indicates the summation:

$$\sum_{i=1}^{3} a_i b_i c_j = a_1 b_1 c_j + a_2 b_2 c_j + a_3 b_3 c_j .$$

Finally, in the case of two indices, both repeated, as for example $a_i b_{ij} c_j$, it is:

$$\begin{aligned} a_i b_{ij} c_j \equiv \sum_{j=1}^{3} \sum_{i=1}^{3} a_i b_{ij} c_j &= \sum_{i=1}^{3} a_i b_{i1} c_1 + \sum_{i=1}^{3} a_i b_{i2} c_2 + \sum_{i=1}^{3} a_i b_{i3} c_3 = \\ &= (a_1 b_{11} c_1 + a_2 b_{21} c_1 + a_3 b_{31} c_1) + (a_1 b_{12} c_2 + a_2 b_{22} c_2 + a_3 b_{33} c_2) \\ &+ (a_1 b_{13} c_3 + a_2 b_{23} c_3 + a_3 b_{33} c_3) . \end{aligned}$$

It has been previously seen that the divergence of a vector **a** is a scalar quantity. Considering the symbolic vector *nabla* defined as $\nabla = (\partial/\partial x, \partial/\partial y, \partial/\partial z)$, the divergence of a vector can be written:

- in vector notation as $\nabla \cdot \mathbf{a}$ being $\nabla \cdot \mathbf{a} = \dfrac{\partial a_x}{\partial x} + \dfrac{\partial a_y}{\partial y} + \dfrac{\partial a_z}{\partial z}$;
- in tensor notation as $\dfrac{\partial a_j}{\partial x_j}$, being $\dfrac{\partial a_j}{\partial x_j} = \sum_{j=1}^{3} \dfrac{\partial a_j}{\partial x_j}$.

The vector (or cross) product of two vectors **a** and **b** can be written:

- in vector notation as $\mathbf{a} \times \mathbf{b}$, being

$$\begin{aligned} \mathbf{c} = \mathbf{a} \times \mathbf{b} = \begin{vmatrix} \mathbf{i} & \mathbf{j} & \mathbf{k} \\ a_x & a_y & a_z \\ b_x & b_y & b_z \end{vmatrix} &= \left(a_y b_z - a_z b_y\right) \mathbf{i} + \left(a_z b_x - a_x b_z\right) \mathbf{j} + \left(a_x b_y - a_y b_x\right) \mathbf{k} \\ = c_1 \mathbf{i} + c_2 \mathbf{j} + c_3 \mathbf{k}; \end{aligned}$$

- in tensor notation as $c_i = \epsilon_{ijk} a_j b_k$, meaning

$$c_1 = \sum_{j=1}^{3}\sum_{k=1}^{3} \epsilon_{1jk} a_j b_k; \qquad c_2 = \sum_{j=1}^{3}\sum_{k=1}^{3} \epsilon_{2jk} a_j b_k; \qquad c_3 = \sum_{j=1}^{3}\sum_{k=1}^{3} \epsilon_{3jk} a_j b_k.$$

Here, the third-order tensor (i.e., a three-dimensional matrix) ϵ_{ijk}—known as the Levi-Civita tensor—has been introduced. Its elements are defined as follows:

$$\varepsilon_{ijk} = \begin{cases} +1 & if(i, j, k)\ is\ an\ even\ permutation,\ i.e.,\ (1, 2, 3), (2, 3, 1), (3, 1, 2) \\ -1 & if(i, j, k)\ is\ an\ odd\ permutation,\ i.e.,\ (3, 2, 1), (1, 3, 2), (2, 1, 3) \\ 0 & if\ the\ two\ indices\ coincide:\ i = j\ and/or\ j = k\ and/or\ k = i \end{cases}$$

Recalling that a permutation is a way of arranging distinct objects in sequence (in this case, numbers), it's worth mentioning that by "even permutations," it is meant permutations obtained with an even number of transpositions, while "odd permutations" refer to those obtained with an odd number of transpositions. By "transposition," it is meant the exchange of two elements that are not necessarily adjacent, and zero is considered even. Considering the permutations of (1, 2, 3), we have:

- '123' is an even permutation because it is obtained with zero transpositions;
- 123 → '213' is an odd permutation because it is obtained with a single transposition: the exchange of 1 and 2;
- 123 → 213 → '231' is an even permutation because it is obtained with two transpositions: the exchange of 1 and 2, and the exchange of 1 and 3.

The curl of a vector can be expressed as the vector symbol nabla cross the vector itself:

- in vector notation

$$\nabla \times \mathbf{a} = \begin{vmatrix} \mathbf{i} & \mathbf{j} & \mathbf{k} \\ \frac{\partial}{\partial x} & \frac{\partial}{\partial y} & \frac{\partial}{\partial z} \\ a_x & a_y & a_z \end{vmatrix} = \left(\frac{\partial a_z}{\partial y} - \frac{\partial a_y}{\partial z}\right)\mathbf{i} + \left(\frac{\partial a_x}{\partial z} - \frac{\partial a_z}{\partial x}\right)\mathbf{j} + \left(\frac{\partial a_y}{\partial x} - \frac{\partial a_x}{\partial y}\right)\mathbf{k};$$

- in tensor notation, the i-th component will be $\epsilon_{ijk}\frac{\partial a_j}{\partial x_k}$, whose explicit meaning is $\sum_{j=1}^{3}\sum_{k=1}^{3} \epsilon_{ijk}\frac{\partial a_j}{\partial x_k}$.

A second-order tensor is represented in vector notation with a bold symbol, such as Γ, while in tensor notation it is represented as Γ_{ij}. The product of a second-order tensor Γ with a vector **a**—resulting in a vector—is represented as follows:

- in vector notation as $\Gamma \cdot \mathbf{a}$;
- in tensor notation as $\Gamma_{ij}a_j$, whose explicit meaning is: $\sum_{j=1}^{3} \Gamma_{ij}a_j$, that indicates the matrix-vector multiplication rule.

1.1.6 Gauss (or Divergence or Ostrogradskij) Theorem

This theorem states that the flux (see Sect. 2.8) of a vector field **a** passing through a closed, piecewise smooth surface ∂V is equal to the integral of the divergence of **a** over the region V bounded by ∂V. In symbols:

$$\oint_{\partial V} \mathbf{a} \cdot d\mathbf{S} = \int_V \nabla \cdot \mathbf{a}\, dV$$

where $d\mathbf{S} = \mathbf{n} dS$ is the product of the unit normal vector to the surface element and the area dS of that element. In other words, the divergence of a vector field **a** for a finite region V in space is equal to the sum of the fluxes leaving the infinitesimal surfaces into which the volume bounding surface ∂V can be divided. To calculate the value of the divergence associated with a specific point, one just needs to consider a very small volume, tending towards zero, to obtain

$$\nabla \cdot \mathbf{a} = \lim_{V \to 0} \frac{1}{V} \oint_{\partial V} \mathbf{a} \cdot d\mathbf{S}.$$

The physical interpretation of this theorem identifies the integral of the divergence of **a** over the region V_P bounded by ∂V_P with the rate of accumulation of the same vector field **a** inside the region V_P. Referring to Fig. 1.2, the divergence of a flow field will quantify the divergent part of the motion, extracting it from the general motion expression. In its generalised form, this theorem is expressed as

$$\oint_{\partial V} \mathbf{n} \bigstar \mathbf{a}\, dS = \int_V \nabla \bigstar \mathbf{a}\, dV$$

where the symbol $\bigstar$ refers to any form of product—scalar $\mathbf{a} \cdot \mathbf{b}$, vector $\mathbf{a} \times \mathbf{b}$, or tensor $\mathbf{ab}$—to which will correspond, in the volume integral the operator $\nabla \cdot$, $\nabla \times$, ∇ respectively.

1.2 Fluid Mechanics

1.2.1 Strain Rate Tensor

Unlike solid mechanics, where strain (displacement per unit length) is a fundamental concept, in fluid mechanics, it is appropriate to use the concept of strain rate rather than strain because there is an equilibrium relationship between fluid stress and strain rate. In continuum mechanics, the *strain rate tensor* is a physical quantity describing the rate of deformation of a material around a point at a given instant. The strain rate tensor can also be interpreted as the time derivative of the strain tensor, or alternatively as the symmetric part of the gradient tensor of the velocity field describing a moving fluid.

The strain rate tensor is a purely kinematic concept that describes the macroscopic motion within a material regardless of its physical state (solid, liquid, or gas), forces or stresses acting on it. The presence of deformation variations, i.e., a non-zero strain rate tensor, leads to the emergence of forces due to friction between adjacent material elements. These forces are described at each point by the strain rate tensor together with certain material-specific quantities.

Given a three-dimensional velocity field $\mathbf{v} = (v_1, v_2, v_3)$, or equivalently $\mathbf{v} = (v_x, v_y, v_z)$, its gradient will be a second-order tensor (see Sect. 1.1.1):

$$\nabla \mathbf{v} = \begin{bmatrix} \frac{\partial v_x}{\partial x} & \frac{\partial v_y}{\partial x} & \frac{\partial v_z}{\partial x} \\ \frac{\partial v_x}{\partial y} & \frac{\partial v_y}{\partial y} & \frac{\partial v_z}{\partial y} \\ \frac{\partial v_x}{\partial z} & \frac{\partial v_y}{\partial z} & \frac{\partial v_z}{\partial z} \end{bmatrix} = \begin{bmatrix} \partial_1 v_1 & \partial_1 v_2 & \partial_1 v_3 \\ \partial_2 v_1 & \partial_2 v_2 & \partial_2 v_3 \\ \partial_3 v_1 & \partial_3 v_2 & \partial_3 v_3 \end{bmatrix}$$

and, consequently, it will be

$$\mathbf{J} = (\nabla \mathbf{v})^T = \begin{bmatrix} \partial_1 v_1 & \partial_2 v_1 & \partial_3 v_1 \\ \partial_1 v_2 & \partial_2 v_2 & \partial_3 v_2 \\ \partial_1 v_3 & \partial_2 v_3 & \partial_3 v_3 \end{bmatrix}$$

the associated Jacobian matrix. To describe the kinematic behaviour of a fluid in the neighbourhood of a point, it is possible to consider an infinitesimal volume element such that the velocity within it can be approximated by a first-order Taylor expansion. Without loss of generality, the centroid of the infinitesimal volume, defined as the origin, can be considered as the point. Subject to the motion field represented by the velocity vector $\mathbf{v}$, its centroid will move with velocity $\mathbf{v}_G$, while a point P, different from G, will move with velocity $\mathbf{v}_P$ because the velocity vector $\mathbf{v}$ is a function of both space and time. In the case where P is in a neighbourhood of G such that a truncated Taylor series expansion to the first order is valid, it can be written

$$\mathbf{v}_P = \mathbf{v}_G + (\nabla \mathbf{v})^T \cdot \mathbf{x}$$

or, equivalently

$$\mathbf{v} = \mathbf{v}_G + \mathbf{J} \cdot \mathbf{x}$$

in which the symbol $\mathbf{x}$ represents the position vector of the generic point P. Like any matrix, the Jacobian matrix can also be decomposed into its symmetric part, defined as

$$\mathbf{E} = \frac{1}{2}\left(\mathbf{J} + \mathbf{J}^T\right) \qquad \text{with} \qquad E_{ij} = \frac{1}{2}\left(\partial_j v_i + \partial_i v_j\right) \tag{1.4}$$

and its antisymmetric (skew) part defined as

$$\mathbf{R} = \frac{1}{2}\left(\mathbf{J} - \mathbf{J}^T\right) \qquad \text{with} \qquad R_{ij} = \frac{1}{2}\left(\partial_j v_i - \partial_i v_j\right)$$

which leads to writing the expression of the velocity of point P as

$$\mathbf{v} = \mathbf{v}_G + \mathbf{E} \cdot \mathbf{x} + \mathbf{R} \cdot \mathbf{x}$$

being

$$\mathbf{J} = \mathbf{E} + \mathbf{R} \qquad \text{with} \qquad J_{ij} = \frac{\partial v_i}{\partial x_j} = \partial_j v_i = E_{ij} + R_{ij}.$$

The antisymmetric part $\mathbf{R} \cdot \mathbf{x}$ represents a rigid rotation of point P around G with angular velocity $\boldsymbol{\omega}$ defined as

$$\boldsymbol{\omega} = \frac{1}{2}\nabla \times \mathbf{v} = \frac{1}{2}\begin{bmatrix} \partial_2 v_3 - \partial_3 v_2 \\ \partial_3 v_1 - \partial_1 v_3 \\ \partial_1 v_2 - \partial_2 v_1 \end{bmatrix}. \tag{1.5}$$

The quantity $\nabla \times \mathbf{v}$ is called the *rotational tensor* of the velocity vector field $\mathbf{v}$, or simply the *vorticity tensor*. Due to the antisymmetry of $\mathbf{R}$, it is $R_{ij} = -R_{ji}$. In the case where the motion is completely described solely by the symmetric tensor, it is referred to as *irrotational motion*. It's worth noting that rigid rotation does not alter the relative positions of points in space, hence the antisymmetric tensor $\mathbf{R}$ does not contribute to deformation. The only contribution to the rate of deformation comes from the symmetric tensor $\mathbf{E}$, also known as the *strain-rate tensor*. Like any other matrix, the strain-rate tensor $\mathbf{E}$ can also be decomposed into its *spherical or hydrostatic part* $\mathbf{S}$ and its *deviatoric part* $\mathbf{D}$:

$$\mathbf{E} = \mathbf{S} + \mathbf{D}.$$

Defining the *trace of a matrix* as the sum of the elements along the main diagonal, the spherical part is defined as

$$\mathbf{S} = \begin{bmatrix} \frac{1}{3}(E_{11}+E_{22}+E_{33}) & 0 & 0 \\ 0 & \frac{1}{3}(E_{11}+E_{22}+E_{33}) & 0 \\ 0 & 0 & \frac{1}{3}(E_{11}+E_{22}E_{33}) \end{bmatrix} = \frac{1}{3} tr(\mathbf{E})\mathbf{I}$$

in which the symbol $tr(\mathbf{E})$ denotes the trace of the strain rate tensor $\mathbf{E}$ and $\mathbf{I}$ represents the identity matrix. Alternatively, in tensor notation

$$S_{i,j} = \frac{1}{3}\left(\sum_k \partial_k v_k\right)\delta_{ij}$$

in which the symbol δ_{ij} ($\delta_{ij} = 1$ when $i = j$, $\delta_{ij} = 0$ when $i \neq j$) is used. It is observed that the deviatoric part can be seen as the scalar sum of the main diagonal elements of $\mathbf{E}$ multiplied by the identity tensor represented by the Kronecker delta; the deviatoric part represents the deformation that causes isotropic volume change (expansion/contraction) and is therefore called the *rate of dilatation tensor*. Notice that the trace of the rate of dilatation tensor coincides with the divergence of the velocity vector field $\mathbf{v}$ and therefore represents the rate of change of volume per unit time, i.e., the rate of volume change. The deviatoric part, called the *rate of shear tensor*, is defined as

$$\mathbf{D} = \mathbf{E} - \mathbf{S} = \mathbf{E} - \frac{1}{3} tr(\mathbf{E})\mathbf{I}$$

or, alternatively

$$D_{i,j} = \frac{1}{2}\left(\partial_i v_j + \partial_j v_i\right) - S_{ij} = \frac{1}{2}\left(\partial_i v_j + \partial_j v_i\right) - \frac{1}{3}\left(\sum_k \partial_k v_k\right)\delta_{ij}.$$

Since $\mathbf{E}$ is a symmetric tensor, so is $\mathbf{D}$, representing deformations that, overall, do not cause volume changes. In conclusion, the motion of point P will be the sum of:

- a translation with the velocity of the particle's centroid G;
- a rigid rotation described by the tensor $\mathbf{R}$;
- a deformation due to isotropic volume change described by the tensor $\mathbf{S}$;
- a deformation due to shear without volume change described by the tensor $\mathbf{D}$.

1.2.2 *Q or Okubo-Weiss Criterion*

This parameter is a scalar used mainly for visualisation purposes and, with reference to Definition 1.5, is defined as

$$Q = \omega_{ij}\omega_{ij} - S_{ij}S_{ij}.$$

The parameter Q is positive in regions where vorticity is greater than the rate of deformation, and vice versa; it approaches zero near the walls.

1.2.3 *Stress Tensor*

By definition, stress is a force per unit area. On any surface where a force acts, an associated stress vector can also be defined. As in the case for the strain-rate tensor, the purpose of the stress-tensor is to uniquely define the stress state at every point in a flow field. As it will be clearer when reading Sect. 2.5, applying Newton's second law to an infinitesimal control volume within the considered continuum medium leads to a second-order tensor called the *stress tensor*, which can be indicated by the symbol $\mathbb{P}$ in vector notation or by the symbol p_{ij} in tensor notation. The stress tensor can also be interpreted as a mathematical operator: considering an infinitesimal oriented surface identified by the vector $\mathbf{dS} = dS\mathbf{n}$, applying the stress tensor to the unit normal unit vector $\mathbf{n}$, results in the stress (force per unit area) $\mathbf{f}$ acting on the surface element considered. In vector notation, this is expressed as $\mathbf{f} = \mathbb{P} \cdot \mathbf{n}$; in tensor notation, it is represented as $f_i = p_{ij}n_j$.

To better understand the importance of the stress tensor, let's consider an infinitesimal oriented surface with its normal coinciding with the $\mathbf{j}$ unit vector of the Cartesian coordinate system. Applying the stress tensor to the $\mathbf{j}$ unit vector yields the stress (with its three components) acting on the infinitesimal surface element belonging to the xz plane with normal $\mathbf{j}$. In formulas:

$$\mathbf{f}^{(y)} = \begin{bmatrix} p_{11} & p_{12} & p_{13} \\ p_{21} & p_{22} & p_{23} \\ p_{31} & p_{32} & p_{33} \end{bmatrix} \begin{bmatrix} 0 \\ 1 \\ 0 \end{bmatrix} = \begin{bmatrix} p_{12} \\ p_{22} \\ p_{32} \end{bmatrix}.$$

Notice that the stress acting on the infinitesimal surface element belonging to the xz plane corresponds to the second column of the stress tensor. In general:

- the stress vector acting on infinitesimal surfaces belonging to a plane with the unit normal vector $\mathbf{n}$ is represented by the n-th column of the stress tensor $\mathbb{P}$;
- the m-th component of the stress acting on infinitesimal surfaces belonging to a plane with the normal vector $\mathbf{n}$ is represented by the element p_{ij} of the stress tensor.

At this point, it is clear that the columns of the stress tensor are an ordered sequence of the stresses acting on infinitesimal surface elements lying on planes having as normal unit vector the unit vectors of the considered Cartesian coordinate system. Even though forces or stresses are measurable on actual solid surfaces, here the interest is on "virtual" surfaces interior to the flow field. For every possible orientation of the virtual surface, there is a different stress vector that describes the forces exerted by the flow on the surface. Aiming at a unique representation of the stress state at a point in a flow, the stress vector is not sufficient to represent it. However, the stresses on three mutually orthogonal differential surfaces can represent the stress state uniquely. The three vectors that describe the stress on these surfaces are represented as a tensor. The nine particular numbers that comprise the tensor depend on the coordinate system in which the tensor is represented.

1.2.4 Constitutive Equations

If, in the case of solids, the constitutive equations represent the relationship between the components of the stress tensor and the deformations with respect to the undeformed configuration, in the case of fluids, the constitutive equations represent the relationship between the components of the stress tensor and the components of the rate of deformation tensor, which coincide with the spatial derivatives of the velocity vector components. Since, generally, for a fluid, when the velocity is zero, the shear stresses are zero while the normal stresses can still be non-zero, we proceed to break down the stress tensor into its spherical part and its deviatoric part as shown below:

$$\begin{bmatrix} p_{11} & p_{12} & p_{13} \\ p_{21} & p_{22} & p_{23} \\ p_{31} & p_{32} & p_{33} \end{bmatrix} = \begin{bmatrix} -p & 0 & 0 \\ 0 & -p & 0 \\ 0 & 0 & -p \end{bmatrix} + \begin{bmatrix} p_{11}+p & p_{12} & p_{13} \\ p_{21} & p_{22}+p & p_{23} \\ p_{31} & p_{32} & p_{33}+p \end{bmatrix} \tag{1.6}$$

In which p is the thermodynamic pressure, defined as the negative of the mean value of the normal stresses in the three coordinate directions: in tensor notation $p = -\frac{p_{kk}}{3}$ (implicit summation notation is used). The negative sign is a matter of convention: a positive pressure is usually understood to be compressive (i.e. inward directed), whereas a positive normal stress is taken to be tensile (i.e. outward directed). Hence the need for the negative sign. The Eq. 1.6 in tensor notation becomes:

$$p_{ij} = -p\delta_{ij} + (p_{ij} + p\delta_{ij}) = -p\delta_{ij} + \tau_{ij}.$$

The tensor $\tau_{ij} = p_{ij} + p\delta_{ij}$ is the deviatoric part of the stress tensor and is called the *viscous stress tensor* by construction, characterized by having zero trace ($\tau_{ii} = 0$). It will be:

$$\tau_{11} = p_{11} - \frac{p_{11} + p_{22} + p_{33}}{3}, \quad \tau_{22} = p_{22} - \frac{p_{11} + p_{22} + p_{33}}{3},$$
$$\tau_{33} = p_{33} - \frac{p_{11} + p_{22} + p_{33}}{3}$$

and therefore

$$\tau_{11} + \tau_{22} + \tau_{33} = 0.$$

τ_{11}, τ_{22} and τ_{33} are the *deviatoric* normal stresses, meaning the fluid-mechanical normal stress plus the thermodynamic pressure. Being related only to fluid motion, the deviatoric stress tensor is zero for a fluid at rest. Referring to Sects. 2.5 and 2.5.1, although viscous stresses are generally tangential in nature, normal viscous stresses ($\tau_{ii} = p_{ii} + p$) can take non-negligible values. For Newtonian fluids, the Stokes hypothesis[1] holds. Based on the Stokes hypothesis, the deviatoric part (i.e., that with zero trace) of the stress tensor is proportional to the deviatoric part of the rate of deformation tensor by a constant of proportionality μ, called *dynamic viscosity*. Referring to Sect. 1.2.1 and using tensor notation, this relationship can be expressed as:

$$p_{ij} - \frac{1}{3}\delta_{ij} p_{kk} = 2\mu \left(E_{ij} - \frac{1}{3}\delta_{ij} E_{kk} \right)$$

equivalent to

$$\tau_{ij} = 2\mu D_{ij}$$

Observing that the trace of the rate of deformation tensor is equal to the divergence of velocity ($E_{kk} = \nabla \cdot \mathbf{v}$), we can write:

$$\tau_{ij} = 2\mu \left(E_{ij} - \frac{1}{3}\delta_{ij} \nabla \cdot \mathbf{v} \right)$$

and, for incompressible flows (see Eq. 1.4)

$$\tau_{ij} = 2\mu E_{ij} = \mu \left(\partial_j v_i + \partial_i v_j \right)$$

This is the three-dimensional generalization of the well-known Newton's law of viscosity, expressing the proportionality, via dynamic viscosity (μ), between the viscous shear stress and the velocity gradient when the velocity has only one component u (along the x-axis) and varies only in the direction orthogonal to the axis along which it is defined (along the y-axis). In formula: $\tau = \mu \partial u / \partial y$.

---

[1] George Gabriel Stokes, *On the theories of the internal friction of fluids in motion and of the equilibrium and motion of elastic solids*, Cambridge, Trans. Cambridge Philos. Soc., 8, 287–319, 1845.

1.3 Differential Equations with Physical Applications

1.3.1 Generalities on Partial Differential Equations

A differential equation is an equation involving one or more derivatives of an unknown function. If all the derivatives are taken with respect to a single independent variable, it is called an ordinary differential equation (ODE), while it is termed a partial differential equation (PDE) when derivatives with respect to multiple independent variables are involved. The differential equation (whether ordinary or partial) has order n if n is the maximum order of the derivatives appearing in it. A partial differential equation generally has an unknown function $u(x_1, x_2, \dots, x_r)$ of r independent variables and establishes a relationship among the independent variables, the function u, and its partial derivatives. Here, we will consider first- and second-order equations. First-order equations are in the form:

$$au_x + bu_y = f$$

while second order equations will be

$$au_{xx} + bu_{xy} + cu_{yy} = f$$

where a, b, c, f are functions of x, y, u as well as they are functions of the first order partial derivatives u_x, u_y, and/or second partial derivatives u_{xx}, u_{yy} in the case of second order equations. In both cases, the equation is called *linear* if the coefficients a and b (respectively a, b, c) depend only on x and y, and f is linear in u (respectively in u, u_x, u_y); if the coefficients depend on x, y, u, u_x, u_y, the equation is termed *quasi-linear*. In the specific problems discussed here, often one variable, either x, is a spatial variable, the other is a temporal variable, and when this occurs, it will be denoted by t instead of y. An integral or solution, in the classical sense, of a n-th order partial differential equation is a function u that satisfies the equation in a given connected open set Ω, where u is continuous with its derivatives up to the n-th order. If boundary or limit conditions are assigned on Γ, the boundary of Ω (or part of it), then u must be continuously differentiable on $\Omega \cup \Gamma$ up to the order required by the conditions. The general integral is the totality of solutions. For ordinary differential equations it makes sense to pose the problem of finding the general integral (i.e., the set of all solutions). For partial differential equations the approach is different, and usually, the more limited objective is to determine any solutions that satisfy some additional conditions, which are generally those imposed at the boundary Γ of Ω (boundary conditions). Being c a positive constant, the following equations are of common interest:

$$u_t + cu_x = 0, \text{ transport equation}$$

$$u_{tt} - c^2 u_{xx} = 0, \text{ waves equation}$$

$$u_{xx} + u_{yy} = 0, \text{ Laplace equation}$$

$$u_t - cu_{xx} = 0, \text{ heat equation.}$$

A general property of the solutions of partial differential equations is that the general integral depends on arbitrary functions rather than arbitrary constants, as is the case of ordinary differential equations.

1.3.2 Mathematical Classification of Linear and Quasi Linear Partial Differential Equations

For simplicity, we consider equations with only two variables.

$$a_{11}(x, y)u_{xx} + 2a_{12}(x, y)u_{xy} + a_{22}(x, y)u_{yy} + b_1(x, y)u_x + b_2(x, y)u_y + \\ + c(x, y)u = d(x, y) \tag{1.7}$$

having the coefficients the necessary regularity.

The classification is based solely on the part containing the second order derivatives:

$$a_{11}(x, y)u_{xx} + 2a_{12}(x, y)u_{xy} + a_{22}(x, y)u_{yy}.$$

This part is termed *principal part of the equation.* The classification of Eq. 1.7 is based on the sign of the discriminant $\delta = a_{12}^2 - a_{11}a_{22}$ of the principal part:

- if $\delta > 0$ the equation is said to be hyperbolic;
- if $\delta = 0$ the equation is said to be parabolic;
- if $\delta < 0$ the equation is said to be elliptic.

Elliptic equations are those having $\nabla^2 u$ as principal part (Laplace,[2] Poisson[3]); the heat equation (and similar ones) are parabolic equations; the transport equation and the wave equation are hyperbolic equations. Each type of equation corresponds to a pair of sets of characteristic curves (see Sect. 1.3.3). In the case of hyperbolic equations, the characteristics are real and distinct, meaning that information propagates with finite velocity along two specific sets of directions in the x-t plane. In the case of parabolic equations, the two sets degenerate into a single set, and thus information propagates with finite velocity along a single direction. In the case of elliptic equations, the characteristic curves are not real, and there are therefore two distinct sets of imaginary curves: there is no preferred direction, and information propagates

[2] P.S. Laplace, *Mémoire sur la théorie de l'anneau de Saturne*, Paris, Mémoires de l'Académie Royale des Sciences de Paris, 1787.

[3] S.D. Poisson, *Remarques sur un equation qui se presente dans la théorie des attractions des spheroides*, Paris, Nouveau bulletin des sciences: par la Société philomat(h)ique (de Paris), 1813.

instantaneously in all directions. The Navier–Stokes equations are second-order nonlinear partial differential equations that possess properties of each of the three types of equations mentioned above. Considering a non-steady, inviscid, and compressible flow, it is possible to observe sound waves and shocks, indicating distinctly the hyperbolic nature of this type of flow. For supersonic steady compressible flows, the nature will be hyperbolic. For subsonic steady compressible flows, the nature of the equations will be mixed hyperbolic and elliptic. For incompressible flows, the properties will resemble those specific to elliptic equations. Typically, flows do not exhibit properties attributable to just one of the types of equations seen before. An illustrative example in this regard is the case of steady transonic flows, which feature both subsonic (elliptic) and supersonic (hyperbolic) regions.

1.3.3 Transport Equation

This equation is also known as the *advection equation*. The term *advection* and the associated phenomenon are distinct from the term *convection* and its related phenomenon, although they are often used interchangeably. Advection refers to the movement of a certain quantity because it is immersed in a moving fluid, while convection refers to the movement of a certain quantity because it is immersed in a moving fluid generated by density gradients caused by thermal gradients. Considered a generic quantity described by a scalar field u(x,y,z,t) and immersed in a velocity field **c**, the corresponding advection equation is the following continuity equation:

$$u_t + \nabla \cdot (u\mathbf{c}) = 0. \tag{1.8}$$

In the case of an incompressible flow, $\nabla \cdot \mathbf{c} = 0$, and the velocity field **c** is said to be *solenoidal*. For incompressible flows, Eq. 1.8 can be written as:

$$u_t + \mathbf{c} \cdot \nabla u = 0. \tag{1.9}$$

Unidimensional transport equation (also known as inviscid Burgers equation[4])

$$u_t + cu_x = 0 \tag{1.10}$$

can be used to describe the pure transport of a quantity $u(x, t)$ carried out by a solenoidal velocity field $\mathbf{c} = (c, 0, 0)$. Below, a possible way to construct such an equation will be illustrated (see also Sects. 2.4 and 2.8).

Figure 1.13 shows the curve representative of a quantity $u(x, t)$ as a function of x for time $t = 0$. Due to the velocity field c, and being this a pure transport phenomenon, the same curve will reappear at time $t = 1$ but translated by an amount $\Delta x = c\Delta t$,

---

[4] J.M. Burgers, *A Mathematical Model Illustrating the Theory of Turbulence*, Amsterdam, Elsevier, 1948.

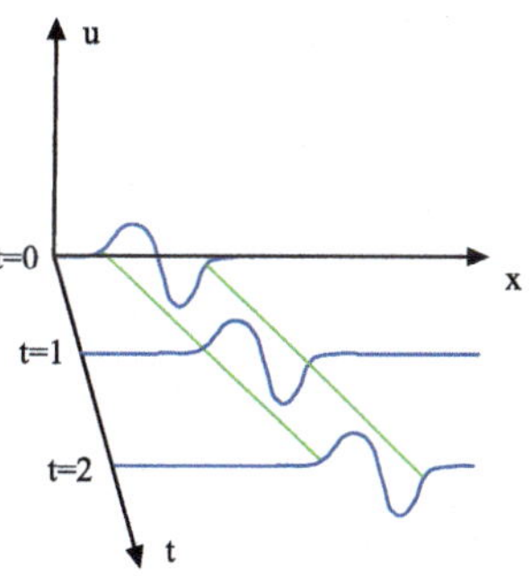

Fig. 1.13 Transport of the quantity u in a velocity field c

where Δt denotes the time interval between $t = 0$ and $t = 1$. The same applies to the transition from $t = 1$ to $t = 2$. It is possible to give an intuitive definition of a wave as a sequence of motion fields resulting from the transport of a generic quantity with a specific velocity. It can be observed that points characterised by the same value of u at different times all lie on a line. For each considered value of u, there will be a different line lying on a plane parallel to the xt plane. The curves obtained from the union of points characterised by the same value of u are called *characteristic curves*. These curves shape depends on the type of differential equation considered; in this case, the characteristic curves are straight parallel lines described by the equation $x - ct = x_0$. These characteristic curves form a *family of curves*. Considering t=0 and x(t=0)=x0, the corresponding u value is called *invariant*. Each invariant identifies a single characteristic curve. The direction of propagation of information will be, in this case, that of increasing x.

Another interpretation of the characteristic curves is as follows. Considering a continuous variation of time, the curves representing the spatial distribution of u will describe a surface, the trace of which at three time values is indicated by the curves in Fig. 1.13. The directional derivative of this surface along the direction indicated by the characteristic curves is always zero, due to the fact that the value of u does not change along the characteristic curves. In formulas, this concept translates into setting equal to zero the dot product between the unit vector indicating the direction of the characteristic curves (in the xt plane) and the gradient of u. Since this dot product is zero, a vector can be considered instead of the unit vector, provided that the direction does not change. In a unit time interval, the distribution of u will be translated by an amount $\Delta x = c\Delta t = c \cdot 1 = c$, and thus, a vector with the same direction as the characteristic curves can be written as $(c, 1)$, while the gradient of u is (u_x, u_t). In formulas:

$$(c, 1) \cdot (u_x, u_t) = 0 \Rightarrow u_t + cu_x = 0$$

namely, the *transport equation*.

The general solution of the transport equation is

$$u(x,t) = F(x - ct)$$

Here, F is a generic function of which it is possible to compute the first derivative considering that

$$u_t = -cu_x, \qquad u_x = \frac{\partial F}{\partial x}.$$

Graphically, $F(x - ct)$ is obtained from $F(x)$ (the configuration at time $t = 0$) by a translation of magnitude ct in the positive direction of the x-axis.

The concept of the inviscid Burgers' equation is also applicable when the velocity field $\mathbf{c}$ in which the scalar field $u(x, y, z, t)$ is immersed is non-solenoidal. In the one-dimensional scenario, the inviscid Burgers' equation can be expressed as:

$$u_t + uu_x = 0 \tag{1.11}$$

where the velocity field $\mathbf{c}$ and the transported quantity $u(x, t)$ coincide. Equation 1.11 represents the one-dimensional motion of an inviscid fluid moving with velocity $u(x, t)$, not subject to external forces. From a physical point of view, it describes the rectilinear uniform motion of individual portions of the fluid, which may have different velocities from point to point. As will be further elucidated in Chapter 2, the Burgers' equation can be considered as a simplification of the conservation equation of momentum, considering only the terms of temporal derivative and advection. Specifically, Burgers' equation is often used to examine the characteristics of numerical discretisation schemes concerning the non-linear advective term uu_x.

To grasp an idea of what Eq. 1.11 represents, one can envision a very narrow corridor traversed by people lined up in a row one by one, each one at a constant speed (no acceleration), and unable to overtake one another. The person who is at position x at time $t = 0$ will move at a constant speed $u(x, 0)$. If $u(x, 0)$ is never decreasing, there will be no collisions; in particular, if in some time interval $u(x, 0)$ is increasing, meaning the people ahead are faster, they will move away from each other (*rarefaction*). If, in some time interval, $u(x, 0)$ is decreasing, i.e., the people behind are faster, collisions will eventually occur. In the case of inviscid Burgers' equation, the characteristic are curves that intersect in the case of collisions and diverge in the case of rarefactions.

1.3.4 Wave Equation

Wave, or vibrating string, equation

$$u_{tt} = c^2 u_{xx} \tag{1.12}$$

has the following physical interpretation: given an elastic string, initially positioned along the x-axis at rest, the configuration is disturbed and the string allowed to vibrate. Then, it can be shown that the normal displacement (along the y-axis) $u(x, t)$ at the instant t and position x is an integral of Eq. 1.12, whose general solution is

$$u(x, t) = F(x + ct) + G(x - ct) \tag{1.13}$$

with F and G being arbitrary functions for which it is possible to write the second derivatives. Equation 1.13 expresses the fact that the motion of the string results from the superposition of two waves travelling in opposite directions with velocity c. Similarly to what was seen with the transport equation, here part of the solution is constant along the family of characteristics $x - ct = \text{const}$ and part along the family $x + ct = \text{const}$. In other words, while in the case of the transport equation, there is only one characteristic passing through each point in the xt-plane, in the case of the wave equation, there will always be two characteristics passing through each point. To better understand, let's consider the following *initial value problem*, or *Cauchy problem*. Given an infinitely long string with initial position and velocity known and respectively given by

$$u(x, 0) = f(x) \quad -\infty < x < \infty$$

and

$$\frac{\partial u(x, 0)}{\partial t} = g(x) \quad -\infty < x < \infty,$$

find the general solution of wave Eq. 1.12. It can be shown that the solution to this problem is given by the D'Alembert[5] formula

$$u(x, t) = \frac{1}{2}[f(x + ct) + f(x - ct)] + \frac{1}{2c}\int_{x-ct}^{x+ct} g(z)dz. \tag{1.14}$$

This formula expresses the dependence on two factors of the value of u at a generic point $P(x^*, t^*)$:

1. the mean value of u in correspondence of the points $A(x^* - ct^*, 0)$ and $B(x^* + ct^*, 0)$: this is the term $\frac{1}{2}[f(x + ct) + f(x - ct)]$ of Formula 1.14 with $x = x^*$ and $t = t^*$;
2. the mean string velocity value in $[A, B]$: this is the term $\frac{1}{2c}\int_{x-ct}^{x+ct} g(z)dz$ of Formula 1.14 with $x = x^*$ and $t = t^*$.

From what has just been said, it follows that, considering a generic point $P(x^*, t^*)$, the solution at P depends only on the values of u in the interval $I = [A, B]$, referred to as the *dependence interval of point* P. Considering the Fig. 1.14, the triangle

[5] Jean-Baptiste Le Rond d'Alembert, *Research on the vibrating strings*, Berlin, History of the Royal Academy of Berlin, 1747.

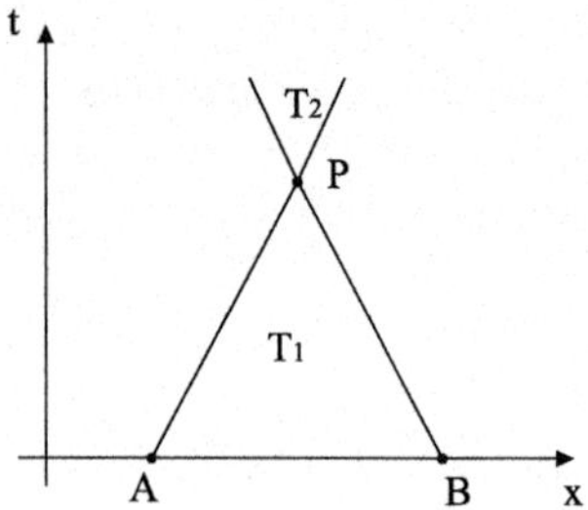

Fig. 1.14 Interval and domain of dependence

bounded by the interval I and by the two characteristic lines passing for P, is called the *continuous dependence domain of point* P. Data on I uniquely determine the solution only in $T1$.

On the other hand, the *influence domain of* P is defined by the triangle $T2$ bounded by characteristic lines passing through P. It is so called because the value of the solution at P influences the solution at all points in $T2$. Figuratively, it can be said that an observer located in x^* at time t^*—the point P—feels the effects of what happened in $T1$, but not of what happens outside of $T1$, and at the same time, the effect of a disturbance at P can only be felt in the domain $T2$.

It is useful to specify here that the term *domain* of dependence or influence refers to the xt plane; when referring only to spatial coordinates, we will speak of *zone* of dependence or influence. In the case of the transport equation seen in Sect. 1.3.3, the zone of dependence is constituted by the points on the x-axis already affected by the passage of the disturbance relative to the considered time; the zone of influence is constituted by the points on the x-axis not yet affected by the passage of the disturbance relative to the considered time. In the case of the wave equation represented in Fig. 1.14, at the time corresponding to point P, the dependence zone will be the interval I while the influence zone will be the entire x-axis.

1.3.5 Heat Equation

The heat equation describes the transport of thermal energy between particles at different temperatures (the same equation also governs, for example, the chemical concentration of different species present in the same domain of interest). The heat equation is also known as *equation of transport by pure diffusion* because it describes the evolution of a generic quantity due solely to the phenomenon of diffusion.

The law of Fourier[6]

$$f = -k\nabla T$$

[6] Jean-Baptiste Joseph Fourier, *The analytical theory of heat*, translation by Alexander Freeman, London, Cambridge University Press, 1878, ed. or.: *Théorie analytique de la chaleur*, Paris, Didot, 1822.

describes the heat flux f as a function of the thermal conductivity coefficient k and the temperature T. The minus sign indicates that thermal energy naturally moves from zones of higher temperature to zones of lower temperature. Here, it is useful to recall the definition of specific heat as the amount of energy (or heat) required to raise or lower the temperature of a given mass or volume of the substance by one unit. When this temperature change is achieved through an isobaric process, it is referred to as the *specific heat at constant pressure*, denoted by the symbol c_p.

Considering a control volume V, the expression of the conservation equation for thermal energy E is given by

$$\int_V \frac{\partial E}{\partial t} dV + \oint_{\partial V} k \nabla T \cdot d\mathbf{S} = 0.$$

From this, considering an infinitesimal one-dimensional element and thermal energy stored solely as enthalpy ($E = \rho c_p T$), we obtain the *differential form of the heat equation*:

$$\frac{\partial T}{\partial t} + \frac{k}{c_p \rho} \frac{\partial^2 T}{\partial x^2} = 0, \tag{1.15}$$

assuming k and c_p not depending on temperature.

In the case where there are no variations in temperature over time, Eq. 1.15 can be written as:

$$\nabla^2 T = \frac{\partial^2 T}{\partial x^2} + \frac{\partial^2 T}{\partial y^2} + \frac{\partial^2 T}{\partial z^2} = 0 \tag{1.16}$$

also named *Laplace equation*.[7] By replacing temperature with velocity in Eq. 1.16, one would obtain the conservation equation for momentum (2.24) in the case where diffusion is the only cause of momentum variation. If there is a source term present in Eq. 1.16, for example due to chemical reactions, we would obtain the *Poisson equation*[8]:

$$\nabla^2 T = s.$$

As shall be elaborated in more detail in Sect. 6.4.1, the Poisson equation, with its elliptic nature, will be used as the pressure correction equation in segregated solution algorithms.

---

[7] Pierre-Simon de Laplace, *Théorie des attractions des sphéroïdes et de la figure des planétes* Paris, Gauthier-Villars, 1782.

[8] Siméon Denis Poisson, *Remarques sur une équation qui se présente dans la théorie des attractions des sphéroïdes*, Nouveau bulletin des sciences: par la Société philomat(h)ique (de Paris), Paris, J. Klostermann fils, t. III, n.75, Dec. 1813, pp. 388–392.

1.4 Gasdynamics

1.4.1 Mechanical Waves

A wave is a disturbance that originates from a source and propagates through time and space, carrying energy or momentum. Mechanical waves are those that propagate exclusively through material media other than vacuum. The medium through which a wave propagates can be thought of as being composed of infinitely small oscillators, each oscillating with its own phase and amplitude to form a wave. Fundamental characteristics of a wave include the following quantities:

- *amplitude*: it is the maximum displacement from the undisturbed position;
- *frequency*: it is the number of complete oscillations per unit time. The unit of measurement in the International System is the hertz (Hz), dimensionally equivalent to the inverse of time (1 Hz = 1 s^{-1}). It is also referred to as the *angular frequency*, denoted by ω and expressed in terms of frequency f as $\omega = 2\pi f$. Angular frequency is measured in radians per second (rad/s).
- *wavelength*: it is the distance between two crests or two troughs of the wave. It is measured in meters and denoted by the symbol λ.
- *wave number*: for wavelength, it is what angular frequency is for frequency. It is denoted by the symbol k and expressed in terms of wavelength λ as $k = 2\pi/\lambda$. The wave number is measured in radians per meter (rad/m) or simply as the inverse of a length, given that radians have no dimension.
- *wave velocity*: The simplest form of wave velocity is the *phase velocity*, defined as the speed of propagation of points of the wave shape characterised by constant phase, such as the velocity at which a crest moves. Wave shape is the profile generated, on a Cartesian plane, by the measurement of a signal with respect to two quantities, for example time and displacement, that characterise it. The time taken by a crest to travel a distance equal to a wavelength λ is $t = \lambda/c$, where c is the *phase velocity of the wave*. In the time interval t, a point on the wave completes one full cycle of oscillation, so $t = 1/f$. Therefore, it can be written that $c = f\lambda$.

Considering the direction of oscillation motion relative to the direction of propagation, and with reference to Figs. 1.15 and 1.16, waves can be distinguished as follows:

- *Transverse waves*: in this type, the oscillatory motion occurs in a direction perpendicular to the overall direction of wave propagation. Examples of transverse waves include those propagating on guitar strings and other stringed instruments;
- *Longitudinal waves*: in this type, the oscillatory motion occurs in a direction parallel to the overall direction of wave propagation. Examples of longitudinal waves include pressure waves in a gas, where the pressure gradient created by the wave passage is parallel to the direction of propagation;
- *Mixed waves*: in this type, the oscillatory motion occurs in all directions. A typical example of this type of wave is ocean waves, which propagate at the interface of

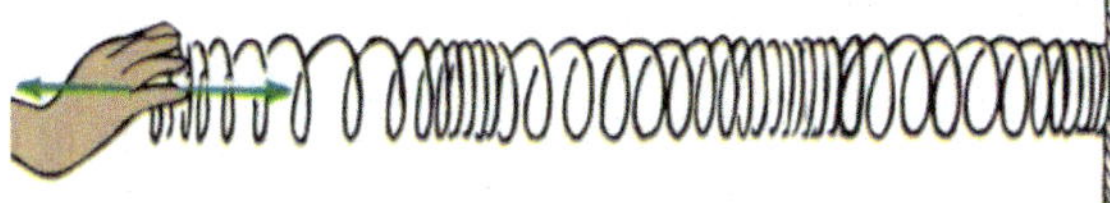

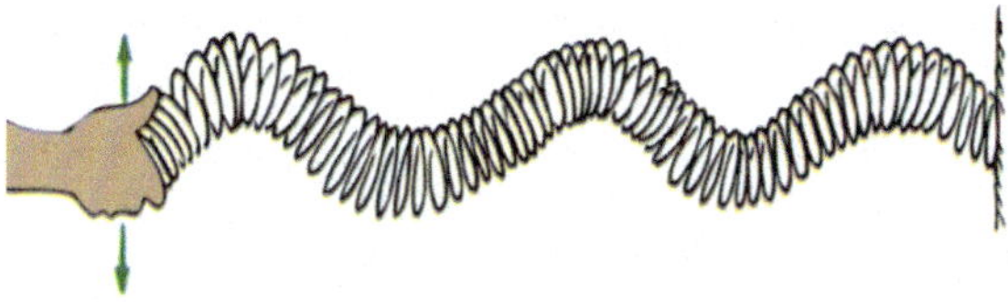

Fig. 1.15 Longitudinal (up) and transverse (bottom) waves

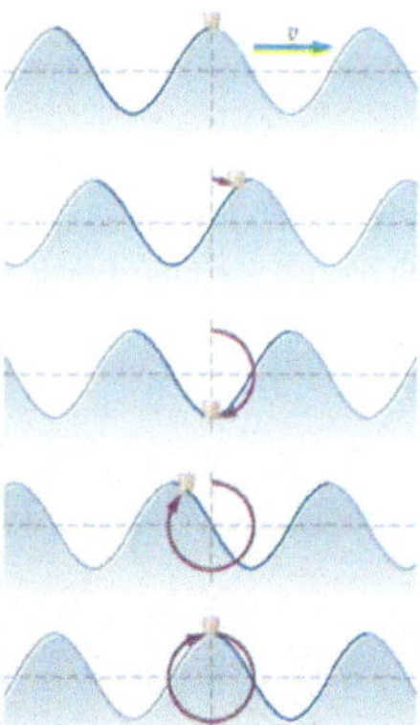

Fig. 1.16 Mixed waves

two fluids with different densities (in this case, water and air). In this case, there is a resultant circulatory motion resulting from the composition of the transverse and longitudinal components of the oscillation.

1.4.2 Acoustic Waves Equation

Acoustic waves are mechanical longitudinal waves and, as such, can be represented as a function of the displacement from the equilibrium position of the medium through which they propagate. Specifically, acoustic waves can also be represented as pressure waves.

1.4.3 One-Dimensional Pressure Waves

For simplicity, let's consider the motion of a fluid element restricted only to the x direction. Figure 1.17 illustrates the variation of absolute pressure P on the two x-normal faces, while Fig. 1.18 illustrates the variation of the x-component u of velocity on the two corresponding faces of the fluid element. Both pressure and velocity are functions of space and time: $P(x, t)$ and $u(x, t)$. The component along the x direction of the total force due to the pressure acting on the fluid element at a given time t is given by the difference in pressure acting on the two x-normal faces. Considering that the variation of pressure along the x direction at a fixed time t is given by the partial derivative of P with respect to x, the pressure difference between the two x-normal faces will be:

$$\delta P = \frac{\partial P}{\partial x}\delta x$$

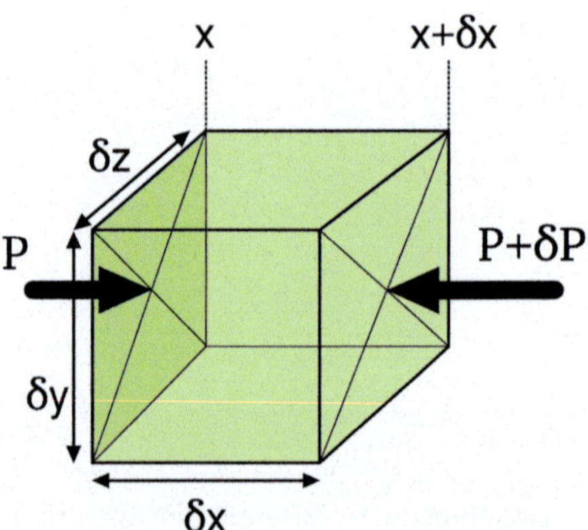

Fig. 1.17 Pressure acting on two fluid element faces

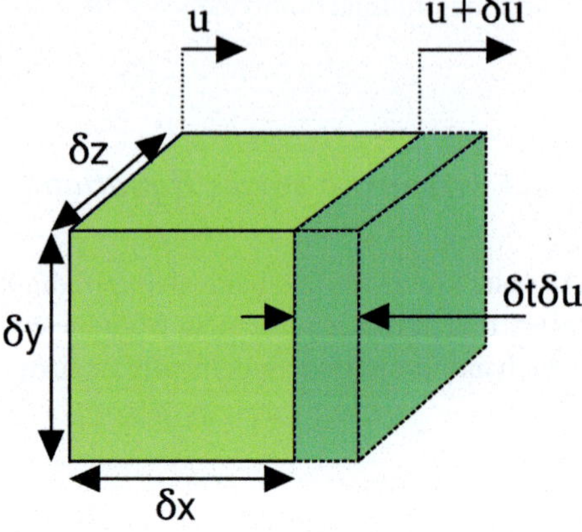

Fig. 1.18 Deformation of a fluid element due to the different velocities of its two surfaces

in which, to indicate that the variations are small, the symbol δ has been used rather than the symbol Δ. Consequently, the component along the x-direction of the total force due to the pressure acting on the fluid element will be

$$\delta F_x = -\frac{\partial P}{\partial x}\delta x \delta y \delta z$$

where the minus sign is introduced to account for the fact that an increase in pressure with increasing x results in a force opposite to the direction of increasing x.

The change in velocity $u(x, t)$ with respect to time at a particular value of x (i.e., $\partial u/\partial t$) is the component along x of the acceleration. Applying the second law of Newton $F_x = ma_x$ to the fluid element, we obtain

$$-\frac{\partial P}{\partial x}\delta x \delta y \delta z = \rho \delta x \delta y \delta z \frac{\partial u}{\partial t}$$

being $\rho \delta x \delta y \delta z$ the mass in the fluid element. Or, equivalently

$$\frac{\partial P}{\partial x} = -\rho \left.\frac{\partial u}{\partial t}\right|_x . \tag{1.17}$$

When the fluid element is crossed by the pressure wave, its two x-normal faces will be characterised by different velocities as shown in Fig. 1.18. The fluid element will therefore undergo a change in volume, for which it is necessary to consider the definition of the modulus of compressibility B (see also Sect. 1.4.5):

$$B = -\frac{\Delta P}{\Delta V / V_0} \tag{1.18}$$

where ΔP is the change in pressure, ΔV is the change in volume, V_0 is the undisturbed volume. Here too, to indicate that the variations are small, the symbol δ will be used instead of the symbol Δ.

The difference in velocity of the two x-normal faces can be expressed as

$$\delta u = \left.\frac{\partial u}{\partial x}\right|_t \delta x$$

that is, the change in velocity along the coordinate x at a fixed moment in time multiplied by the thickness δx of the fluid element. This difference in velocity corresponds to a change in length over the time interval δt, which is equal to $\delta u \delta t$, and consequently a change in volume equal to

$$\delta V = \left.\frac{\partial u}{\partial x}\right|_t \delta x \delta y \delta z \delta t.$$

Given $V_0 = \delta x \delta y \delta z$, we have

$$\frac{\delta V}{V_0} = \left.\frac{\partial u}{\partial x}\right|_t \delta t.$$

Now, we can revisit Formula 1.18 to write

$$\delta P = -B\frac{\delta V}{V_0} = -B\left.\frac{\partial u}{\partial x}\right|_t \delta t$$

thus

$$\frac{\delta P}{\delta t} = -B\left.\frac{\partial u}{\partial x}\right|_t.$$

For δt approaching zero

$$\frac{\partial P}{\partial t} = -B\left.\frac{\partial u}{\partial x}\right|_t. \tag{1.19}$$

The partial derivative symbol is used because pressure P is a function of both x and t, and in this case, we are considering the variation over time for a fixed value of x. Equations 1.17 and 1.19 are two expressions that relate the variation in pressure to the variation in velocity. To obtain a single equation from these, we can differentiate Eq. 1.17 with respect to x to get:

$$\frac{\partial^2 P}{\partial x^2} = -\rho\frac{\partial u}{\partial x \partial t}$$

and differentiate Eq. 1.19 with respect to t to get

$$\frac{\partial^2 P}{\partial t^2} = -B\frac{\partial u}{\partial t \partial x}.$$

Remembering that

$$\frac{\partial u}{\partial x \partial t} = \frac{\partial u}{\partial t \partial x},$$

the *pressure equation in terms of absolute pressure* P is:

$$\frac{\partial^2 P}{\partial x^2} = \frac{\rho}{B}\frac{\partial^2 P}{\partial t^2}.$$

This relationship also holds when considering the difference $p(x, t)$ between the absolute pressure P and the pressure P_0 of the medium under undisturbed conditions:

$$p(x, t) = P(x, t) - P_0.$$

In this case

$$\frac{\partial^2 p}{\partial x^2} = \frac{\rho}{B}\frac{\partial^2 p}{\partial t^2}.$$

1.4.4 Acoustic Waves Described by Displacement from the Equilibrium Position of the Transmitting Medium

For simplicity, we consider the motion of a fluid element restricted to the x direction. Referring to Fig. 1.17 and denoting the displacement of the transmitting medium in time and space by the symbol $\phi(x, t)$, the velocity of each point of the transmitting medium can be expressed as:

$$u(x, t) = \frac{\partial \phi(x, t)}{\partial t}.$$

Remembering the Formula 1.17, it is possible to write

$$\frac{\partial P}{\partial x} = -\rho\frac{\partial u}{\partial t} = -\rho\frac{\partial^2 \phi}{\partial t^2} = \frac{\partial p}{\partial x}. \tag{1.20}$$

Referring to Fig. 1.19, the change in volume of the fluid element as a function of displacement can be expressed as:

$$\delta V = \delta y \delta z \left[\phi(x + \delta x, t) - \phi(x, t)\right].$$

Consequently,

$$\frac{\delta V}{V_0} = \frac{[\phi(x + \delta x, t) - \phi(x, t)]}{\delta x}$$

being $V_0 = \delta x \delta y \delta z$ the initial volume of the fluid element. Considering the definition of bulk modulus expressed by Eq. 1.18, it is clear that we are observing the change

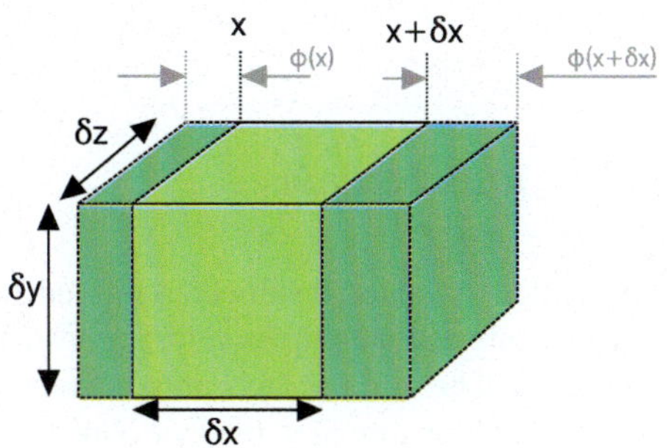

Fig. 1.19 Fluid element volume variation as a function of displacement

of volume (relative to the undisturbed value) when varying the pressure from the undisturbed value (P_0) to the value $p + P_0$. In other words we are observing the change of volume caused by the pressure change $\Delta P = p$. Consequently, from Eq. 1.18, we can write:

$$p = -B\frac{\delta V}{V_0} = -B\frac{[\phi(x + \delta x, t) - \phi(x, t)]}{\delta x}$$

and, considering $\delta x \to 0$

$$p = -B\frac{\partial \phi}{\partial x}$$

which, derived with respect to x gives

$$\frac{\partial p}{\partial x} = -B\frac{\partial^2 \phi}{\partial x^2}.$$

Remembering Eq. 1.20, it is possible to delete the pressure term to write the *equation of displacement wave*

$$\frac{\partial^2 \phi}{\partial x^2} = \frac{\rho}{B}\frac{\partial^2 \phi}{\partial t^2}.$$

It is noted that the displacement wave has a phase velocity $\sqrt{B/\rho}$ equal to that of the pressure wave. The solution of the displacement wave equation turns out to be:

$$\phi(x, t) = Ae^{i(kx - \omega t)}$$

as a consequence, the solution of the pressure equation is

$$p(x, t) = -B\frac{\partial \phi}{\partial x} = -ikBAe^{i(kx - \omega t)} = kBAe^{i(kx - \omega t) - \pi/2}.$$

Therefore, the pressure is observed to have a phase shift of $\pi/2$ compared to the displacement, which means that the pressure reaches its maximum when the displacement of the transmissive medium is zero. This behaviour is counterintuitive when compared to the behavior of a simple mass-spring system, where the force is maximum when the displacement from the equilibrium position is maximum.

1.4.5 Bulk Modulus

An element of material subject to limited stresses of compression, tension, or shear, strains while maintaining its volume constant. When immersed in a fluid, the material element will be subjected to a pressure field acting on each of its faces (see Fig. 1.20), exerting a force normal to the face. Due to these stresses (also known as *bulk stress*),

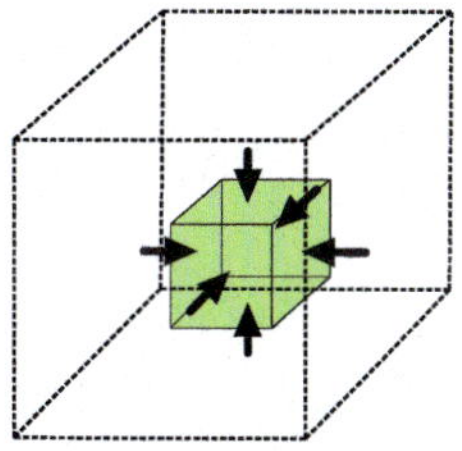

Fig. 1.20 Fluid element volume variation due to pressure

the material element will deform uniformly (*bulk strain*) in every direction, resulting in a change in volume of the material element itself. Denoting the initial volume of the element by the symbol V_0, and the volume change by the symbol ΔV, the volume strain can be defined as:

$$Bulk\ strain = \frac{\Delta V}{V_0}.$$

The bulk modulus B can be defined as the proportionality coefficient between pressure variation Δp the corresponding volume variation:

$$B = -\frac{\Delta p}{\Delta V / V_0}.$$

Here, the minus sign is necessary to obtain a positive value of B in case of positive pressure variation associated with volume decrease. Alternatively, it is possible to define the bulk modulus as

$$B = \rho \frac{\partial p}{\partial \rho}$$

in which ρ is the density and p the pressure.

Bulk modulus is also a thermodynamic quantity. Specifically, we define the *isothermal compressibility modulus* B_T (in the case of constant temperature transformation) and the *isentropic compressibility modulus* B_S (in the case of isentropic transformation). In practice, this distinction is relevant only for gases, to a lesser extent for liquids, and even less for solids. For an ideal gas, we define

$$B_S = \gamma p$$

where γ is the *adiabatic expansion coefficient* (ratio between constant pressure and constant volume heats).

1.5 Numerical Calculus

1.5.1 Taylor Series Expansion and Accuracy

A function $\phi(x)$ that passes through a point x_0 and has all necessary derivatives at that point, can be approximated, at the point x_0 by a polynomial (*Taylor polynomial*) defined as follows:

$$\phi_k(x) = \phi(x_0) + \frac{1}{1!}\phi'(x_0)(x - x_0) + \frac{1}{2!}\phi''(x_0)(x - x_0)^2 + \frac{1}{3!}\phi'''(x_0)(x - x_0)^3 + \ldots + \frac{1}{k!}\phi^{(k)}(x_0)(x - x_0)^k.$$

Considering one point at a distance Δx from point x_0, it is possible to write

$$\phi(x_0 + \Delta x) = \phi(x_0) + \Delta x \left(\frac{\partial \phi}{\partial x}\right)_{x_0} + \frac{(\Delta x)^2}{2!}\left(\frac{\partial^2 \phi}{\partial x^2}\right)_{x_0} + \frac{(\Delta x)^3}{3!}\left(\frac{\partial^3 \phi}{\partial x^3}\right)_{x_0} + HOT \tag{1.21}$$

having indicated by the symbol HOT the higher-order terms.

The error incurred when the HOT are not considered is called the *truncation error*. Neglecting the HOT, the error incurred is not greater than the value of the first derivative that is neglected. In the case where the first neglected derivative is the second-order derivative, the obtained value is said to be *approximated to the second order*. Considering Fig. 3.2, given the value of ϕ and its corresponding derivative at point P, we can use the Taylor series expansion to obtain an approximate value of the function ϕ at point E:

$$\phi_E = \phi_P + \Delta_{PE}\left(\frac{\partial \phi}{\partial x}\right)_P + HOT.$$

Equivalently, it is possible to use the Taylor series expansion to approximate the function ϕ value in e:

$$\phi_e = \phi_P + \Delta_{Pe}\left(\frac{\partial \phi}{\partial x}\right)_P + HOT. \tag{1.22}$$

By neglecting these last two terms of higher order, we obtain an expression accurate to the second order for the behaviour of the considered quantity. In general, the term "*accuracy*" refers to the difference between the exact and the calculated solution. Since, in most practical cases, the exact solution is not available, we settle for considering the order of magnitude of the truncation error as a measure of accuracy. In the case of discretisation of equations, the order of magnitude of the truncation error of the discretisation scheme is the highest among the orders of magnitude of the truncation errors associated with each term of the equation. Note that in this case, accuracy differs from truncation error providing only a measure of how much the truncation error decreases as the size of the cells used for discretisation of the computational domain decreases. Specifically, considering the one-dimensional case of Eq. 1.22, we have an expression for ϕ_e accurate to the second order, which means that the truncation error will decrease by four times if the value of Δ_{Pe} is halved. It is therefore clear that schemes with a higher order of accuracy can produce errors of smaller order of magnitude for the same distances.

1.5.2 Mean Value Approximation

In the one-dimensional case, considering Eq. 1.21, the variation of a quantity $\phi(x)$ passing through a point P contained in a control volume V_P can be expressed as:

$$\phi(x) = \phi_P + (x - x_P)\left(\frac{\partial \phi}{\partial x}\right)_P + \frac{(x - x_P)^2}{2!}\left(\frac{\partial^2 \phi}{\partial x^2}\right)_P + \frac{(x - x_P)^3}{3!}\left(\frac{\partial^3 \phi}{\partial x^3}\right)_P + HOT.$$

Integrating over the control volume, one gets

$$\int_{V_P} \phi(x) dV = \int_{V_P} \left[\phi_P + (x - x_P)\left(\frac{\partial \phi}{\partial x}\right)_P + \frac{(x - x_P)^2}{2!}\left(\frac{\partial^2 \phi}{\partial x^2}\right)_P \right. \\ \left. + \frac{(x - x_P)^3}{3!}\left(\frac{\partial^3 \phi}{\partial x^3}\right)_P + HOT\right] dV.$$

Assuming that the variation law $\phi(x)$ within the control volume is linear, all derivatives of order higher than the second vanish:

$$\int_{V_P} \phi(x) dV = \int_{V_P} \left[\phi_P + (x - x_P)\left(\frac{\partial \phi}{\partial x}\right)_P\right] dV.$$

If P is the control volume centroid, it is

$$\left(\frac{\partial \phi}{\partial x}\right)_P \int_{V_P} (x - x_P) dV = 0$$

and so

$$\int_{V_P} \phi(x) dV = \int_{V_P} \phi_P dV$$

namely

$$\int_{V_P} \phi(x)dV = \phi_P V_P.$$

Lastly, considering the mean value definition, it is

$$\overline{\phi} = \frac{1}{V_P}\int_{V_P} \phi(x)dV = \phi_P$$

The statement asserts that the average value of the quantity $\phi(x)$ within the control volume V_P equals the value of the same quantity at the centroid of the control volume if the variation of $\phi(x)$ is linear or constant within the control volume V_P. If the variation of $\phi(x)$ is not linear or constant, a second-order accurate approximation is obtained.

1.5.3 Derivatives Approximation

Given a function $f : [a, b] \rightarrow \Re$ continuously differentiable in an interval $[a, b]$, we want to approximate its first derivative at a generic point $\overline{x}$ in (a, b). With reference to Fig. 1.21, by definition, the first derivative at $\overline{x}$ is:

$$f'(\overline{x}) = \lim_{h\rightarrow 0} \frac{f(\overline{x}+h) - f(\overline{x})}{h}.$$

The value $f'(\overline{x})$ provides the slope of the tangent to f in$\overline{x}$. For values of h sufficiently small and positive, the quantity

$$(\delta_+ f)(\overline{x}) = \frac{f(\overline{x}+h) - f(\overline{x})}{h} \tag{1.23}$$

is called *forward finite difference* and it represents an approximation of $f'(\overline{x})$.

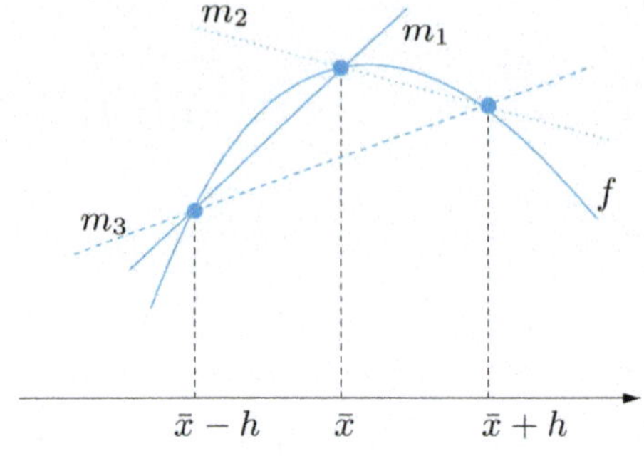

Fig. 1.21 Finite difference approximation $f'(\overline{x})$: backward (continuous line), forward (dotted line) and central (dashed line). $m_1 = (\delta_- f)(\overline{x})$, $m_2 = (\delta_+ f)(\overline{x})$ and $m_3 = (\delta_f)(\overline{x})$ represent the gradient of the straight lines

Considering the Taylor series expansion

$$f(\overline{x}+h) = f(\overline{x}) - hf'(\overline{x}) + HOT,$$

a first order accurate approximation of $f'(\overline{x})$ is:

$$(\delta_- f)(\overline{x}) = \frac{f(\overline{x}) - f(\overline{x}-h)}{h} \tag{1.24}$$

named *backward finite difference*. Lastly, the *central difference* is defined as

$$(\delta_f)(\overline{x}) = \frac{f(\overline{x}+h) - f(\overline{x}-h)}{2h}$$

and it is second order accurate.

1.5.4 Explicit and Implicit Methods

This subsection deals with the resolution of the so-called *Cauchy problems*, namely problems of the form: find $y : I \subset \Re \to \Re$ so that

$$\begin{cases} y'(t) = f(t, y(t)) & \forall t \in I \\ y(t_0) = y_0 \end{cases} \tag{1.25}$$

where I is an interval, $f : I \times \Re \to \Re$ is a function, and y' indicates the derivative of y with respect to t. Lastly, t_0 is a value in I and y_0 is an assigned value called *initial value*. The problem of Cauchy (1.25) is termed *linear* if the function $f(t, y)$ is linear with respect to the variable y. Only a limited number of ordinary differential equations admit explicit solutions. Therefore, numerical methods are sought to approximate the solution for every class of ordinary differential equations that admit a solution. The general strategy of such methods involves dividing the integration interval $I = [t_0, T]$, with $T < +\infty$, into N_h sub-intervals of width $h = (T - t_0)/N_h$; h is referred to as the discretisation step. For each node $t_n = t_0 + nh$ (for $n = 1, \ldots, N_h$), the unknown value u_n approximating $y_n = y(t_n)$ is sought. The set of values $\{u_0 = y_0, u_1, \ldots, u_{N_h}\}$ forms the *numerical solution*.

A classic method is the *forward Euler method*, which generates the following sequence:

$$u_{n+1} = u_n + hf(t_n, u_n), \qquad n = 0, \ldots, N_h - 1.$$

This method is derived from the differential equation in the Cauchy problem (1.25), considered at each node t_n with $n = 1, \ldots, N_h$, where the exact derivative $y'(t_n)$ is approximated by the difference quotient (1.23). Similarly, by using the difference quotient (1.24) to approximate $y(t_{n+1})$, we obtain the *backward Euler method*:

$$u_{n+1} = u_n + hf(t_{n+1}, u_{n+1}), \qquad n = 0, \ldots, N_h - 1.$$

Summing up each step of the forward Euler and backward Euler methods yields another implicit one-step method termed the Crank-Nicolson method:

$$u_{n+1} = u_n + \frac{h}{2}\left[f(t_n, u_n) + f(t_{n+1}, u_{n+1})\right], \qquad n = 0, \ldots, N_h - 1.$$

They are three examples of *one-step methods*, so called because to compute the numerical solution at the node t_{n+1}, only the information related to the previous node t_n is necessary. More precisely, while in the Forward Euler method, the numerical solution u_{n+1} depends solely on the previously computed value u_n, in the Backward Euler and Crank-Nicolson methods it also depends, through $f(t_{n+1}, u_{n+1})$, on itself. For this reason, the former method is called *explicit*, while the latter two are called *implicit* (the Forward Euler and Backward Euler methods are also known respectively as *explicit Euler* and *implicit Euler*).

Implicit methods are computationally more expensive than explicit ones because if the function f in the Cauchy problem (1.25) is non-linear in y, they require solving a non-linear problem at each time level t_{n+1} to compute u_{n+1}. On the other hand, implicit methods manifest better stability properties than explicit schemes.

As an illustrative example, consider the ordinary differential equation

$$\frac{dy}{dt} = -y^2, \; t \in [0, a]$$

with the initial condition $y(0) = 1$. Considering the discretisation $t_k = a\frac{k}{n}$ with $0 \leqslant k \leqslant N_h$, i.e. $h = a/n$, being y_k the value $y(t_k)$. Using the Euler method, one gets

$$\left(\frac{dy}{dt}\right)_k \approx \frac{y_{k+1} - y_k}{h} = -y_k^2$$

from which one can derive the explicit formula

$$y_{k+1} = y_k - hy_k^2$$

valid for $k = 0, \ldots, N_h - 1$. Using the backward Euler method, it is

$$\frac{y_{k+1} - y_k}{h} = -y_{k+1}^2$$

from which one can derive the implicit formula for y_{k+1}

$$y_{k+1} + hy_{k+1}^2 = y_k.$$

Using the Crank-Nicolson method, it is

$$\frac{y_{k+1} - y_k}{h} = -\frac{1}{2}y_{k+1}^2 - \frac{1}{2}y_k^2$$

from which one can derive the implicit formula for y_{k+1}

$$y_{k+1} + \frac{1}{2}hy_{k+1}^2 = y_k - \frac{1}{2}hy_k^2.$$

Both implicit formulas can be numerically solved to compute the value of y_{k+1} using, for example, the Newton's algorithm.

1.5.5 Fixed Point Iteration

With a calculator, it's easy to verify that repeatedly applying the cosine function starting from the number 1 generates the following sequence of real numbers.

$$\begin{aligned}
x^{(1)} &= \cos(1) = 0.54030230586814,\\
x^{(2)} &= \cos(x^{(1)}) = 0.85755321584639,\\
&\vdots\\
x^{(10)} &= \cos(x^{(9)}) = 0.74423735490056,\\
&\vdots\\
x^{(20)} &= \cos(x^{(19)}) = 0.73918439977149,
\end{aligned}$$

which tends to $\alpha = 0.73908513\ldots$ Since by construction, $x^{(k+1)} = \cos(x^{(k)})$ for $k = 0, 1, \ldots$ (with $x(0) = 1$), α is such that $\cos(\alpha) = \alpha$: for this reason it is called *fixed point of the cosine function*. The interest in a method that exploits iterations of this type is evident: if α is a fixed point for the cosine function, then it is a zero of the function $f(x) = x - \cos(x)$ and the method just proposed could be used to compute the zeros of f (only one in this case).

To better specify this intuitive idea, consider the following problem: given a function $\phi : [a, b] \rightarrow \mathbb{R}$, find $\alpha \in [a, b]$ such that $\alpha = \phi(\alpha)$. If such an α exists, it is called a *fixed point of* ϕ and it can be determined as the limit of the following sequence:

$$x^{(k+1)} = \phi(x^{(k)}), \quad k \geqslant 0$$

where $x(0)$ is the initial value. This algorithm is called the *fixed-point iteration method*, and ϕ is called its *iteration function*. The introductory example is thus a fixed-point iteration algorithm for the function $\phi(x) = cos(x)$.

Chapter 2
Governing Equations of Fluid Dynamics

The governing equations of fluid dynamics represent the mathematical formulation of three fundamental principles of physics:

1. conservation of mass;
2. Newton's second law of motion;
3. conservation of energy.

2.1 Control Volume

In applying these fundamental principles of physics to a moving fluid, it is useful to resort to one of the two models described below. Figure 2.1 shows a finite-sized region within a generic flow. This region is called the *control volume*, denoted as V, and it is bounded by the *control surface*, denoted as S. The control volume can be fixed in space, with the fluid passing through it, as shown in the left-hand side of Fig. 2.1, or it can be moving with the fluid so as to always contain the same mass of fluid, as shown in the right-hand side of Fig. 2.1. The equations obtained by applying the three fundamental principles of physics to the finite control volume—whether fixed or moving—are said to be in *integral form*. With appropriate mathematical procedures, the corresponding differential equations can be derived. In the case of a control volume fixed in space, we refer to the governing equations in *conservative form*, whether they are in integral or differential form. In the case where the conservation principles are applied to the control volume moving with the fluid, this is known as governing equations in *non-conservative form*, whether they are in integral or differential form.

The control volume may also have infinitesimal dimensions, denoted as dV, while still containing a sufficient number of molecules to allow the fluid to be treated as a continuous medium. As with the finite control volume, the infinitesimal control volume may be either fixed or in motion. The equations obtained by applying the three

G. Caramia and E. Distaso, *A Practical Approach to Computational Fluid Dynamics Using OpenFOAM®*, https://doi.org/10.1007/978-3-031-88957-8_2

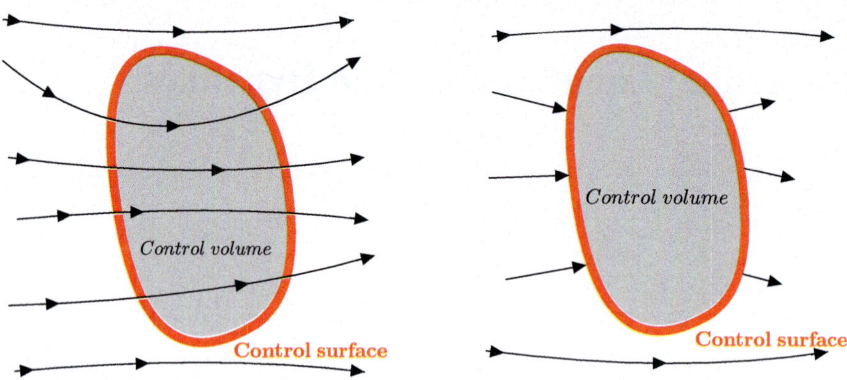

Fig. 2.1 On the left, a finite control volume fixed in space. On the right, a finite control volume moving with the fluid

fundamental principles of physics to the infinitesimal control volume, whether stationary or in motion, are said to be in *differential form*. These differential equations are in *conservative form* if derived by applying the three principles to a stationary infinitesimal control volume. They are in *non-conservative form* if derived by considering the infinitesimal control volume moving with the fluid.

2.2 Substantial Derivative

Here we aim to emphasise the physical interpretation of the concept of the substantial derivative. To this end, we consider the motion of an infinitesimal control volume in Cartesian space.

Indicating with **i**, **j** and **k** the unit vectors of the three coordinate axes x, y and z, it is possible to express the velocity vector as $\mathbf{V} = u\mathbf{i} + v\mathbf{j} + w\mathbf{k}$. In the case of non-stationary motion, the three components of the velocity vector are functions of both space and time.

$$u = u(x, y, z, t),$$

$$v = v(x, y, z, t),$$

$$w = w(x, y, z, t).$$

In addition to this, the motion is characterised by a scalar field of density also dependent on both space and time

$$\rho = \rho(x, y, z, t).$$

As shown in Fig. 2.2, the infinitesimal control volume at time t_1 is located at point 1 where the density has the value $\rho_1 = \rho(x_1, y_1, z_1, t_1)$.

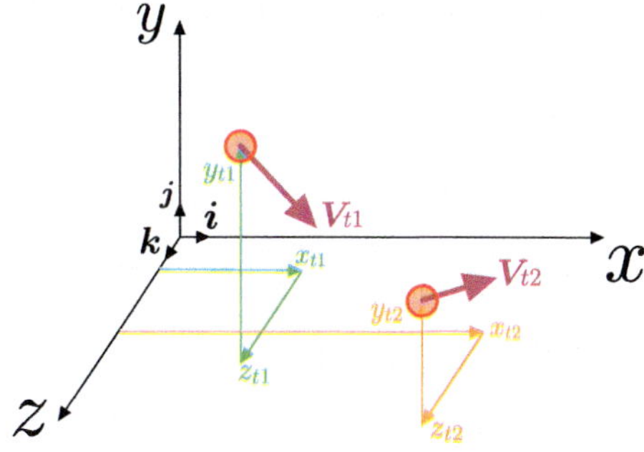

Fig. 2.2 Infinitesimal control volume at two successive moments in time

At a later time t_2, the *same* infinitesimal control volume is located at point 2 where the density has the value $\rho_2 = \rho(x_2, y_2, z_2, t_2)$. Since the density ρ is a scalar function, its Taylor series expansion can be written as

$$\rho_2 = \rho_1 + \left(\frac{\partial \rho}{\partial x}\right)_1 (x_2 - x_1) + \left(\frac{\partial \rho}{\partial y}\right)_1 (y_2 - y_1) + \left(\frac{\partial \rho}{\partial z}\right)_1 (z_2 - z_1) + \left(\frac{\partial \rho}{\partial t}\right)_1 (t_2 - t_1) + HOT.$$

By dividing it by $(t_2 - t_1)$ and neglecting the terms of higher order, we obtain

$$\frac{\rho_2 - \rho_1}{t_2 - t_1} = \left(\frac{\partial \rho}{\partial x}\right)_1 \left(\frac{x_2 - x_1}{t_2 - t_1}\right) + \left(\frac{\partial \rho}{\partial y}\right)_1 \left(\frac{y_2 - y_1}{t_2 - t_1}\right) + \left(\frac{\partial \rho}{\partial z}\right)_1 \left(\frac{z_2 - z_1}{t_2 - t_1}\right) + \left(\frac{\partial \rho}{\partial t}\right)_1 \tag{2.1}$$

which, looking at the left-hand side, represents the average variation of density over time that the infinitesimal control volume undergoes moving from point 1 to point 2. The limit of the quotient on the left-hand side of Eq. 2.1 as t_2 approaches t_1 is indicated by the symbol $D\rho/Dt$:

$$\lim_{t_2 \to t_1} \left(\frac{\rho_2 - \rho_1}{t_2 - t_1}\right) \equiv \frac{D\rho}{Dt}$$

and represents the instantaneous variation of density that the infinitesimal control volume undergoes moving from point 1 to point 2. The symbol D/Dt is called *the substantial derivative*. The substantial derivative is completely different from the partial derivative of the density with respect to time at point 1 $(\partial \rho/\partial t)_1$ because the latter represents the instantaneous variation of the density at the fixed point 1: in other words, the partial derivative measures the variations at a point fixed in both space and time due to the sole dependence on the time variable t. The substantial derivative measures the instantaneous variations of density when the infinitesimal volume moves in space from one position to another.

Returning to Eq. 2.1, we note that by taking the limit as $t_2 \rightarrow t_1$, we obtain

$$\frac{D\rho}{Dt} = u\frac{\partial \rho}{\partial x} + v\frac{\partial \rho}{\partial y} + w\frac{\partial \rho}{\partial z} + \frac{\partial \rho}{\partial t},$$

noting that

$$\lim_{t_2 \rightarrow t_1} \left(\frac{x_2 - x_1}{t_2 - t_1} \right) \equiv u,$$

$$\lim_{t_2 \rightarrow t_1} \left(\frac{y_2 - y_1}{t_2 - t_1} \right) \equiv v,$$

$$\lim_{t_2 \rightarrow t_1} \left(\frac{z_2 - z_1}{t_2 - t_1} \right) \equiv w.$$

Therefore, the substantial derivative in Cartesian coordinates can be written as

$$\frac{D}{Dt} \equiv u\frac{\partial}{\partial x} + v\frac{\partial}{\partial y} + w\frac{\partial}{\partial z} + \frac{\partial}{\partial t}.$$

In Cartesian coordinates, the operator ∇ can be written as

$$\nabla \equiv \mathbf{i}\frac{\partial}{\partial x} + \mathbf{j}\frac{\partial}{\partial y} + \mathbf{k}\frac{\partial}{\partial z}$$

from which

$$\frac{D}{Dt} \equiv \frac{\partial}{\partial t} + (\mathbf{V} \cdot \nabla) \tag{2.2}$$

which is the vector form of the substantial derivative and it is therefore valid in any reference system. To better clarify the concepts expressed by Definition 2.2, consider a domain of interest characterised by a specific velocity field that at each point takes the value of the vector **V**. In the same domain of interest, there is a generic property (for example, temperature) that varies both with respect to space and time. The instantaneous variation of the generic property associated with the mass contained in an infinitesimal moving volume—the substantial derivative—is equal to the sum of two contributions:

1. *the local derivative* $\partial/\partial t$, i.e., the instantaneous variation at the considered point due to the time dependence of the property;
2. *the convective derivative* $\mathbf{V} \cdot \nabla$, i.e., the variation due to the displacement of the infinitesimal element which, as time passes, moves, due to the aforementioned velocity field, to a point where the property has a different value.

In other words, the generic property associated with the mass contained in an infinitesimal element has changed over time both because the property itself varies over time and because, due to the velocity field **V**, the infinitesimal element has moved to a point in the domain where the value of the property is different. As mentioned, the substantial derivative can be applied to any property of the fluid such as temperature, pressure, etc. Considering the temperature, it can be written

$$\frac{DT}{Dt} \equiv \frac{\partial T}{\partial t} + (\mathbf{V} \cdot \nabla)\, T = u\frac{\partial T}{\partial x} + v\frac{\partial T}{\partial y} + w\frac{\partial T}{\partial z} + \frac{\partial T}{\partial t}$$

in which the temperature variation of an infinitesimal fluid element, when it moves from one point to another, is due to the temperature variation with time (term $\partial T/\partial t$) and to the infinitesimal fluid element movement to points characterised by different temperature values.

2.3 The Physical Meaning of the Velocity Divergence

As in the previous section, a control volume moving with the fluid is considered. Although such a control volume always contains the same mass, its dimension $\mathscr{V}$ and its bounding surface S vary as the position changes since the density ρ of the fluid varies from point to point. In Fig. 2.3 such a control volume is shown at a generic instant of time. Also in the same figure, an infinitesimal element dS of the control surface S is shown. The infinitesimal element dS will cover a space $V\Delta t$ in a time Δt due to the velocity **V** with which the fluid moves. Therefore, considering the volume increase $\Delta\mathscr{V}$ due to the motion of the only element dS, this will be equal to the volume of the cylinder having as base the element dS and as height the value $(\mathbf{V}\Delta t) \cdot \mathbf{n}$ being $\mathbf{n}$ the unit vector normal to the infinitesimal element dS. In formulas

$$\Delta\mathscr{V} = [(\mathbf{V}\Delta t) \cdot \mathbf{n}]\, dS = (\mathbf{V}\Delta t) \cdot d\mathbf{S}$$

having indicated with $d\mathbf{S}$ the product $\mathbf{n}\, dS$. To obtain the total volume variation, it will be necessary to sum the contributions due to all the infinitesimal surface elements, that is, it will be necessary to calculate the following surface integral:

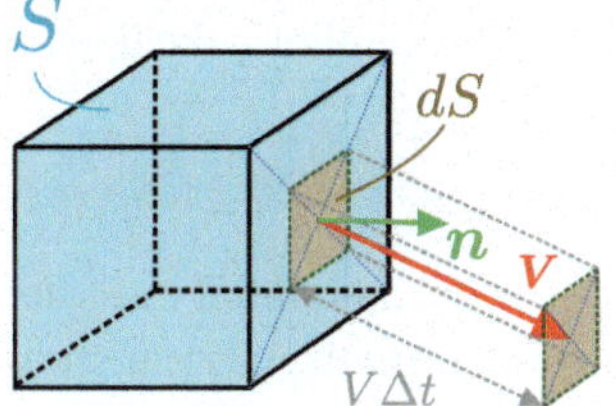

Fig. 2.3 Control volume moving with the fluid

$$\oint_S (\mathbf{V}\Delta t) \cdot d\mathbf{S}.$$

Dividing this integral by the time interval Δt, we obtain the instantaneous variation of the volume considered due to the displacement. This is precisely the definition of substantial derivative applied to the volume $\mathscr{V}$. In formulas

$$\frac{D\mathscr{V}}{Dt} = \frac{1}{\Delta t}\oint_S (\mathbf{V}\Delta t) \cdot d\mathbf{S} = \oint_S \mathbf{V} \cdot d\mathbf{S}.$$

Applying the divergence theorem, we get

$$\frac{D\mathscr{V}}{Dt} = \oint_S \mathbf{V} \cdot d\mathbf{S} = \int_{\mathscr{V}} (\nabla \cdot \mathbf{V})\, d\mathscr{V}.$$

If instead of the volume $\mathscr{V}$, we were to consider an infinitesimal volume $\delta\mathscr{V}$ and if this were so small as to be able to consider the value of $\nabla \cdot \mathbf{V}$ constant within it, then we could write

$$\frac{D\,(\delta\mathscr{V})}{Dt} = \int_{\delta\mathscr{V}} (\nabla \cdot \mathbf{V})\, d\mathscr{V} = (\nabla \cdot \mathbf{V})\, \delta\mathscr{V}$$

that is

$$\nabla \cdot \mathbf{V} = \frac{1}{\delta\mathscr{V}} \frac{D\,(\delta\mathscr{V})}{Dt} \tag{2.3}$$

which says that the divergence of the velocity is equal to the substantial derivative applied to the infinitesimal volume $\delta\mathscr{V}$ divided by the measure of the infinitesimal volume itself. In other words, the divergence of the velocity is equal to the time variation, *per unit of volume*, of the measure of the volume occupied by an infinitesimal mass of fluid due to the movement of the fluid itself (see also Sect. 1.1.2).

2.4 The Continuity Equation

If we denote by *infinitesimal fluid element* the infinitesimal control volume moving with the fluid, then the mass contained in it will be constant and equal to δm and its volume can be indicated by the symbol $\delta\mathscr{V}$ as in the previous section. Indicating with the symbol ρ the density, we can write

$$\delta m = \rho\delta\mathscr{V}.$$

Since the amount of mass contained in the fluid element is always the same, its variation due to the movement of the fluid will be null. As a consequence, its substantial derivative will also be null:

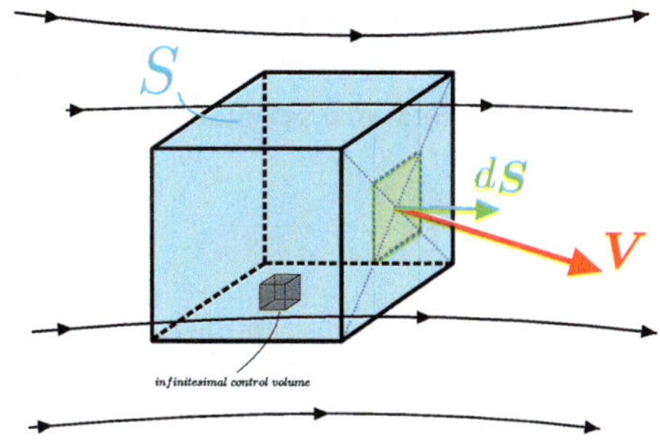

Fig. 2.4 Finite control volume and infinitesimal control volume fixed in space

$$\frac{D(\delta m)}{Dt} = \frac{D(\rho \delta \mathcal{V})}{Dt} = \delta \mathcal{V} \frac{D\rho}{Dt} + \rho \frac{D(\delta \mathcal{V})}{Dt} = 0,$$

that is

$$\frac{D\rho}{Dt} + \rho \left[\frac{1}{\delta \mathcal{V}} \frac{D(\delta \mathcal{V})}{Dt} \right] = 0$$

which, recalling Eq. 2.3, becomes

$$\frac{D\rho}{Dt} + \rho \nabla \cdot \mathbf{V} = 0$$

which is the continuity equation in *non-conservative differential form*. In order to write the continuity equation in *integral conservative form*, we now consider a finite control volume $\mathcal{V}$ fixed and stationary in space as shown in Fig. 2.4. It will be bounded by the control surface S on which it will be possible to define an infinitesimal element identified by the vector $d\mathbf{S}$ and characterised by the velocity $\mathbf{V}$ as in the previous section. Within the control volume, it will be possible to define an infinitesimal volume element $d\mathcal{V}$.

The principle of mass conservation applied to this control volume implies the balance between the net mass flow rate exiting through the surface S and the decrease in mass contained in the control volume itself.

In general, the mass flow rate through a surface is equal to the product of density, surface area, and the component of the velocity normal to the surface. In the case of the infinitesimal element dS, it will be:

$$\rho V_n dS = \rho \mathbf{V} \cdot d\mathbf{S}.$$

Considering that the vector $d\mathbf{S}$ points out of the control volume, the product $\rho \mathbf{V} \cdot d\mathbf{S}$ will be positive when the mass is leaving the control volume and negative otherwise. The net flux that crosses the entire surface S will be the sum of the contributions of all the infinitesimal elements that make it up, namely

$$\oint_S \rho \mathbf{V} \cdot d\mathbf{S}.$$

To quantify the variation of the mass contained in the control volume, one can note that the mass contained in an infinitesimal volume element $d\mathscr{V}$ is equal to $\rho d\mathscr{V}$, and therefore, the total mass contained in the control volume is

$$\int_{\mathscr{V}} \rho d\mathscr{V}.$$

Considering the decrease in mass contained in the control volume equal to

$$-\frac{\partial}{\partial t}\int_{\mathscr{V}} \rho d\mathscr{V},$$

the principle of mass conservation applied to this control volume $\mathscr{V}$ can be expressed as

$$\oint_S \rho \mathbf{V} \cdot d\mathbf{S} = -\frac{\partial}{\partial t}\int_{\mathscr{V}} \rho d\mathscr{V}$$

or

$$\oint_S \rho \mathbf{V} \cdot d\mathbf{S} + \frac{\partial}{\partial t}\int_{\mathscr{V}} \rho d\mathscr{V} = 0 \tag{2.4}$$

which is the continuity equation in *integral conservative form*. This form is called *conservative* because, as will be seen more clearly later, it only contains *conserved* variables (in this case ρ and $\rho\mathbf{V}$). Given that the control volume is fixed in space—both in terms of position and in terms of shape and size—in Eq. 2.4, it is possible to bring the partial derivative inside the integral sign. Applying also the divergence theorem to the first term on the left-hand side of Eq. 2.4 one gets

$$\oint_S \rho \mathbf{V} \cdot d\mathbf{S} = \int_{\mathscr{V}} \nabla \cdot (\rho \mathbf{V})\, d\mathscr{V}.$$

At this point, Eq. 2.4 can be rewritten as

$$\int_{\mathscr{V}} \frac{\partial \rho}{\partial t} d\mathscr{V} + \int_{\mathscr{V}} \nabla \cdot (\rho \mathbf{V})\, d\mathscr{V} = 0$$

that is

$$\int_{\mathscr{V}} \frac{\partial \rho}{\partial t} + \nabla \cdot (\rho \mathbf{V})\, d\mathscr{V} = 0.$$

Given the arbitrariness of the choice of the control volume, the cancellation of this integral is equivalent to the cancellation of the integrand, namely:

$$\frac{\partial \rho}{\partial t} + \nabla \cdot (\rho \mathbf{V}) = 0 \tag{2.5}$$

which is the continuity equation in *differential conservative form*.

2.5 Conservation of Momentum

To determine this equation, Newton's second law will be applied to an infinitesimal fluid element in motion, as shown in Fig. 2.5. When applied to a moving fluid element, Newton's second law states that the resultant of the forces **F** applied on the element is equivalent to its total mass m multiplied by its acceleration **a**. Newton's second law is a vector relation, of which, for now, only the component in the direction of the x-axis is considered:

$$F_x = ma_x \tag{2.6}$$

so, the component along the x-axis of all the forces F_x acting on the element is equal to its mass m multiplied by its acceleration a_x along the x-axis. The forces acting on the element can be of two types:

1. *Mass forces*, such as gravitational, electrical, or magnetic forces.
2. *Surface forces*, which are those forces that act on the surface that delimits the considered element and that can be further divided into two classes: pressure forces and friction forces.

Indicating with the symbol **f** the resultant of the mass forces per unit of mass acting on the element, f_x will be the component along the x-axis. If we denote with ρ the density and considering the volume of the infinitesimal element equal to the product $dxdydz$, it will be

$$component\ along\ x\ of\ the\ resultant\ mass\ forces = \rho f_x\, dx\, dy\, dz. \tag{2.7}$$

Figures 2.6 and 2.7 show the two friction-related stresses for the xy plane only. It is useful to recall that stress is defined as a force per unit area. The shear stress, denoted by the symbol τ_{yx}, determines the deformation of the fluid element over time; the normal stress, denoted by the symbol τ_{xx}, determines the change in volume that the fluid element undergoes over time. Both the shear stress and the normal stress are due to the velocity gradients present in the fluid, although the normal stress is generally negligible compared to the shear stress. They are an exception in cases

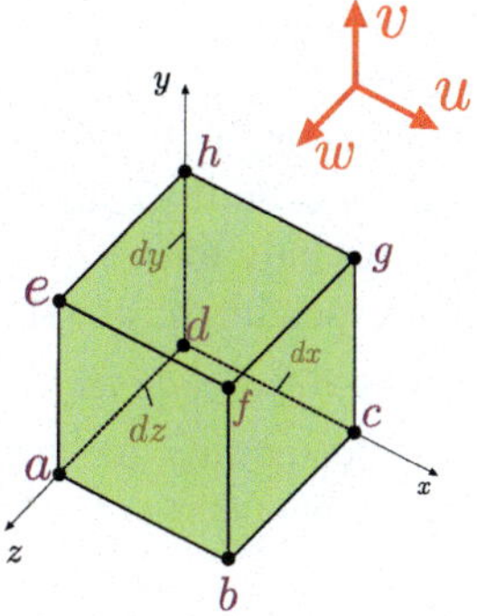

Fig. 2.5 Infinitesimal fluid element in motion

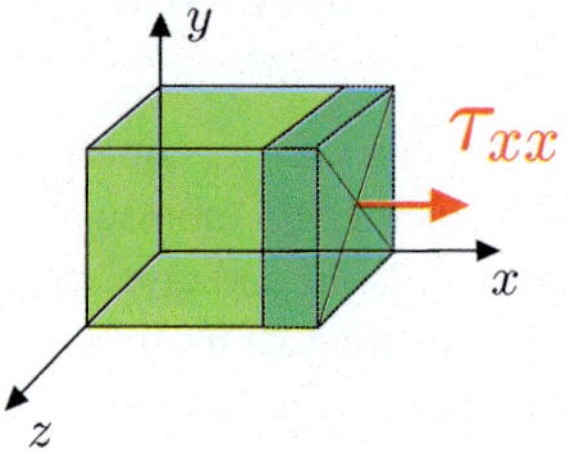

Fig. 2.6 Stresses due to friction: normal stress and corresponding deformation

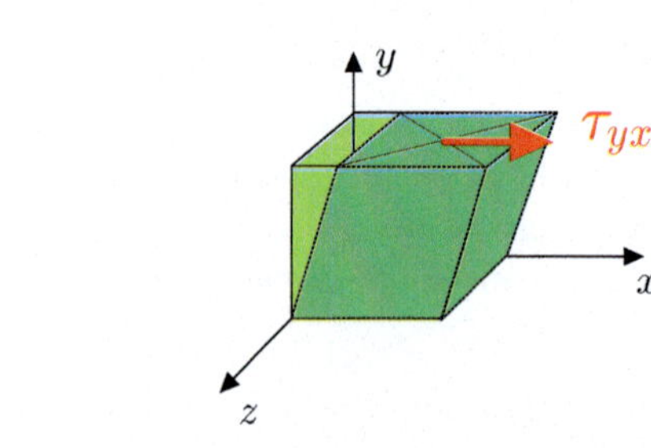

Fig. 2.7 Stresses due to friction: shear stress and corresponding deformation

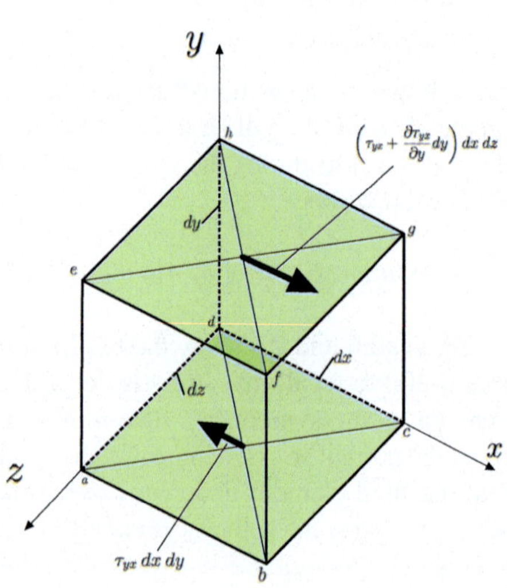

Fig. 2.8 Shear stress on the faces normal to the y-axis

where there are strong velocity gradients in the same direction as the main flow (i.e. shocks).

Referring to Fig. 2.5, with the symbol τ_{ij}, we will denote the stress in direction j exerted on the plane having as normal the direction i. Looking at Figs. 2.5 and 2.8, the only force $\tau_{yx} dx\, dz$ acting on face $abcd$ is that due to the shear stress τ_{yx}. Correspondingly, on face $efgh$, distant dy from $abcd$, the only acting force will be $\left[\tau_{yx} + \frac{\partial \tau_{yx}}{\partial y} dy\right] dx\, dz$.

To determine the direction of application of these forces, it is assumed that the three components of velocity $u,\ v,\ w$ increase in the positive direction of the three

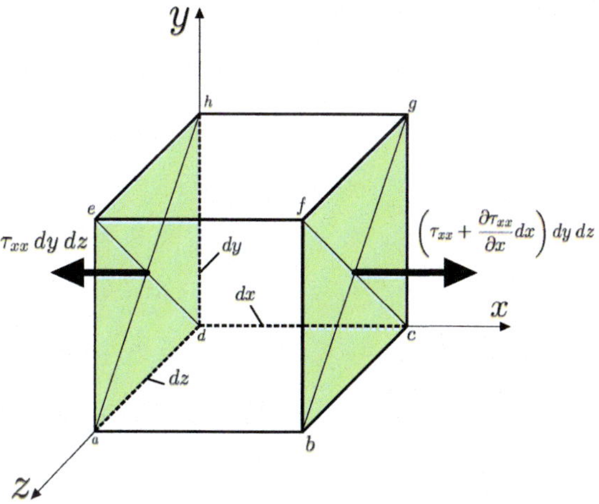

Fig. 2.9 Normal stress due to the velocity gradient

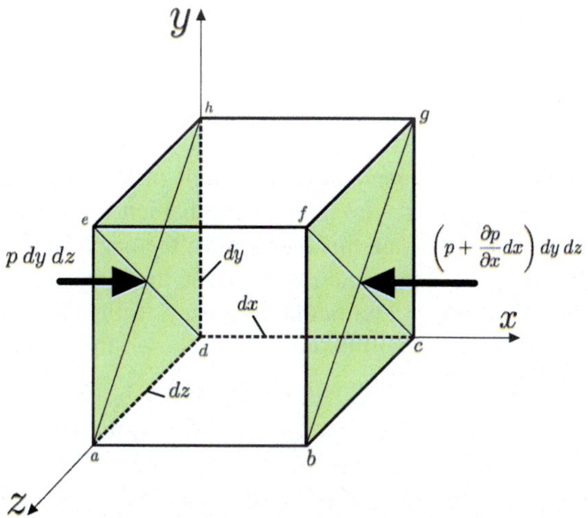

Fig. 2.10 Contribution of pressure

axes (see Fig. 2.5). Considering the component u, it will increase in the positive direction of the x-axis, and in the xy plane it will increase in the positive direction of the y-axis, and in the xz plane it will increase in the positive direction of the z-axis. Therefore, considering the face $efgh$, the u component immediately above it will be greater than that on the face itself, causing it to feel a *dragging* effect that tends to increase velocity: the direction of the shear stress on this face will be that of the positive x-axis. If, on the other hand, the face $abcd$ is considered, the component u immediately below it will be less than that on the face itself, causing it to feel a dragging effect that tends to reduce the velocity: the direction of the shear stress on this face will be in the negative direction of the x-axis.

Looking at Figs. 2.9 and 2.10, the u component immediately to the right of the face $bcgf$ will be greater than that on the face itself, causing it to feel a dragging

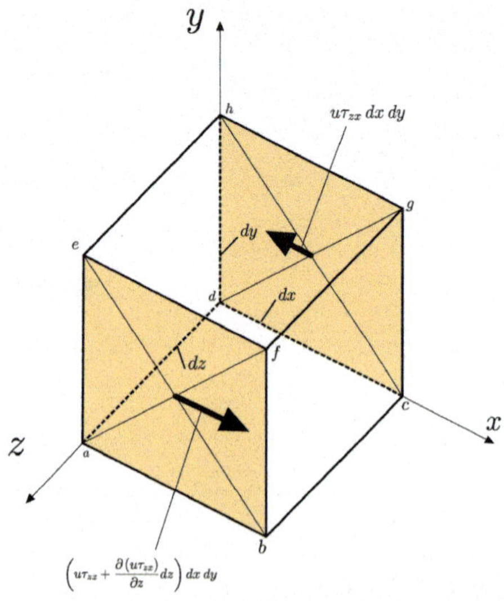

Fig. 2.11 Shear stress on the faces normal to the z-axis

effect that tends to increase the velocity: the direction of the normal stress on this face will be the positive direction of the x-axis. If we consider the face $adhe$, the component u immediately to the left of it will be less than that on the face itself, causing it to feel a *dragging* effect that tends to reduce the velocity: the direction of the normal stress on this face will be the negative direction of the x-axis. Following this logic, it is possible to determine the directions of application of all the forces due to friction. In particular, referring to Fig. 2.11, on the face $dcgh$ the direction of the stress τ_{zy} will be negative while on the face $abfe$ the direction of the stress $\left(\tau_{zx} + \frac{\partial \tau_{zx}}{\partial z}dz\right) dx\,dy$ will be positive.

As for the pressure forces, we refer to the Figs. 2.5 and 2.10. Since we are considering only the component along x of all the forces acting on the element, the only faces for which the contribution of the pressure is not null are $bcgf$ and $adhe$. Considering the contribution of the pressure always directed towards the inside of the element, on the face $adhe$ the force due to the pressure will be directed in the positive direction of the x-axis and equal to $p\,dy\,dz$. On the face $bcgf$ the force due to the pressure will be directed in the negative direction of the x-axis and equal to $\left[p + \frac{\partial p}{\partial x}dx\right] dy\,dz$. Adding up the contributions described so far, we obtain F_x, which is the component in the direction of the x-axis of the total force acting on the element:

$$F_x = \left[p - \left(p + \frac{\partial p}{\partial x}dx\right)\right] dy\,dz$$
$$+ \left[\left(\tau_{xx} + \frac{\partial \tau_{xx}}{\partial x}dx\right) - \tau_{xx}\right] dy\,dz$$

$$+\left[\left(\tau_{yx}+\frac{\partial \tau_{yx}}{\partial y}dy\right)-\tau_{yx}\right]dx\,dz$$
$$+\left[\left(\tau_{zx}+\frac{\partial \tau_{zx}}{\partial z}dz\right)-\tau_{zx}\right]dx\,dy$$
$$+\rho f_x\,dx\,dy\,dz$$

from which

$$F_x=\left(-\frac{\partial p}{\partial x}+\frac{\partial \tau_{xx}}{\partial x}+\frac{\partial \tau_{yx}}{\partial y}+\frac{\partial \tau_{zx}}{\partial z}\right)dx\,dy\,dz+\rho f_x\,dx\,dy\,dz$$

represents the element on the left-hand side of Eq. 2.6. As for the remaining part of Eq. 2.6, the mass m of the element will be

$$m=\rho\,dx\,dy\,dz$$

while its acceleration a_x is the variation over time of the velocity u of a moving fluid element.

By definition, a_x will therefore be equal to the substantial derivative applied to u:

$$a_x=\frac{Du}{Dt}.$$

It is now possible to write Eq. 2.6 as

$$\rho\frac{Du}{Dt}=-\frac{\partial p}{\partial x}+\frac{\partial \tau_{xx}}{\partial x}+\frac{\partial \tau_{yx}}{\partial y}+\frac{\partial \tau_{zx}}{\partial z}+\rho f_x \tag{2.8}$$

and, considering the remaining two coordinate directions

$$\rho\frac{Dv}{Dt}=-\frac{\partial p}{\partial y}+\frac{\partial \tau_{xy}}{\partial x}+\frac{\partial \tau_{yy}}{\partial y}+\frac{\partial \tau_{zy}}{\partial z}+\rho f_y, \tag{2.9}$$

$$\rho\frac{Dw}{Dt}=-\frac{\partial p}{\partial z}+\frac{\partial \tau_{xz}}{\partial x}+\frac{\partial \tau_{yz}}{\partial y}+\frac{\partial \tau_{zz}}{\partial z}+\rho f_z. \tag{2.10}$$

The Eqs. 2.8, 2.9 and 2.10 are scalar equations named *Navier–Stokes equations*. They are in *non-conservative form* because they are obtained from the direct application of Newton's second law to an infinitesimal fluid element in motion. To obtain the corresponding conservative form, it can be noted that by definition of substantial derivative, it is

$$\rho\frac{Du}{Dt}=\rho\frac{\partial u}{\partial t}+\rho\mathbf{V}\cdot\nabla u. \tag{2.11}$$

Moreover, considering the variation over time of the only component in x of the momentum, it will be

$$\frac{\partial(\rho u)}{\partial t} = \rho\frac{\partial u}{\partial t} + u\frac{\partial \rho}{\partial t},$$

that is

$$\rho\frac{\partial u}{\partial t} = \frac{\partial(\rho u)}{\partial t} - u\frac{\partial \rho}{\partial t}$$

corresponding to the first term on the right-hand side in Eq. 2.11. Again, remembering the rule for calculating the divergence of the product of a scalar times a vector,

$$\nabla \cdot (\rho u \mathbf{V}) = u\nabla \cdot (\rho \mathbf{V}) + (\rho \mathbf{V}) \cdot \nabla u,$$

from which

$$\rho \mathbf{V} \cdot \nabla u = \nabla \cdot (\rho u \mathbf{V}) - u\nabla \cdot (\rho \mathbf{V})$$

corresponding to the second term on the right-hand side in Eq. 2.11 which can now be written as

$$\rho\frac{Du}{Dt} = \frac{\partial(\rho u)}{\partial t} - u\frac{\partial \rho}{\partial t} + \nabla \cdot (\rho u \mathbf{V}) - u\nabla \cdot (\rho \mathbf{V}).$$

which, when rearranged, becomes

$$\rho\frac{Du}{Dt} = \frac{\partial(\rho u)}{\partial t} - u\left[\frac{\partial \rho}{\partial t} + \nabla \cdot (\rho \mathbf{V})\right] + \nabla \cdot (\rho u \mathbf{V}).$$

The terms within the square brackets are nothing more than the left-hand side of the continuity equation in conservative differential form Eq. 2.5 so they can be eliminated leading to write

$$\rho\frac{Du}{Dt} = \frac{\partial(\rho u)}{\partial t} + \nabla \cdot (\rho u \mathbf{V}).$$

Substituting this into the first of the three Navier-Stokes equations, we get

$$\frac{\partial(\rho u)}{\partial t} + \nabla \cdot (\rho u \mathbf{V}) = -\frac{\partial p}{\partial x} + \frac{\partial \tau_{xx}}{\partial x} + \frac{\partial \tau_{yx}}{\partial y} + \frac{\partial \tau_{zx}}{\partial z} + \rho f_x. \tag{2.12}$$

Recalling Eq. 1.1, it can be written

$$div\ \mathbf{V} = \nabla \cdot \mathbf{V} = \sum_{i=1}^{3} \frac{\partial}{\partial x_i} V_i = \frac{\partial V_1}{\partial x_1} + \frac{\partial V_2}{\partial x_2} + \frac{\partial V_3}{\partial x_3} = \frac{\partial u}{\partial x} + \frac{\partial v}{\partial y} + \frac{\partial w}{\partial z}$$

and, consequently

$$div\ (\rho u \mathbf{V}) = \nabla \cdot (\rho u \mathbf{V}) = \frac{\partial}{\partial x}(\rho u u) + \frac{\partial}{\partial y}(\rho u v) + \frac{\partial}{\partial z}(\rho u w).$$

At this stage, Eq. 2.12 may be expressed in its extended form as

$$\frac{\partial(\rho u)}{\partial t}+\frac{\partial}{\partial x}(\rho uu)+\frac{\partial}{\partial y}(\rho uv)+\frac{\partial}{\partial z}(\rho uw)=-\frac{\partial p}{\partial x}+\frac{\partial \tau_{xx}}{\partial x}+\frac{\partial \tau_{yx}}{\partial y}+\frac{\partial \tau_{zx}}{\partial z}+\rho f_x. \tag{2.13}$$

In a similar manner, taking into account the remaining two Navier–Stokes equations, we have

$$\frac{\partial(\rho v)}{\partial t}+\nabla\cdot(\rho v\mathbf{V})=-\frac{\partial p}{\partial y}+\frac{\partial \tau_{xy}}{\partial x}+\frac{\partial \tau_{yy}}{\partial y}+\frac{\partial \tau_{zy}}{\partial z}+\rho f_y, \tag{2.14}$$

$$\frac{\partial(\rho w)}{\partial t}+\nabla\cdot(\rho w\mathbf{V})=-\frac{\partial p}{\partial z}+\frac{\partial \tau_{xz}}{\partial x}+\frac{\partial \tau_{yz}}{\partial y}+\frac{\partial \tau_{zz}}{\partial z}+\rho f_z \tag{2.15}$$

and, in extended form

$$\frac{\partial(\rho v)}{\partial t}+\frac{\partial}{\partial x}(\rho vu)+\frac{\partial}{\partial y}(\rho vv)+\frac{\partial}{\partial z}(\rho vw)=-\frac{\partial p}{\partial y}+\frac{\partial \tau_{xy}}{\partial x}+\frac{\partial \tau_{yy}}{\partial y}+\frac{\partial \tau_{zy}}{\partial z}+\rho f_y, \tag{2.16}$$

$$\frac{\partial(\rho w)}{\partial t}+\frac{\partial}{\partial x}(\rho wu)+\frac{\partial}{\partial y}(\rho wv)+\frac{\partial}{\partial z}(\rho ww)=-\frac{\partial p}{\partial z}+\frac{\partial \tau_{xz}}{\partial x}+\frac{\partial \tau_{yz}}{\partial y}+\frac{\partial \tau_{zz}}{\partial z}+\rho f_z. \tag{2.17}$$

Equations 2.12, 2.14 and 2.15 or, equivalently, Eqs. 2.13, 2.16 and 2.17, together constitute the Navier–Stokes equations in *conservative differential form*. Using vector notation, it is possible to express these same equations in a more compact form as

$$\frac{\partial(\rho\mathbf{V})}{\partial t}+\nabla\cdot(\rho\mathbf{V}\mathbf{V})=-\nabla p+\nabla\cdot\boldsymbol{\tau}+\mathbf{f}_b$$

where the symbol $\boldsymbol{\tau}$ represents the viscous stress tensor as defined:

$$\boldsymbol{\tau}=\begin{bmatrix}\tau_{xx} & \tau_{xy} & \tau_{xz}\\ \tau_{yx} & \tau_{yy} & \tau_{yz}\\ \tau_{zx} & \tau_{zy} & \tau_{zz}\end{bmatrix}.$$

2.5.1 Newtonian Fluids

To summarise what was initially presented in Sects. 1.2.3 and 1.2.4, *Newtonian* are those fluids for which

- the shear stress is zero when the fluid is still or with a null velocity gradient;
- it is possible to define a proportionality constant—called *dynamic viscosity* and indicated by the symbol μ—between shear stress and velocity gradient. In other words, it is possible to define a linear relationship between stresses and strain rates (see Sect. 1.2.1);

- the value of dynamic viscosity does not depend on the direction considered for the gradient calculation (i.e., the fluid is *isotropic*).

For such fluids (see also Sect. 1.2.4), it is possible to express the tensor of viscous stresses as

$$\boldsymbol{\tau} = \mu\left[\nabla\mathbf{V} + (\nabla\mathbf{V})^T\right] + \lambda\,(\nabla\cdot\mathbf{V})\,\mathbf{I} \tag{2.18}$$

where the symbol **I** represents the identity matrix. The symbol λ represents the *volume or dilatational viscosity*, which has the dimensions of dynamic viscosity. Volume viscosity is null for incompressible fluids, and it measures the viscous resistance of a (compressible) fluid to volume variation. Volume viscosity is important only when the fluid is rapidly compressed or expanded, as in the case of sound or shock waves (see also Sect. 1.4.2). Volume viscosity explains the energy loss of these types of waves, as described by Stokes' law on sound attenuation. Stokes himself hypothesised to be

$$\lambda = -\frac{2}{3}\mu.$$

In a Cartesian reference frame, Eq. 2.18 can be expressed as

$$\boldsymbol{\tau} = \begin{bmatrix} 2\mu\dfrac{\partial u}{\partial x} + \lambda\nabla\cdot\mathbf{V} & \mu\left(\dfrac{\partial v}{\partial x} + \dfrac{\partial u}{\partial y}\right) & \mu\left(\dfrac{\partial u}{\partial z} + \dfrac{\partial w}{\partial x}\right) \\ \mu\left(\dfrac{\partial v}{\partial x} + \dfrac{\partial u}{\partial y}\right) & 2\mu\dfrac{\partial v}{\partial x} + \lambda\nabla\cdot\mathbf{V} & \mu\left(\dfrac{\partial w}{\partial y} + \dfrac{\partial v}{\partial z}\right) \\ \mu\left(\dfrac{\partial u}{\partial z} + \dfrac{\partial w}{\partial x}\right) & \mu\left(\dfrac{\partial w}{\partial y} + \dfrac{\partial v}{\partial z}\right) & 2\mu\dfrac{\partial z}{\partial x} + \lambda\nabla\cdot\mathbf{V} \end{bmatrix} \tag{2.19}$$

in which the viscous stresses have been expressed as

$$\begin{aligned} \tau_{xx} &= \lambda\nabla\cdot\mathbf{V} + 2\mu\frac{\partial u}{\partial x}, \\ \tau_{yy} &= \lambda\nabla\cdot\mathbf{V} + 2\mu\frac{\partial v}{\partial y}, \\ \tau_{zz} &= \lambda\nabla\cdot\mathbf{V} + 2\mu\frac{\partial w}{\partial z}, \\ \tau_{xy} = \tau_{yx} &= \mu\left(\frac{\partial v}{\partial x} + \frac{\partial u}{\partial y}\right), \\ \tau_{xz} = \tau_{zx} &= \mu\left(\frac{\partial u}{\partial z} + \frac{\partial w}{\partial x}\right), \end{aligned} \tag{2.20}$$

$$\tau_{zy} = \tau_{yz} = \mu\left(\frac{\partial w}{\partial y} + \frac{\partial v}{\partial z}\right). \tag{2.21}$$

The dynamic viscosity μ is, in other words, the resistance of a fluid to flow. It is worth mentioning that the assumption of an isotropic fluid has allowed the setting of $\tau_{xy} = \tau_{yx}, \tau_{xz} = \tau_{zx}, \tau_{zy} = \tau_{yz}$, which results in the symmetry of the stress tensor $\boldsymbol{\tau}$. In the case of an incompressible flow, the divergence of the velocity will be null, and Eq. 2.19 can be expressed as

$$\boldsymbol{\tau} = \begin{bmatrix} 2\mu\frac{\partial u}{\partial x} & \mu\left(\frac{\partial v}{\partial x} + \frac{\partial u}{\partial y}\right) & \mu\left(\frac{\partial u}{\partial z} + \frac{\partial w}{\partial x}\right) \\ \mu\left(\frac{\partial v}{\partial x} + \frac{\partial u}{\partial y}\right) & 2\mu\frac{\partial v}{\partial x} & \mu\left(\frac{\partial w}{\partial y} + \frac{\partial v}{\partial z}\right) \\ \mu\left(\frac{\partial u}{\partial z} + \frac{\partial w}{\partial x}\right) & \mu\left(\frac{\partial w}{\partial y} + \frac{\partial v}{\partial z}\right) & 2\mu\frac{\partial z}{\partial x} \end{bmatrix}. \tag{2.22}$$

Using the notation based on indices, it is possible to express the elements of Eq. 2.22 as

$$\tau_{ij} = \mu s_{ij} = \mu\left(\frac{\partial u_i}{\partial x_j} + \frac{\partial u_j}{\partial x_i}\right). \tag{2.23}$$

The convention used for this notation assumes that i or $j = 1$ correspond to the first coordinate direction (x), i or $j = 2$ correspond to the second coordinate direction (y), and i or $j = 3$ correspond to the third coordinate direction (z). As an example

$$\tau_{12} = \tau_{xy} = \mu\left(\frac{\partial u_1}{\partial x_2} + \frac{\partial u_2}{\partial x_1}\right) = \mu\left(\frac{\partial u}{\partial y} + \frac{\partial v}{\partial x}\right).$$

In conclusion, for a Newtonian fluid, the equation of momentum conservation can be expressed as

$$\frac{\partial(\rho\mathbf{V})}{\partial t} + \nabla\cdot(\rho\mathbf{V}\mathbf{V}) = -\nabla p + \nabla\cdot\left\{\mu\left[\nabla\mathbf{V} + (\nabla\mathbf{V})^T\right]\right\} + \nabla\left(\lambda\nabla\cdot\mathbf{V}\right) + \mathbf{f}_b$$

and, in the case of incompressible flow

$$\frac{\partial(\rho\mathbf{V})}{\partial t} + \nabla\cdot(\rho\mathbf{V}\mathbf{V}) = -\nabla p + \nabla\cdot\mu\left[\nabla\mathbf{V} + (\nabla\mathbf{V})^T\right] + \mathbf{f}_b.$$

Ultimately, in the case where the viscosity is also constant, there is a further simplification:

$$\frac{\partial(\rho\mathbf{V})}{\partial t} + \nabla\cdot(\rho\mathbf{V}\mathbf{V}) = -\nabla p + \mu\nabla^2\mathbf{V} + \mathbf{f}_b. \tag{2.24}$$

2.6 Energy Conservation Equation

To determine this equation, the first law of thermodynamics will be applied to the infinitesimal fluid element in motion shown in Fig. 2.5. Indicating with the symbol A the variation of energy within the fluid element, with the symbol B the net inflow of thermal energy into the fluid element, and with C the work done by the mass forces and the surface forces, the first law of thermodynamics applied to an infinitesimal fluid element in motion can be expressed as

$$A = B + C. \tag{2.25}$$

Initially, considering the term C, it can be noted that the work done by a force on a moving body is equal to the product of the force itself and the component of the body's velocity in the direction of application of the force. Taking into account Eq. 2.7, the work done by the mass forces can be expressed as

$$\rho \mathbf{f} \cdot \mathbf{V}(dx\, dy\, dz).$$

As for the contribution of surface forces (pressure, shear, and normal stress), initially only the component of such forces in the x-direction is considered. Figures 2.12, 2.13, 2.14, 2.15 show the work done by the component in the x-direction of such forces to be equal to their product with the u component in the same direction of the velocity $\mathbf{V}$. Assuming the u component of the velocity is oriented in the positive direction of the x-axis, the work done by surface forces will be positive if they are also oriented in the positive direction of the x-axis, and it will be negative if oriented in the negative direction of the x-axis.

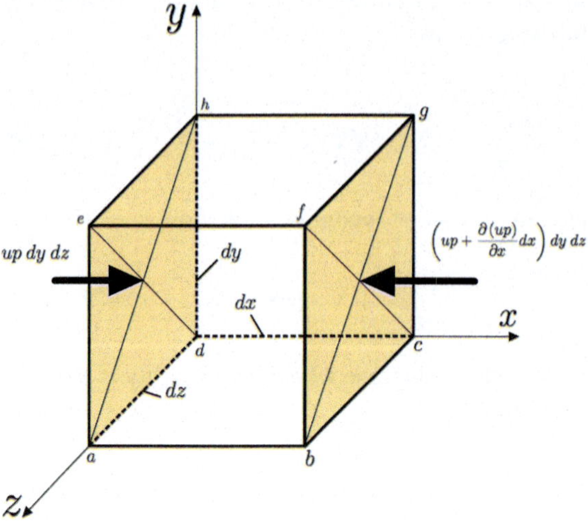

Fig. 2.12 Work done by pressure forces in the x direction

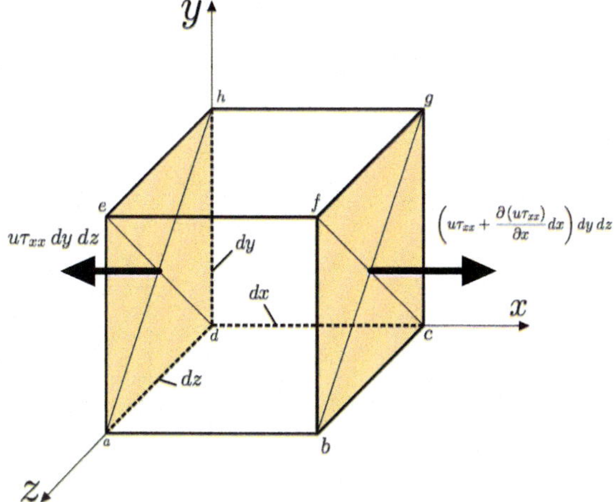

Fig. 2.13 Work done considering only the x direction and only the faces *adhe* and *bcgf*

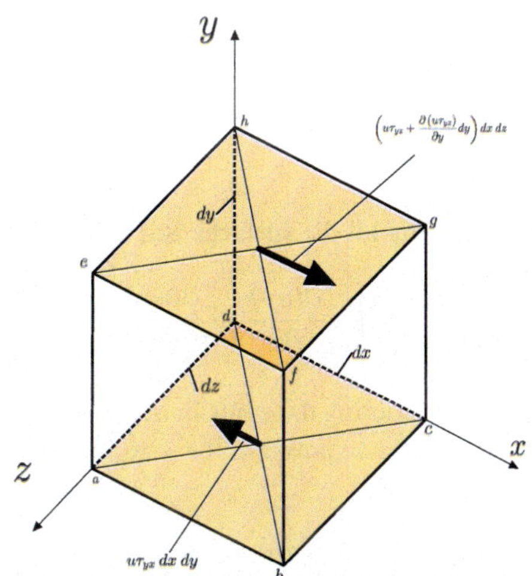

Fig. 2.14 Work done considering only the x direction and only the faces *abcd* and *efgh*

In the case of pressure forces, the resultant of the work done by them, considering only the x-direction, will be

$$\left[up - \left(up + \frac{\partial(up)}{\partial x}dx\right)\right]dy\,dz = -\frac{\partial(up)}{\partial x}dx\,dy\,dz.$$

In the case of shear stress, the resultant of the work done, considering only the x-direction and only the faces $abcd$ and $efgh$, will be

$$\left[u\tau_{yx} - \left(u\tau_{yx} + \frac{\partial(u\tau_{yx})}{\partial y}dy\right)\right]dx\,dz = \frac{\partial(u\tau_{yx})}{\partial y}dx\,dy\,dz.$$

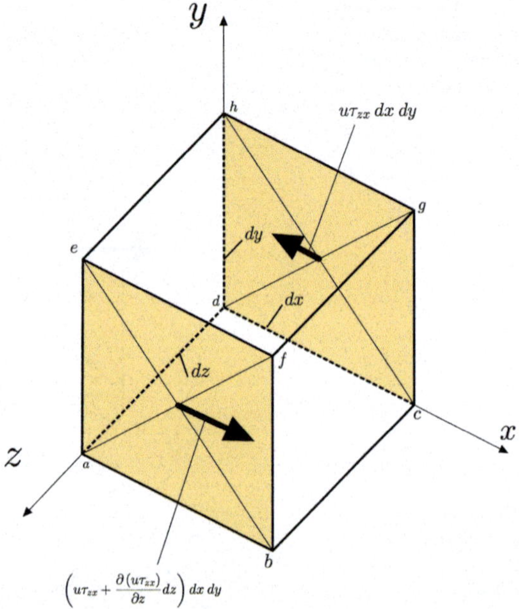

Fig. 2.15 Work done considering only the x direction and only the faces $abfe$ and $cdhg$

Considering all the surfaces and all the surface forces, we derive:

$$\left[-\frac{\partial(up)}{\partial x} + \frac{\partial(u\tau_{xx})}{\partial x} + \frac{\partial(u\tau_{yx})}{\partial y} + \frac{\partial(u\tau_{zx})}{\partial z}\right] dx\, dy\, dz.$$

Now, considering the contributions of all the forces in the three coordinate directions, we derive the expression for term C of Eq. 2.25:

$$\begin{aligned} C = \Bigg[& -\left(\frac{\partial(up)}{\partial x} + \frac{\partial(vp)}{\partial y} + \frac{\partial(wp)}{\partial z}\right) \\ & + \frac{\partial(u\tau_{xx})}{\partial x} + \frac{\partial(u\tau_{yx})}{\partial y} + \frac{\partial(u\tau_{zx})}{\partial z} \\ & + \frac{\partial(v\tau_{xy})}{\partial x} + \frac{\partial(v\tau_{yy})}{\partial y} + \frac{\partial(v\tau_{zy})}{\partial z} \\ & + \frac{\partial(w\tau_{xz})}{\partial x} + \frac{\partial(w\tau_{yz})}{\partial y} + \frac{\partial(w\tau_{zz})}{\partial z}\Bigg] dx\, dy\, dz + \rho \mathbf{f} \cdot \mathbf{V}(dx\, dy\, dz). \end{aligned}$$

We now consider term B of Eq. 2.25. The flux of thermal energy that affects the fluid element can be of a volumetric type, that is, due to the presence inside it of sources/sinks of energy (i.e., chemical reactions, nuclear processes, electromagnetic irradiation), or it can be due to conductive thermal phenomena present on the faces that delimit it. Let $\dot{q}$ as the rate of volumetric heat addition per unit mass. The total heat generated inside the fluid element is

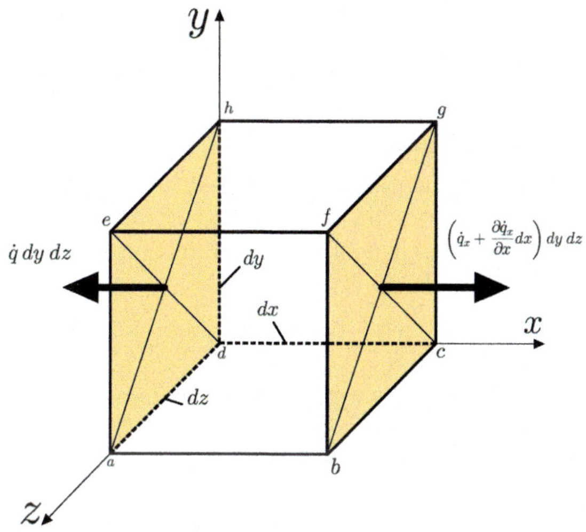

Fig. 2.16 Total thermal flux that crosses the fluid element by conduction in the direction of the x-axis

$$\rho\,\dot{q}\,dx\,dy\,dz.$$

Initially, considering only the direction of the x-axis and assuming the thermal flux due to conduction is always oriented in the positive direction of the same axis, the heat $\dot{q}_x\,dy\,dz$ passes across the face $aedh$. Let $\dot{q}_x$ be the heat flux per unit of time and per unit of surface transmitted by conduction. The total heat flux that crosses the fluid element by conduction in the direction of the x-axis will be (Fig. 2.16)

$$\left[\dot{q}_x - \left(\dot{q}_x + \frac{\partial \dot{q}_x}{\partial x} dx\right)\right] dy\,dz = -\frac{\partial \dot{q}_x}{\partial x} dx\,dy\,dz.$$

Now, considering the contributions in the three coordinate directions, one derives the expression for term B of Eq. 2.25:

$$B = \left[\rho\dot{q} - \left(\dot{q}_x + \frac{\partial \dot{q}_x}{\partial x} + \frac{\partial \dot{q}_y}{\partial y} + \frac{\partial \dot{q}_z}{\partial z}\right)\right] dx\,dy\,dz.$$

If now the terms $\dot{q}_x, \dot{q}_y, \dot{q}_z$ are considered proportional through the thermal conductivity coefficient k to the local gradient of temperature, then we have

$$\dot{q}_x = -k\frac{\partial T}{\partial x}; \qquad \dot{q}_y = -k\frac{\partial T}{\partial y}; \qquad \dot{q}_z = -k\frac{\partial T}{\partial z}$$

and therefore

$$B = \left[\rho\dot{q} + \frac{\partial}{\partial x}\left(k\frac{\partial T}{\partial x}\right) + \frac{\partial}{\partial y}\left(k\frac{\partial T}{\partial y}\right) + \frac{\partial}{\partial z}\left(k\frac{\partial T}{\partial z}\right) \right] dx\, dy\, dz.$$

Finally, term A of Eq. 2.25 is considered. The total energy per unit of mass of the moving fluid element is equal to the sum of its internal energy per unit of mass, indicated by the symbol e, and its kinetic energy per unit of mass, $\frac{V^2}{2}$. Since we are considering a moving fluid element, the time variation of its total energy is expressed by a substantial derivative. Therefore, considering the mass of the fluid element to be equal to $\rho\, dx\, dy\, dz$, it will be

$$A = \rho \frac{D}{Dt}\left(e + \frac{V^2}{2}\right) dx\, dy\, dz.$$

Now, Eq. 2.25 can be expressed as

$$\begin{aligned} \rho \frac{D}{Dt}\left(e + \frac{V^2}{2}\right) = {} & \rho\dot{q} + \frac{\partial}{\partial x}\left(k\frac{\partial T}{\partial x}\right) + \frac{\partial}{\partial y}\left(k\frac{\partial T}{\partial y}\right) + \frac{\partial}{\partial z}\left(k\frac{\partial T}{\partial z}\right) \\ & - \frac{\partial (up)}{\partial x} - \frac{\partial (vp)}{\partial y} - \frac{\partial (wp)}{\partial z} \\ & + \frac{\partial (u\tau_{xx})}{\partial x} + \frac{\partial (u\tau_{yx})}{\partial y} + \frac{\partial (u\tau_{zx})}{\partial z} \\ & + \frac{\partial (v\tau_{xy})}{\partial x} + \frac{\partial (v\tau_{yy})}{\partial y} + \frac{\partial (v\tau_{zy})}{\partial z} \\ & + \frac{\partial (w\tau_{xz})}{\partial x} + \frac{\partial (w\tau_{yz})}{\partial y} + \frac{\partial (w\tau_{zz})}{\partial z} + \rho \mathbf{f} \cdot \mathbf{V}. \end{aligned} \tag{2.26}$$

Equation 2.26 is *the energy conservation equation in non-conservative form in terms of total energy* $e + \frac{V^2}{2}$. To obtain the same equation in terms of the solely internal energy, one can multiply the Navier–Stokes Eqs. 2.8, 2.9, and 2.10 respectively by u, v, and w, resulting in

$$\rho \frac{D\left(\frac{u^2}{2}\right)}{Dt} = -u\frac{\partial p}{\partial x} + u\frac{\partial \tau_{xx}}{\partial x} + u\frac{\partial \tau_{yx}}{\partial y} + u\frac{\partial \tau_{zx}}{\partial z} + u\rho f_x, \tag{2.27}$$

$$\rho \frac{D\left(\frac{v^2}{2}\right)}{Dt} = -v\frac{\partial p}{\partial y} + v\frac{\partial \tau_{xy}}{\partial x} + v\frac{\partial \tau_{yy}}{\partial y} + v\frac{\partial \tau_{zy}}{\partial z} + v\rho f_y, \tag{2.28}$$

$$\rho \frac{D\left(\frac{w^2}{2}\right)}{Dt} = -w\frac{\partial p}{\partial z} + w\frac{\partial \tau_{xz}}{\partial x} + w\frac{\partial \tau_{yz}}{\partial y} + w\frac{\partial \tau_{zz}}{\partial z} + w\rho f_z. \tag{2.29}$$

Keeping in mind that $V^2 = u^2 + v^2 + w^2$, Eqs. 2.27, 2.28, and 2.29, summed together, result in the expression

$$
\begin{aligned}
\rho \frac{D\left(\frac{V^2}{2}\right)}{Dt} = & -u\frac{\partial p}{\partial x} - v\frac{\partial p}{\partial y} - w\frac{\partial p}{\partial z} \\
& + u\left(\frac{\partial \tau_{xx}}{\partial x} + \frac{\partial \tau_{yx}}{\partial y} + \frac{\partial \tau_{zx}}{\partial z}\right) \\
& + v\left(\frac{\partial \tau_{xy}}{\partial x} + \frac{\partial \tau_{yy}}{\partial y} + \frac{\partial \tau_{zy}}{\partial z}\right) \\
& + w\left(\frac{\partial \tau_{xz}}{\partial x} + \frac{\partial \tau_{yz}}{\partial y} + \frac{\partial \tau_{zz}}{\partial z}\right) + \rho \mathbf{f} \cdot \mathbf{V}
\end{aligned}
$$

Remembering that $\rho \mathbf{f} \cdot \mathbf{V} = \rho\left(uf_x + vf_y + wf_z\right)$, substituting in Eq. 2.26 yields

$$
\begin{aligned}
\rho \frac{De}{Dt} = & \rho \dot{q} + \frac{\partial}{\partial x}\left(k\frac{\partial T}{\partial x}\right) + \frac{\partial}{\partial y}\left(k\frac{\partial T}{\partial y}\right) + \frac{\partial}{\partial z}\left(k\frac{\partial T}{\partial z}\right) \\
& - p\left(\frac{\partial u}{\partial x} + \frac{\partial v}{\partial y} + \frac{\partial w}{\partial z}\right) \\
& + \tau_{xx}\frac{\partial u}{\partial x} + \tau_{yx}\frac{\partial u}{\partial y} + \tau_{zx}\frac{\partial u}{\partial z} \\
& + \tau_{xy}\frac{\partial v}{\partial x} + \tau_{yy}\frac{\partial v}{\partial y} + \tau_{zy}\frac{\partial v}{\partial z} \\
& + \tau_{xz}\frac{\partial w}{\partial x} + \tau_{yz}\frac{\partial w}{\partial y} + \tau_{zz}\frac{\partial w}{\partial z}
\end{aligned}
\tag{2.30}
$$

which is *the energy conservation equation in non-conservative form in terms of internal energy e*. A careful observer will note the absence of the term due to volume forces from this form of the energy conservation equation. Remembering the expressions from (2.20) to (2.21), and in particular setting $\tau_{xy} = \tau_{yx}$, $\tau_{xz} = \tau_{zx}$, and $\tau_{zy} = \tau_{yz}$, we can express

$$
\begin{aligned}
\rho \frac{De}{Dt} = & \rho \dot{q} + \frac{\partial}{\partial x}\left(k\frac{\partial T}{\partial x}\right) + \frac{\partial}{\partial y}\left(k\frac{\partial T}{\partial y}\right) + \frac{\partial}{\partial z}\left(k\frac{\partial T}{\partial z}\right) \\
& - p\left(\frac{\partial u}{\partial x} + \frac{\partial v}{\partial y} + \frac{\partial w}{\partial z}\right) + \tau_{xx}\frac{\partial u}{\partial x} + \tau_{yy}\frac{\partial v}{\partial y} + \tau_{zz}\frac{\partial w}{\partial z} \\
& + \tau_{xy}\left(\frac{\partial u}{\partial y} + \frac{\partial v}{\partial x}\right) + \tau_{zx}\left(\frac{\partial u}{\partial z} + \frac{\partial w}{\partial x}\right) + \tau_{zy}\left(\frac{\partial v}{\partial z} + \frac{\partial w}{\partial y}\right)
\end{aligned}
$$

and again

$$
\begin{aligned}
\rho\frac{De}{Dt} &= \rho\dot{q} + \frac{\partial}{\partial x}\left(k\frac{\partial T}{\partial x}\right) + \frac{\partial}{\partial y}\left(k\frac{\partial T}{\partial y}\right) + \frac{\partial}{\partial z}\left(k\frac{\partial T}{\partial z}\right) \\
&- p\left(\frac{\partial u}{\partial x} + \frac{\partial v}{\partial y} + \frac{\partial w}{\partial z}\right) + \lambda\left(\frac{\partial u}{\partial x} + \frac{\partial v}{\partial y} + \frac{\partial w}{\partial z}\right)^2 \\
&+ \mu\left[2\left(\frac{\partial u}{\partial x}\right)^2 + 2\left(\frac{\partial v}{\partial y}\right)^2 + 2\left(\frac{\partial w}{\partial z}\right)^2\right. \\
&\left. + \left(\frac{\partial u}{\partial y} + \frac{\partial v}{\partial x}\right)^2 + \left(\frac{\partial u}{\partial z} + \frac{\partial w}{\partial x}\right)^2 + \left(\frac{\partial v}{\partial z} + \frac{\partial w}{\partial y}\right)^2\right]
\end{aligned} \tag{2.31}
$$

which is still the energy conservation equation in non-conservative form in terms of the internal energy e only, in which only flow variables appear. In order to obtain the conservative form of the energy conservation equation in terms of the internal energy e, the expression for the substantial derivative is considered:

$$
\rho\frac{De}{Dt} = \rho\frac{\partial e}{\partial t} + \rho\mathbf{V}\cdot\nabla e.
$$

Recalling that

$$
\frac{\partial(\rho e)}{\partial t} = \rho\frac{\partial e}{\partial t} + e\frac{\partial \rho}{\partial t} \quad \Rightarrow \quad \rho\frac{\partial e}{\partial t} = \frac{\partial(\rho e)}{\partial t} - e\frac{\partial \rho}{\partial t}
$$

and that

$$
\nabla\cdot(\rho e\mathbf{V}) = e\nabla\cdot(\rho\mathbf{V}) + \rho\mathbf{V}\cdot\nabla e \quad \Rightarrow \quad \rho\mathbf{V}\cdot\nabla e = \nabla\cdot(\rho e\mathbf{V}) - e\nabla\cdot(\rho\mathbf{V}),
$$

it is

$$
\rho\frac{De}{Dt} = \frac{\partial(\rho e)}{\partial t} - e\left[\frac{\partial\rho}{\partial t} + \nabla\cdot(\rho\mathbf{V})\right] + \nabla\cdot(\rho e\mathbf{V}). \tag{2.32}
$$

Noticing that the term in square brackets in Eq. 2.32 is null due to the continuity equation, we can express

$$
\rho\frac{De}{Dt} = \frac{\partial(\rho e)}{\partial t} + \nabla\cdot(\rho e\mathbf{V}) \tag{2.33}
$$

which, substituted into Eq. 2.31, yields the energy conservation equation in *conservative form* in terms of the internal energy e:

$$
\begin{aligned}
\frac{\partial(\rho e)}{\partial t} + \nabla \cdot (\rho e \mathbf{V}) = \rho \dot{q} &+ \frac{\partial}{\partial x}\left(k\frac{\partial T}{\partial x}\right) + \frac{\partial}{\partial y}\left(k\frac{\partial T}{\partial y}\right) + \frac{\partial}{\partial z}\left(k\frac{\partial T}{\partial z}\right) \\
&- p\left(\frac{\partial u}{\partial x} + \frac{\partial v}{\partial y} + \frac{\partial w}{\partial z}\right) + \lambda\left(\frac{\partial u}{\partial x} + \frac{\partial v}{\partial y} + \frac{\partial w}{\partial z}\right)^2 \\
&+ \mu\left[2\left(\frac{\partial u}{\partial x}\right)^2 + 2\left(\frac{\partial v}{\partial y}\right)^2 + 2\left(\frac{\partial w}{\partial z}\right)^2\right. \\
&\left.+ \left(\frac{\partial u}{\partial y} + \frac{\partial v}{\partial x}\right)^2 + \left(\frac{\partial u}{\partial z} + \frac{\partial w}{\partial x}\right)^2 + \left(\frac{\partial v}{\partial z} + \frac{\partial w}{\partial y}\right)^2\right].
\end{aligned}
\tag{2.34}
$$

The same procedure, applied to the total energy $e + \frac{V^2}{2}$, results in

$$
\rho \frac{D\left(e + \frac{V^2}{2}\right)}{Dt} = \frac{\partial}{\partial t}\left[\rho\left(e + \frac{V^2}{2}\right)\right] + \nabla \cdot \left[\rho\left(e + \frac{V^2}{2}\right)\mathbf{V}\right] \tag{2.35}
$$

which, substituted into Eq. 2.26, yields the energy conservation equation in *conservative form* in terms of the total internal energy $e + \frac{V^2}{2}$:

$$
\begin{aligned}
\frac{\partial}{\partial t}\left[\rho\left(e + \frac{V^2}{2}\right)\right] + \nabla \cdot \left[\rho\left(e + \frac{V^2}{2}\right)\mathbf{V}\right] = \rho \dot{q} & \\
&+ \frac{\partial}{\partial x}\left(k\frac{\partial T}{\partial x}\right) + \frac{\partial}{\partial y}\left(k\frac{\partial T}{\partial y}\right) + \frac{\partial}{\partial z}\left(k\frac{\partial T}{\partial z}\right) \\
&- \frac{\partial(up)}{\partial x} - \frac{\partial(vp)}{\partial y} - \frac{\partial(wp)}{\partial z} \\
&+ \frac{\partial(u\tau_{xx})}{\partial x} + \frac{\partial(u\tau_{yx})}{\partial y} + \frac{\partial(u\tau_{zx})}{\partial z} \\
&+ \frac{\partial(v\tau_{xy})}{\partial x} + \frac{\partial(v\tau_{yy})}{\partial y} + \frac{\partial(v\tau_{zy})}{\partial z} \\
&+ \frac{\partial(w\tau_{xz})}{\partial x} + \frac{\partial(w\tau_{yz})}{\partial y} + \frac{\partial(w\tau_{zz})}{\partial z} \\
&+ \rho \mathbf{f} \cdot \mathbf{V}.
\end{aligned}
\tag{2.36}
$$

Those shown here are not the only possible expressions of the equation of energy: an example is the representation as a function of enthalpy or total enthalpy, which is not reported here for brevity.

2.7 Considerations on the Governing Equations

The equations considered so far account for dissipative phenomena due to viscosity and to thermal conductivity. The fluid considered is homogeneous, and there are no chemical reactions; otherwise, it would be necessary to consider also additional conservation equations of mass and momentum related to the various chemical species present. The energy equation, too, would, in this case, show the presence of additional terms related to transport and diffusion of various chemical species. When it is possible to ignore the phenomena related to viscosity, the flow is said to be *non-viscous* or *inviscid*. Table 2.1 summarises the governing equations for compressible, three-dimensional, non-viscous, and unsteady flows: it is evident that such equations are derivable from those seen in the previous sections by eliminating the terms due to viscosity.

Some observations.

- The governing equations analysed so far constitute a system of differential non-linear equations for which there is no analytical solution.
- In the case of momentum and energy conservation equation the two conservative and non-conservative forms differ only in the terms on the left hand side.

Table 2.1 Summary table of governing equations for non-viscous flows

Equation	Non-conservative form	Conservative form
Conservation of mass	$\frac{D\rho}{Dt} + \rho\nabla\cdot\mathbf{V} = 0$	$\frac{\partial\rho}{\partial t} + \nabla\cdot(\rho\mathbf{V}) = 0$
Conservation of momentum (in x)	$\rho\frac{Du}{Dt} = -\frac{\partial p}{\partial x} + \rho f_x$	$\frac{\partial(\rho u)}{\partial t} + \nabla\cdot(\rho u\mathbf{V}) = -\frac{\partial p}{\partial x} + \rho f_x$
Conservation of momentum (in y)	$\rho\frac{Dv}{Dt} = -\frac{\partial p}{\partial y} + \rho f_y$	$\frac{\partial(\rho v)}{\partial t} + \nabla\cdot(\rho v\mathbf{V}) = -\frac{\partial p}{\partial y} + \rho f_y$
Conservation of momentum (in z)	$\rho\frac{Dw}{Dt} = -\frac{\partial p}{\partial z} + \rho f_z$	$\frac{\partial(\rho w)}{\partial t} + \nabla\cdot(\rho w\mathbf{V}) = -\frac{\partial p}{\partial z} + \rho f_z$
Energy conservation	$\rho\frac{D}{Dt}\left(e + \frac{V^2}{2}\right) =$	$\frac{\partial}{\partial t}\left[\rho\left(e + \frac{V^2}{2}\right)\right] + \nabla\cdot\left[\rho\left(e + \frac{V^2}{2}\right)\mathbf{V}\right] =$
	$\rho\dot{q} - \frac{\partial(up)}{\partial x} - \frac{\partial(vp)}{\partial y} - \frac{\partial(wp)}{\partial z} + \rho\mathbf{f}\cdot\mathbf{V}$	$\rho\dot{q} - \frac{\partial(up)}{\partial x} - \frac{\partial(vp)}{\partial y} - \frac{\partial(wp)}{\partial z} + \rho\mathbf{f}\cdot\mathbf{V}$

- The conservative form always contains a divergence term among those on the left hand side—for example $\nabla \cdot (\rho \mathbf{V})$ or $\nabla \cdot (\rho u \mathbf{V})$. For this reason, the equations in conservative form are also called *equations in divergence form.*
- Normal and shear stresses depend on the velocity gradient.
- The system constituted by the governing equations consists of five equations in six unknowns ($\rho,\ p,\ \rho u,\ \rho v,\ \rho w,\ e$). The additional equation to consider is the state equation which, in the case of a perfect gas is

$$p = \rho R T,$$

where R is the constant of the considered gas. This relationship introduces a further unknown T which corresponds to a further equation that closes the system and which is represented by a thermodynamic relationship between the state variables:

$$e = c_v T$$

where c_v is the specific heat at constant volume for the considered gas.
- In Sect. 2.5 the momentum conservation equations for non-stationary compressible viscous flows are defined as the Navier–Stokes equations. Although this is historically correct, in modern literature, when talking about numerical solution of Navier–Stokes equations, it means the solution of the system formed by all the governing equations including the continuity equation and conservation of energy for an unsteady compressible viscous flow.

2.8 Further Insights on the Conservative Form

It is worth recalling that, by definition, the *flux of a certain quantity* is the measure of that quantity that crosses a unit of surface area in the unit of time. For example, the mass flux will be dimensionally equivalent to a mass per unit of time and surface area:

$$[mass \quad flux] = \frac{kg}{m^2 s}.$$

It is immediately clear that the mass flux does not coincide with the *mass flow rate.* Multiplying the mass flux by a surface area gives the mass flow rate, which can therefore be interpreted as the flux through a non-unitary surface. Proceeding with the dimensional analysis of the term ρu, we obtain

$$\rho u = \frac{kg}{m^3}\frac{m}{s} = \frac{kg}{m^2 s} = [mass \quad flux]$$

that is, the momentum per unit volume can be interpreted as a mass flux. The dimensional analysis of the term ρu^2 leads to

$$\rho u^2 = \rho u u = \frac{kg}{m^3}\frac{m}{s}\frac{m}{s} = \frac{kg\frac{m}{s}}{m^2 s} = [flux \quad of \quad momentum].$$

As for the dimensional analysis of the pressure term p, we obtain

$$\begin{aligned} p &= \frac{force}{surface} = \frac{mass \cdot acceleration}{surface} \\ &= kg\frac{m}{s^2}\frac{1}{m^2} = \frac{kg\frac{m}{s}}{m^2 s} = \left[flux \quad of \quad momentum\right]. \end{aligned}$$

This is an opportunity to highlight how a force can be interpreted as a flow rate of momentum:

$$\left[force\right] = \left[pressure\right] \cdot \left[surface\right] = \left[flux \quad of \quad momentum\right] \cdot \left[surface\right].$$

With these premises, it is observed once again that, in the conservative form of the governing equations, the divergence of the flux of a certain quantity always appears:

- the mass conservation equation contains the divergence of the mass flux $\rho\mathbf{V}$;
- the momentum conservation equation along x contains the divergence of the flux $\rho u\mathbf{V}$ of the component along x of the momentum;
- the conservation equation of momentum along y contains the divergence of the flux $\rho v\mathbf{V}$ of the component along x of the momentum;
- the conservation equation of momentum along z contains the divergence of the flux $\rho w\mathbf{V}$ of the component along x of the momentum;
- the energy conservation equation such flux contains the divergence of the flux $\rho\left(e + \frac{V^2}{2}\right)\mathbf{V}$ of the energy.

It is recalled here that only by considering a fixed control volume in space the conservative form of the governing equations can be obtained from the application of the concepts of conservation of mass, momentum (Newton's second law), and energy (first law of thermodynamics). Therefore, in the case where a fixed control volume in space is considered, the governing equations will have as dependent variables the fluxes rather than the primitive variables such as pressure, density, velocity, etc. Based on this last consideration, the governing equations can be more clearly understood if written in conservative form (so-called because only the conserved variables appear, i.e., the fluxes rather than the primitive variables). The governing equations are typically written as

$$\frac{\partial \mathbf{U}}{\partial t} + \frac{\partial \mathbf{F}}{\partial x} + \frac{\partial \mathbf{G}}{\partial y} + \frac{\partial \mathbf{H}}{\partial z} = \mathbf{J}. \tag{2.37}$$

This equation is able to represent the entire system of Navier-Stokes equations setting

$$\mathbf{U} = \begin{bmatrix} \rho \\ \rho u \\ \rho v \\ \rho w \\ \rho\left(e + \frac{V^2}{2}\right) \end{bmatrix};$$

$$\mathbf{F} = \begin{bmatrix} \rho u \\ \rho u^2 + p - \tau_{xx} \\ \rho v u - \tau_{xy} \\ \rho w u - \tau_{xz} \\ \rho\left(e + \frac{V^2}{2}\right) u + p u - k\frac{\partial T}{\partial x} - u\tau_{xx} - v\tau_{xy} - w\tau_{xz} \end{bmatrix};$$

$$\mathbf{G} = \begin{bmatrix} \rho v \\ \rho u v - \tau_{yx} \\ \rho u^2 + p \\ \rho u v \\ \rho u w \\ \rho\left(e + \frac{V^2}{2}\right) u + p u \end{bmatrix};$$

$$\mathbf{G} = \begin{bmatrix} \rho v \\ \rho u v \\ \rho v^2 + p \\ \rho v w \\ \rho\left(e + \frac{V^2}{2}\right) v + p v \end{bmatrix};$$

$$\mathbf{H} = \begin{bmatrix} \rho w \\ \rho u w \\ \rho v w \\ \rho w^2 + p \\ \rho\left(e + \frac{V^2}{2}\right) w + p w \end{bmatrix};$$

$$\mathbf{J} = \begin{bmatrix} 0 \\ \rho f_x \\ \rho f_y \\ \rho f_z \\ \rho\left(u f_x + v f_y + w f_z\right) + \rho\dot{q} \end{bmatrix}.$$

The terms **F**, **G**, **H** in Eq. 2.37 are referred to as *flux terms or flux vectors*. The term **J** is called *source term* (which is zero if the volume forces are negligible). The term **U** is referred to as *solution vector* because its elements are the dependent variables whose values are the result of numerical iterative solution methods. Once the elements of **U**—also called *conserved variables*—are known, the value of the primitive variables can be obtained using the following relationships:

$$\rho = \rho; \qquad u = \frac{\rho u}{\rho}; \qquad v = \frac{\rho v}{\rho}; \qquad w = \frac{\rho w}{\rho}; \qquad e = \frac{\rho\left(e + V^2/2\right)}{\rho} - \frac{u^2 + v^2 + w^2}{2}.$$

In the case of inviscid flows, the terms in Eq. 2.37 become

$$\mathbf{U} = \begin{bmatrix} \rho \\ \rho u \\ \rho v \\ \rho w \\ \rho\left(e + \frac{V^2}{2}\right) \end{bmatrix};$$

$$\mathbf{F} = \begin{bmatrix} \rho u \\ \rho u^2 + p \\ \rho v u \\ \rho w u \\ \rho\left(e + \frac{V^2}{2}\right) u + p u \end{bmatrix}; \quad \mathbf{G} = \begin{bmatrix} \rho v \\ \rho u v \\ \rho v^2 + p \\ \rho w v \\ \rho\left(e + \frac{V^2}{2}\right) v + p v \end{bmatrix};$$

$$\mathbf{H} = \begin{bmatrix} \rho w \\ \rho u w \\ \rho v w \\ \rho w^2 + p \\ \rho\left(e + \frac{V^2}{2}\right) w + p w \end{bmatrix}; \quad \mathbf{J} = \begin{bmatrix} 0 \\ \rho f_x \\ \rho f_y \\ \rho f_z \\ \rho\left(u f_x + v f_y + w f_z\right) + \rho \dot{q} \end{bmatrix}.$$

The conservative form of the Navier-Stokes equations shown with the Eq. 2.37 is referred to as *strong* because all the flux variables always appear solely as an argument of the derivative sign. Conversely, the other conservative forms discussed are called *weak* because the flux variables also appear outside the derivative sign. The representation in strong conservative form is particularly important in the case of compressible flows which are characterised by the presence of discontinuities in primitive variables such as shocks. In the case where such discontinuities manifest in the computational domain without specific detection techniques being incorporated into the general calculation algorithm, we speak of *shock-capturing methods*. An alternative approach involves the explicit introduction in the computational domain of discontinuities and the use of Rankine-Hugoniot relations to calculate the value of the primitive quantities upstream and downstream of the shock; the Navier-

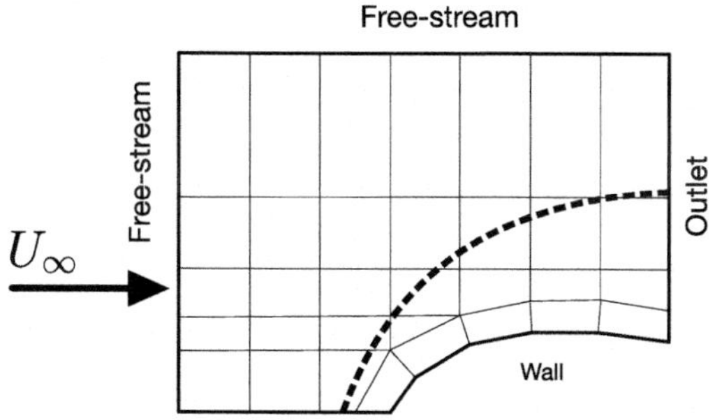

Fig. 2.17 Computational domain for the case of shock-capturing methods

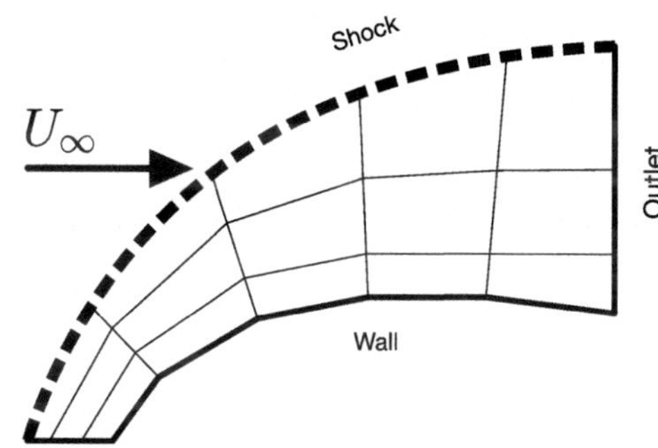

Fig. 2.18 Computational domain for the case of shock-fitting methods

Stokes equations are then used to determine the solution of the flow in the remaining parts of the computational domain. We speak in this case of *shock-fitting* methods. Figures 2.17 and 2.18 illustrate the computational domains corresponding to the two approaches.

The use of the conservative form proves to be of fundamental importance in shock-capturing methods. Considering a non-viscous flow in which a shock is present and referring to Fig. 2.19, it can be observed that, across the shock, the primitive variables, such as density or pressure, exhibit a discontinuity. If one were to consider, using the non-conservative form, the primitive variables as dependent variables, the presence of discontinuity would lead to "unstable" calculations (the iterative algorithm would diverge) as well as to incorrect results. On the contrary, the conserved variables such as the fluxes ρu or $(p + \rho u^2)$ remain unchanged across the shock: the conservative form of the Navier-Stokes equations does not recognise discontinuity of the dependent variables thereby significantly improving the stability of the iterative process and the accuracy of the results.

2.9 General Transport Equation

Referring to Sect. 2.7, it is noted that, for fluids consisting of a single substance, the governing equations in conservative form can be written as the following balance

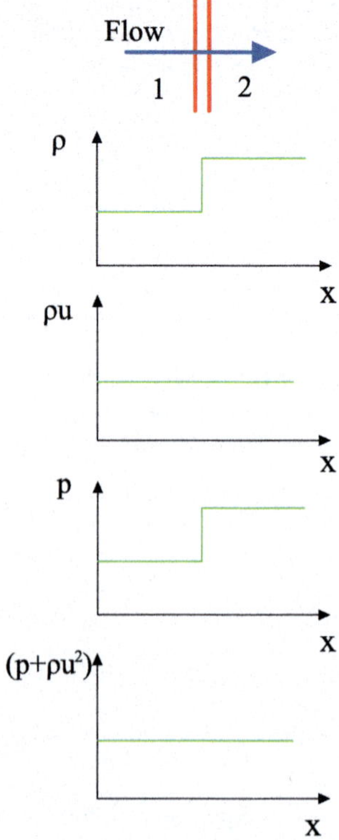

Fig. 2.19 Variation of properties across a shock

equation known as *general transport equation in differential form*. Introducing the symbol ϕ to represent the generic quantity being transported, the general transport equation in differential form is

$$\underbrace{\frac{\partial \rho \phi}{\partial t}}_{\textit{Temporal variation}} + \overbrace{\nabla \cdot (\rho \mathbf{u} \phi)}^{\textit{Convection}} = \overbrace{\nabla \cdot (\rho \Gamma_{\phi} \nabla \phi)}^{\textit{Diffusion}} + \overbrace{S_{\phi}(\phi)}^{\textit{Source}} \tag{2.38}$$

where ρ is the density, $\mathbf{u}$ is the velocity of the fluid that carries the quantity ϕ with its three components u, v, w, Γ_{ϕ} is the diffusion coefficient (viscosity μ or thermal diffusivity α) of ϕ, S_{ϕ}, with its three components S_u, S_v, S_w, is the generation/destruction of ϕ within the fluid element. Equation 2.38 represents the fact that the sum of the time variation of ϕ within the fluid element and the net outgoing/incoming flux due to conduction from/into the element of fluid, is equal to the variation of ϕ within the fluid element due to the net outgoing/incoming flux due to diffusion from/into the element of fluid plus the variation, caused by the presence of sources or sinks, of ϕ. By setting $\phi = 1$, $\Gamma_{\phi} = 0$, $S_{\phi} = 0$ it is possible to derive the continuity equation:

$$\frac{\partial \rho}{\partial t} + \nabla \cdot (\rho \mathbf{u}) = 0.$$

By setting $\phi = u$, $\Gamma_\phi = \mu$, $S_\phi = S_u - \frac{\partial p}{\partial x}$ it is possible to derive the first component of the momentum conservation equation:

$$\frac{\partial \rho u}{\partial t} + \nabla \cdot (\rho \mathbf{u} u) = \nabla \cdot (\mu \nabla u) - \frac{\partial p}{\partial x} + S_u. \tag{2.39}$$

By setting $\phi = v$, $\Gamma_\phi = \mu$, $S_\phi = S_v - \frac{\partial p}{\partial y}$ it is possible to derive the second component of the momentum conservation equation:

$$\frac{\partial \rho v}{\partial t} + \nabla \cdot (\rho \mathbf{u} v) = \nabla \cdot (\mu \nabla v) - \frac{\partial p}{\partial y} + S_v. \tag{2.40}$$

By setting $\phi = w$, $\Gamma_\phi = \mu$, $S_\phi = S_w - \frac{\partial p}{\partial z}$ it is possible to derive the third component of the momentum conservation equation:

$$\frac{\partial \rho w}{\partial t} + \nabla \cdot (\rho \mathbf{u} w) = \nabla \cdot (\mu \nabla w) - \frac{\partial p}{\partial z} + S_w. \tag{2.41}$$

In Eqs. 2.39, 2.40, 2.41 the symbol μ denotes the dynamic viscosity which is related to the *kinematic viscosity* ν by the relation $\nu = \mu/\rho$. The expression of the three components of momentum conservation equation can be compacted using the vector notation:

$$\frac{\partial \rho \mathbf{u}}{\partial t} + \nabla \cdot (\rho \mathbf{u} \mathbf{u}) = \nabla \cdot (\mu \nabla \mathbf{u}) - \nabla p + \mathbf{S}. \tag{2.42}$$

In the case where the density can be considered constant and the term $\mathbf{S}$ is absent, Eq. 2.42 becomes:

$$\frac{\partial \mathbf{u}}{\partial t} + \nabla \cdot (\mathbf{u} \mathbf{u}) = \nabla \cdot (\nu \nabla \mathbf{u}) - \nabla \left(\frac{p}{\rho} \right). \tag{2.43}$$

The term $\frac{p}{\rho}$ is referred to as *kinematic pressure*. Setting $\phi = h$, $\Gamma_\phi = \frac{k}{C_p}$, $S_\phi = S_h$ it is possible to derive the energy conservation equation:

$$\frac{\partial \rho h}{\partial t} + \nabla \cdot (\rho \mathbf{u} h) = \nabla \cdot \left(\frac{k}{C_p} \nabla T \right) + S_h,$$

having indicated with h the specific enthalpy, with k the thermal conductivity, with C_p the specific heat at constant pressure, with T the absolute temperature. The term S_h contains in itself contributions such as viscous dissipation, the time and spatial variation of pressure, all terms derived from the application of the conservation of total energy to the control volume considered and not explicitly reported here. To bet-

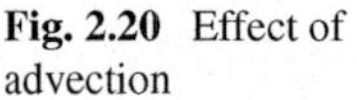

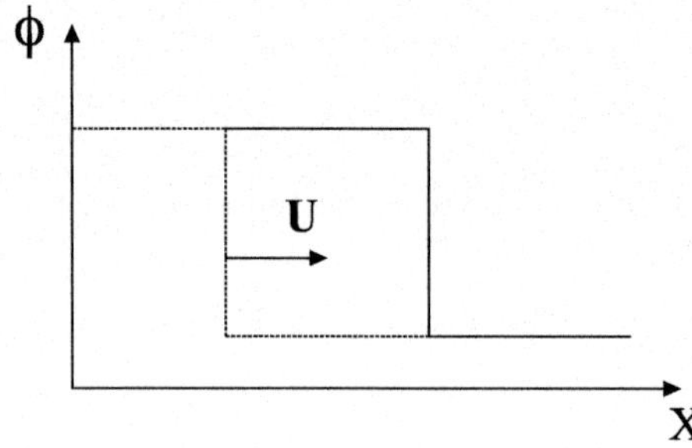

Fig. 2.20 Effect of advection

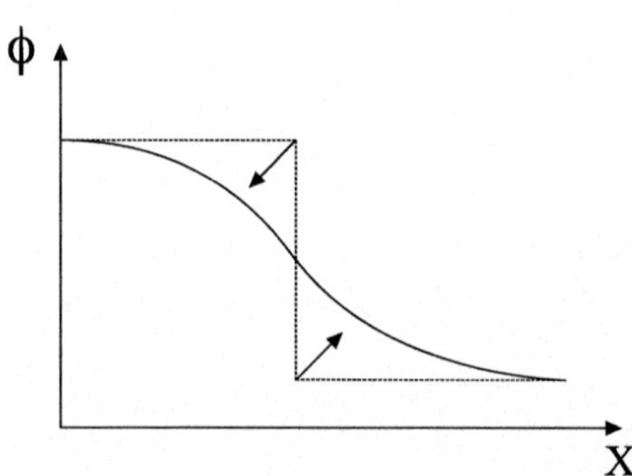

Fig. 2.21 Effect of diffusion

ter understand the phenomena described by the various terms of Eq. 2.38, Figs. 2.20 and 2.21 illustrate the effects of advection (the term $\nabla \cdot (\rho \mathbf{u} \phi)$) and diffusion (the term $\nabla \cdot (\rho \Gamma_\phi \nabla \phi)$) on a quantity ϕ whose initial distribution is represented by the dashed line:

- in the case of advection the profile is not modified in shape but only in position by a quantity equal to the distance travelled in time t due to the effect of the velocity field $\mathbf{u}$;
- in the case of diffusion the profile does not undergo any displacements but only deformations related to the decrease in gradients.

See Sect. 1.3.3 for the definitions of advection and convection.

Chapter 3
The Finite Volume Method

Given a computational domain with known initial and boundary conditions, the objective is to solve the general transport equation for a quantity ϕ, which is presented here for simplicity.

$$\underbrace{\frac{\partial \rho \phi}{\partial t}}_{Temporal\ variation} + \overbrace{\nabla \cdot (\rho \mathbf{u} \phi)}^{Convection} = \overbrace{\nabla \cdot (\rho \Gamma_\phi \nabla \phi)}^{Diffusion} + \overbrace{S_\phi(\phi)}^{Source} . \tag{3.1}$$

This is a second-order partial differential equation. To achieve an acceptable degree of accuracy in its numerical resolution, the employed discretisation scheme must have a degree of accuracy equal to or higher than that of the equation. To satisfy this constraint, it is necessary to assume a linear variation of the quantity ϕ in space and time in the vicinity of the generic point P and time t under consideration, respectively. By considering a Taylor series expansion (see Sect. 1.5.1) in the vicinity of point P, accurate to the second order, we obtain

$$\phi(\mathbf{x}) = \phi_P + (\mathbf{x} - \mathbf{x}_P) \cdot (\nabla \phi)_P \qquad with \qquad \phi_P = \phi(\mathbf{x}_P)$$

$$\phi(t + \delta t) = \phi^t + \delta t \left(\frac{\partial \phi}{\partial t} \right)^t \qquad with \qquad \phi^t = \phi(t).$$

In the finite volume method, the computational domain is divided (*discretised*) into an arbitrary and finite number of *control volumes* or *cells*. These control volumes can have any shape, with the only constraint that the surfaces (*faces*) delimiting them must be flat. Information such as the position of the centroid of each control volume, the position of the centroid of each face of the surface delimiting the control volume, the volume of each cell, the area of each face, and the cells to which each face belongs will also be known. Therefore, all the necessary geometric information will be available. All variables will be calculated and stored at the centroids (*centres*)

G. Caramia and E. Distaso, *A Practical Approach to Computational Fluid Dynamics Using OpenFOAM®*, https://doi.org/10.1007/978-3-031-88957-8_3

of the cells, within which the values of the calculated variables will be considered constant.

The finite volume method[1] applies the conservation principles discussed in Sects. 2.4, 2.5, and 2.6 to each of the cells into which the computational domain has been discretised. Denoting by V_P the volume of the generic cell centred at P, it is possible to integrate Eq. 3.1 with respect to V_P to obtain

$$\int_{V_P} \frac{\partial \rho\phi}{\partial t} dV + \int_{V_P} \nabla \cdot (\rho \mathbf{u} \phi) \, dV = \int_{V_P} \nabla \cdot (\rho \Gamma_\phi \nabla \phi) \, dV + \int_{V_P} S_\phi(\phi) \, dV.$$

Then, using the divergence theorem (see Sect. 1.1.6), it is possible to transform the volume integrals into surface integrals (see also Sect. 2.8).

$$\frac{\partial}{\partial t} \int_{V_P} (\rho\phi) \, dV + \oint_{\partial V_P} d\mathbf{S} \cdot (\rho \mathbf{u} \phi) - \oint_{\partial V_P} d\mathbf{S} \cdot (\rho \Gamma_\phi \nabla \phi) = \int_{V_P} S_\phi(\phi) \, dV.$$

From this, a second-order approximate expression can be derived using the midpoint integration rule (see Sect. 1.5.2). Regarding the convective term, we have

$$\begin{aligned} \int_{V_P} \nabla \cdot (\rho \mathbf{u} \phi) \, dV &= \oint_{\partial V_P} d\mathbf{S} \cdot \underbrace{(\rho \mathbf{u} \phi)}_{convective\ \ flux} \\ &= \sum_f \int_f d\mathbf{S} \cdot (\rho \mathbf{u} \phi)_f \approx \sum_f \mathbf{S}_f \cdot \overline{(\rho \mathbf{u} \phi)}_f \\ &= \sum_f \mathbf{S}_f \cdot (\rho \mathbf{u} \phi)_f \end{aligned} \tag{3.2}$$

where $(\rho \mathbf{u} \phi)_f$ represents the value of the quantity $\rho \mathbf{u} \phi$ at the centre of face f. For the diffusive term, on the other hand,

$$\begin{aligned} \int_{V_P} \nabla \cdot (\rho \Gamma_\phi \nabla \phi) \, dV &= \oint_{\partial V_P} d\mathbf{S} \cdot \underbrace{(\rho \Gamma_\phi \nabla \phi)}_{diffusive\ \ flux} \\ &= \sum_f \int_f d\mathbf{S} \cdot (\rho \Gamma_\phi \nabla \phi)_f \approx \sum_f \mathbf{S}_f \cdot \overline{(\rho \Gamma_\phi \nabla \phi)}_f \\ &= \sum_f \mathbf{S}_f \cdot (\rho \Gamma_\phi \nabla \phi)_f \end{aligned} \tag{3.3}$$

[1] Akshai Runchal, Transfer Processes in Steady Two-Dimensional Separated Flows, Ph.D. thesis, Faculty of Engineering, University of London, 1969.

where $(\rho\Gamma_\phi\nabla\phi)_f$ represents the value of the quantity $\rho\Gamma_\phi\nabla\phi$ at the centre of face f. For the source term, one can distinguish between the constant contribution S_c and the linear contribution S_p.

$$\int_{V_P} S_\phi(\phi)\,dV = \overline{S}_\phi \Delta V = S_c V_P + S_p V_P \phi_P. \tag{3.4}$$

Observing Eqs. 3.2 and 3.3, it is clear that it is necessary to determine the convective and diffusive fluxes at the centroids of the cell faces as functions of the values assumed at the centroids of the cells to which the faces belong. Below, specific interpolation methods used for this purpose are illustrated.

3.1 Convective-Diffusive Fluxes

The equation describing the transport of a quantity ϕ in a steady flow, accounting for both diffusive and convective phenomena, can be derived from the general transport Eq. 3.1 by neglecting the time derivative term:

$$\nabla \cdot (\rho \mathbf{u} \phi) = \nabla \cdot (\Gamma_\phi \nabla \phi) + S_\phi \tag{3.5}$$

which, in the one-dimensional case with flow velocity u, takes the form

$$\frac{d}{dx}(\rho u \phi) = \frac{d}{dx}\left(\Gamma_\phi \frac{d\phi}{dx}\right) + S_{\phi_u}. \tag{3.6}$$

The component in the x-axis direction, S_{ϕ_u}, of the source term is a function of the variable ϕ. Specifically, a linear approximation can be considered, as shown in Eq. 3.4. In the absence of sources, the source term S_{ϕ_u} is zero. In addition to the transport equation, the conservation of mass equation must also be satisfied:

$$\frac{d\,(\rho u)}{dx} = 0. \tag{3.7}$$

Referring to Fig. 3.1 and integrating equation 3.6 over the cell centred at P, we obtain

$$(\rho u \phi A)_e - (\rho u \phi A)_w = \left(\Gamma_\phi \frac{d\phi}{dx} A\right)_e - \left(\Gamma_\phi \frac{d\phi}{dx} A\right)_w + (S_u + S_P \phi_P) \tag{3.8}$$

where the subscripts e and w indicate the values of the quantity at the right and left borders, respectively, of the cell centred at P. Although there are no sources of the quantity ϕ in the analysed problem, the terms related to the linearised source term

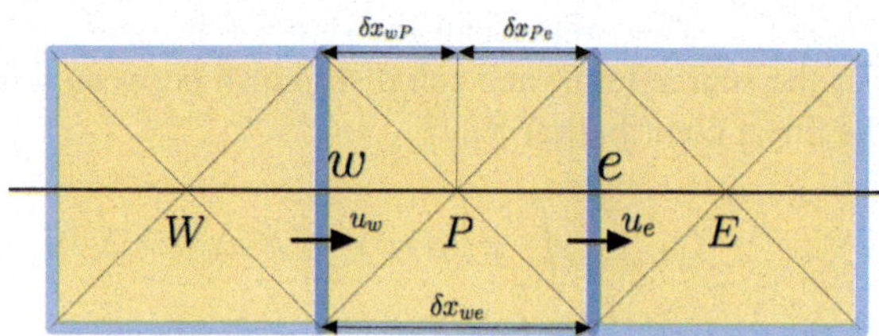

Fig. 3.1 Cell P centre

$(S_u + S_P\phi_P)$ are included in Eq. 3.8 to allow, as shown in the numerical example below, the correct assignment of boundary conditions. Integrating equation 3.7, we obtain

$$(\rho u A)_e - (\rho u A)_w = 0. \tag{3.9}$$

We can define the flux (flow rate per unit surface area, see Sect. 2.8) of convective mass as

$$F = \rho u$$

and the diffusive mass flux (diffusive conductance) as

$$D = \frac{\Gamma}{\delta x}$$

in which the subscript ϕ for the quantity Γ has been omitted to simplify the notation. Therefore, at the left border of the cell centred at P, it will be

$$F_w = (\rho u)_w \qquad D_w = \frac{\Gamma_w}{\delta x_{WP}}.$$

whereas at the right border, it will be

$$F_e = (\rho u)_e \qquad D_e = \frac{\Gamma_e}{\delta x_{PE}}.$$

Equations 3.8 and 3.9 can be discretised by assuming, for simplicity, that $A = A_w = A_e$ and using the centred difference approach to determine the contribution of the diffusive terms. Note that the elimination of the term related to the area of the faces from Eq. 3.8 was possible only because there are no source terms, terms derived from the discretisation of time derivatives, or terms arising from a discretisation that does not involve the calculation of surface integrals. Therefore, the discretised form of Eq. 3.8 becomes

$$F_e\phi_e - F_w\phi_w = D_e(\phi_E - \phi_P) - D_w(\phi_P - \phi_W) + (S_u + S_P\phi_P) \tag{3.10}$$

and the following for Eq. 3.9

$$F_e - F_w = 0. \tag{3.11}$$

The advection velocity values u at the cell boundaries are generally calculated using simple linear interpolation between the values at the two adjacent cell centres. It is now clear that, to solve Eq. 3.10, it is necessary to compute the values of the transported quantity, ϕ_w and ϕ_e, at the cell boundaries. Various methods for determining these values will be illustrated below.

3.1.1 Linear Interpolation or Central Differencing

Initially, only the value ϕ_e is considered, which can be calculated by performing a simple linear interpolation starting from the values at the two cell centres, P and E (see Fig. 3.2).

In practice, it will be

$$\phi_e = f_x \phi_P + (1 - f_x)\phi_E \qquad with \qquad f_x = \frac{eE}{PE} = \frac{|\,\mathbf{x}_e - \mathbf{x}_E\,|}{|\,\mathbf{d}\,|}.$$

Considering, for simplicity, a grid with uniform spacing, it will be

$$\phi_e = \frac{\phi_P + \phi_E}{2} \quad , \quad \phi_w = \frac{\phi_W + \phi_P}{2}.$$

Substituting these values into Eq. 3.10 leads to writing

$$\frac{F_e}{2}(\phi_P + \phi_E) - \frac{F_w}{2}(\phi_W + \phi_P) = D_e(\phi_E - \phi_P) - D_w(\phi_P - \phi_W) + (S_u + S_P\phi_P)$$

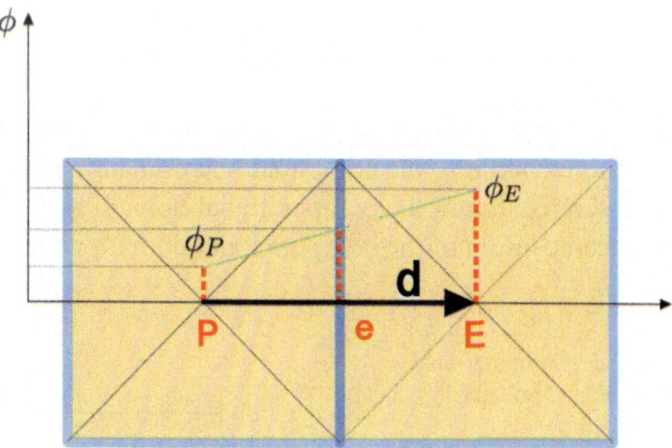

Fig. 3.2 Linear interpolation or central difference

that is,

$$\left[\left(D_w + \frac{F_w}{2}\right) + \left(D_e - \frac{F_e}{2}\right) + (F_e - F_w) - S_P\right]\phi_P$$
$$= \left(D_w + \frac{F_w}{2}\right)\phi_W + \left(D_e - \frac{F_e}{2}\right)\phi_E + S_u. \tag{3.12}$$

Setting

$$a_P = \left[\left(D_w + \frac{F_w}{2}\right) + \left(D_e - \frac{F_e}{2}\right) + (F_e - F_w) - S_P\right],$$
$$a_W = \left(D_w + \frac{F_w}{2}\right),$$
$$a_E = \left(D_e - \frac{F_e}{2}\right),$$

it will be possible to express Eq. 3.12 in a more compact form as

$$a_P\phi_P = a_W\phi_W + a_E\phi_E + S_u. \tag{3.13}$$

When written for each cell of the considered one-dimensional domain, Eq. 3.13 results in a system of algebraic equations, the solution of which represents the distribution of the transported quantity ϕ in terms of the value at the centre of each discretisation cell.

3.1.1.1 Numerical Example

Consider the one-dimensional domain shown in Fig. 3.3, within which a scalar quantity ϕ is transported with velocity u in the presence of both convective and diffusive phenomena.

The governing equation is Eq. 3.6, and the boundary conditions to be set are $\phi_0 = 0$ for $x = 0$ and $\phi_L = 1$ for $x = L$. The task is to compute the distribution of ϕ as a function of the coordinate x using the central differencing scheme in the following cases:

1. discretisation of the computational domain with 5 nodes and $u = 0.1$ m/s;
2. discretisation of the computational domain with 5 nodes and $u = 2.5$ m/s;
3. discretisation of the computational domain with 20 nodes and $u = 2.5$ m/s;

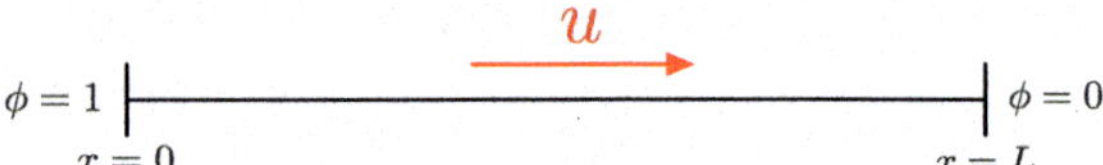

Fig. 3.3 Computational domain and boundary conditions

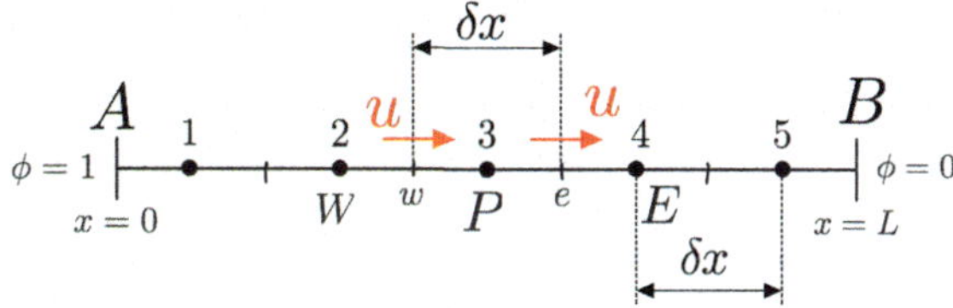

Fig. 3.4 Discretised computational domain

4. compare the results obtained in each of the previous cases with the analytical solution given by

$$\frac{\phi - \phi_0}{\phi_L - \phi_0} = \frac{\exp(\rho u x/\Gamma) - 1}{\exp(\rho u L/\Gamma) - 1}. \tag{3.14}$$

Further data necessary for the solution are: $\rho = 1\,\text{kg/m}^3, L = 1\,\text{m}, \Gamma = 0.1\,\text{kg/(ms)}$. It should be noted that, although it is constant, the density term is still maintained in the equations considered below to avoid complicating aspects related to dimensional analysis. Considering the provided data and observing Fig. 3.4, it can be stated that all cells will have the same values for the following quantities: $F_e = F_w = F = \rho u$, $D_e = D_w = D = \Gamma/\delta x$.

Figure 3.4 represents the case of a computational domain discretised into five cells. Equation 3.13 can be applied to the internal cells 2, 3, and 4, while the boundary cells 1 and 5 require a slightly different approach. Considering cell 1, it is noted that the face w corresponds to the boundary on which the value of the quantity ϕ is specified; therefore, no calculation is necessary to determine its value. The same applies to the face e of cell 5. Therefore, for cell 1, it can be written

$$\frac{F_e}{2}(\phi_P + \phi_E) - F_A\phi_A = D_e(\phi_E + \phi_P) - D_A(\phi_P - \phi_A) \tag{3.15}$$

and for cell 5

$$F_B\phi_B - \frac{F_w}{2}(\phi_P + \phi_W) = D_B(\phi_B + \phi_P) - D_w(\phi_P - \phi_W). \tag{3.16}$$

Considering that $D_A = D_B = 2\Gamma/\delta x = 2D$ and that $F_A = F_B = F$ Eqs. 3.15 and 3.16 can be written in a compact form as

$$a_P\phi_P = a_W\phi_W + a_E\phi_E + S_u \tag{3.17}$$

with

$$a_P = a_W + a_E + (F_E - F_W) - S_P. \tag{3.18}$$

Notice that the expression for a_P differs depending on whether the cell is on the boundary or not.

Table 3.1 summarises the expression of the coefficients for the cells.

Table 3.2 shows the numerical values of the coefficients for the cells in case 1 (symbol $C1$) for which is $u = 0.1m/s$, $F = \rho u = 0.1$, $D = \Gamma/\delta x = 0.1/0.2 = 0.5$.

Table 3.1 Expression of the coefficients for the cells

Node	a_W	a_E	S_P	S_u
1	0	$D - F/2$	$-(2D + F)$	$(2D + F)\phi_A$
2, 3, 4	$D + F/2$	$D - F/2$	0	0
5	$D + F/2$	0	$-(2D - F)$	$(2D - F)\phi_B$

Table 3.2 Value of the coefficients for the cells

Node	a_W		a_E		S_u		S_P		a_P	
	$C1$	$C2$	$C1$	$C2$	$C1$	$C2$	$C1$	$C2$	$C1$	$C2$
1	0		0.45	−0.75	$1.1\phi_A$	$3.5\phi_A$	−1.1	−3.5	1.55	2.75
2	1.75	−0.75	0.45	−0.75	0		0		1	
3	1.75	−0.75	0.45	−0.75	0		0		1	
4	1.75	−0.75	0.45	−0.75	0		0		1	
5	1.75	0	0		$0.9\phi_B$	$-1.5\phi_B$	−0.9	1.5	1.45	0.25

Therefore, applying Eq. 3.17 to each of the five cells, starting from cell 1 up to cell 5, the following system of algebraic equations can be written:

$$\begin{cases} 1.55\phi_1 - 0.45\phi_2 + 0\phi_3 + 0\phi_4 + 0\phi_5 & = 1.1 \\ -0.55\phi_1 + 1\phi_2 - 0.45\phi_3 + 0\phi_4 + 0\phi_5 & = 0 \\ 0\phi_1 - 0.55\phi_2 + 1\phi_3 - 0.45\phi_4 + 0\phi_5 & = 0 \\ 0\phi_1 + 0\phi_2 - 0.550\phi_3 + 1\phi_4 - 0.45\phi_5 & = 0 \\ 0\phi_1 + 0\phi_2 + 0\phi_3 - 0.55\phi_4 + 1.45\phi_5 & = 0 \end{cases}$$

which, in matrix equation form, $\mathbf{A}\phi = \mathbf{b}$ becomes

$$\begin{bmatrix} 1.55 & -0.45 & 0 & 0 & 0 \\ -0.55 & 1 & -0.45 & 0 & 0 \\ 0 & -0.55 & 1 & -0.45 & 0 \\ 0 & 0 & -0.55 & 1 & -0.45 \\ 0 & 0 & 0 & -0.55 & 1.45 \end{bmatrix} \begin{bmatrix} \phi_1 \\ \phi_2 \\ \phi_3 \\ \phi_4 \\ \phi_5 \end{bmatrix} = \begin{bmatrix} 1.1 \\ 0 \\ 0 \\ 0 \\ 0 \end{bmatrix}$$

whose solution is

$$\begin{bmatrix} \phi_1 \\ \phi_2 \\ \phi_3 \\ \phi_4 \\ \phi_5 \end{bmatrix} = \begin{bmatrix} 0.9421 \\ 0.8006 \\ 0.6276 \\ 0.4163 \\ 0.1579 \end{bmatrix}.$$

Table 3.3 Comparison between numerical and analytical solution

Node	Distance	Numerical solution		Analytical solution		Percentage error	
		Case 1	Case 2	Case 1	Case 2	Case 1	Case 2
1	0.1	0.9421	1.0356	0.9387	1.0000	−0.36	−3.56
2	0.3	0.8006	0.8694	0.7963	0.9999	−0.53	13.05
3	0.5	0.6276	1.2573	0.6224	0.9999	−0.83	−25.74
4	0.7	0.4163	0.3521	0.4100	0.9994	−1.53	64.70
5	0.9	0.1579	2.4644	0.1505	0.9179	−4.91	−168.48

Notice that each row of the coefficient matrix refers to the conservation equation written for the corresponding cell centre, and that, in correspondence with the elements of the main diagonal, there are always the values related to the cell centre, while the off-diagonal elements are the values related to cells adjacent to the one considered. Given the data of the problem and considering Eq. 3.14, the analytical solution is given by

$$\phi(x) = \frac{2.7183 - \exp(x)}{1.7183}.$$

Table 3.3 and Fig. 3.5 summarise the comparison between the analytical and numerical solutions obtained by applying the finite volume method to case 1.

Table 3.2 shows the numerical values of the coefficients for the cells for case 2 (symbol $C2$) for which $u = 2.5$ m/s, $F = \rho u = 2.5$, $D = \Gamma/\delta x = 0.1/0.2 = 0.5$. Considering the data of the problem and the Eq. 3.14, the analytical solution is given by

$$\phi(x) = 1 + \frac{1 - \exp(25x)}{7.20 \cdot 10^{10}}.$$

Table 3.3 and Fig. 3.6 summarise the comparison between the analytical solution and the numerical solution obtained by applying the finite volume method to case 2. The

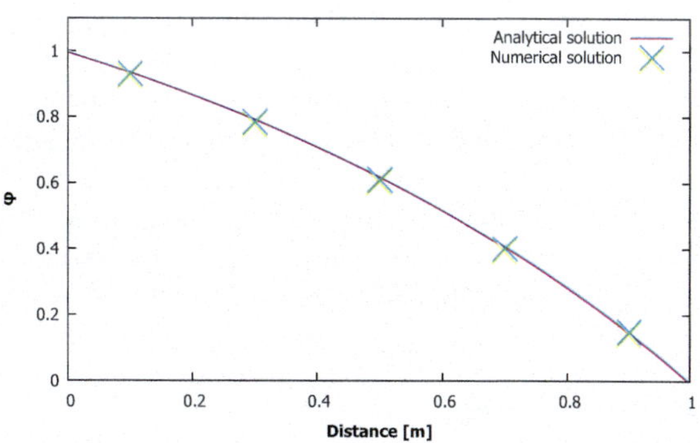

Fig. 3.5 Graphical comparison between numerical and analytical solution for case 1

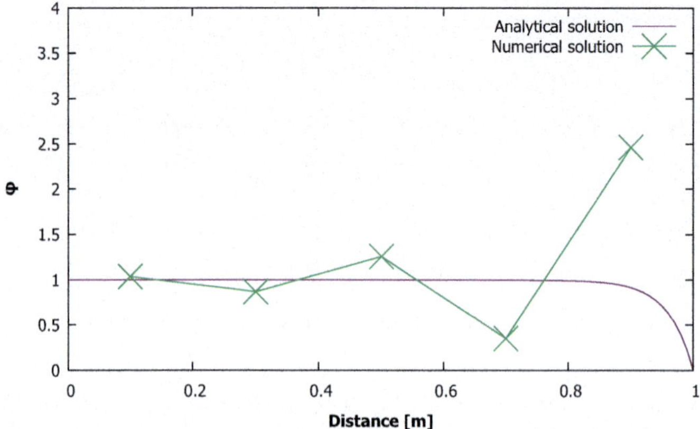

Fig. 3.6 Graphical comparison between numerical and analytical solution related to case 2

Table 3.4 Value of the coefficients for the cells in case 3

Node	a_W	a_E	S_u	S_P	a_P
1	0	0.75	$6.5\phi_A$	−6.5	7.25
2–19	3.25	0.75	0	0	4
20	3.25	0	$1.5\phi_B$	−1.5	4.75

discrepancy between the two solutions is clear in this case. The oscillations present in the numerical solution are known in the literature as "wiggles".

Referring to case 3, for which it is $u = 2.5$ m/s, $F = \rho u = 2.5$, $\delta x = 0.05$, $D = \Gamma/\delta x = 0.1/0.05 = 2$, Table 3.4 summarises the values of the coefficients for the cells. It can be seen that, by increasing the number of intervals, the numerical solution no longer presents significant errors. This is due to the decrease in the value of the F/D ratio, which with twenty intervals is 1.25, while with five intervals it is 5.

3.1.2 *Properties of Discretisation Schemes*

As previously shown, the onset of oscillatory phenomena in the numerical solution can be avoided by increasing the number of cells. However, in most common use cases, the level of refinement required is not acceptable in terms of computational resources and the time required for calculation. It is therefore necessary to analyse some properties that can provide indications about the behaviour of numerical schemes when dealing with computational grids that have a reduced number of cells. Among these properties, the most important are:

- conservativeness;
- boundedness;
- transportiveness.

3.1.2.1 Conservativeness

Considering that each face internal to the computational domain is shared by only two cells, the corresponding flux can be calculated by considering the face as belonging alternately to each of the two cells. The numerical scheme is said to have *the property of conservativeness* if it provides the same flux value—except for the sign—both when the face is considered to belong to one cell and when the same face is considered to belong to the other cell.

To better understand this concept, an example of a scheme that possesses this property is now illustrated. Consider the case of a stationary one-dimensional problem of pure diffusion in the absence of source terms, as illustrated in Fig. 3.7. Considering cell 2 and applying linear interpolation, the flux crossing the left face will be $\Gamma_{w_2}(\phi_2 - \phi_1)/\delta x$, while the flux crossing the right face will be $\Gamma_{e_2}(\phi_3 - \phi_2)/\delta x$. Considering the remaining cells, it is:

$$\left[\Gamma_{e_1}\frac{(\phi_2-\phi_1)}{\delta x} - q_A\right] + \left[\Gamma_{e_2}\frac{(\phi_3-\phi_2)}{\delta x} - \Gamma_{w_2}\frac{(\phi_2-\phi_1)}{\delta x}\right] +$$
$$\left[\Gamma_{e_3}\frac{(\phi_4-\phi_3)}{\delta x} - \Gamma_{w_3}\frac{(\phi_3-\phi_2)}{\delta x}\right] + \left[q_B - \Gamma_{w_4}\frac{(\phi_4-\phi_3)}{\delta x}\right] = q_B - q_A. \quad (3.19)$$

Given that $\Gamma_{e_1} = \Gamma_{w_2}$, $\Gamma_{e_2} = \Gamma_{w_3}$, and $\Gamma_{e_3} = \Gamma_{w_4}$, the fluxes at the interfaces cancel out (it is said in this case that they are expressed in a *consistent manner*), and therefore Eq. 3.19 is satisfied. The consistency of the flux expression resulting from the application of linear interpolation determines the *conservation* of ϕ throughout the computational domain (i.e., the flux that crosses all the internal faces has, except for the sign, always the same value regardless of whether one or the other of the two cells to which the face belongs).

Inconsistent interpolation laws give rise to schemes that do not *conserve* the transported quantity ϕ. This is the case illustrated in Fig. 3.8, where a quadratic interpolation curve is used for the calculation of the fluxes at the interface: it can be noted that the value of ϕ on the face that separates cell 2 from cell 3 is different depending on the values used to determine the quadratic curve. The quadratic curve using values ϕ_1, ϕ_2, ϕ_3 differs from that using the values ϕ_2, ϕ_3, ϕ_4. In other words, the flux exiting from cell 2 through the face is not equal—*it does not conserve*—in modulus to the one entering cell 3 through the same face.

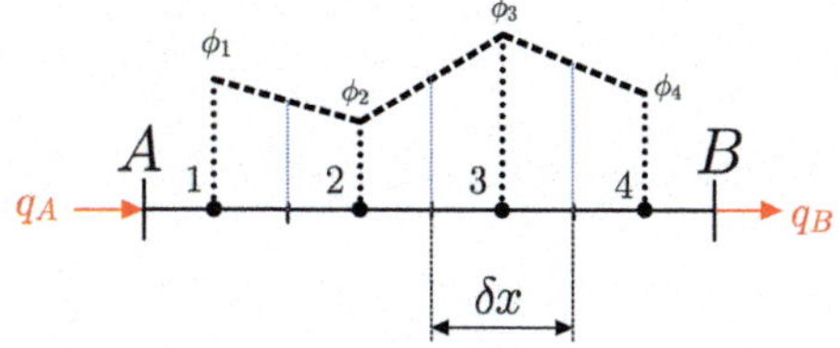

Fig. 3.7 Application of the central differencing scheme for the calculation of flows at the interface

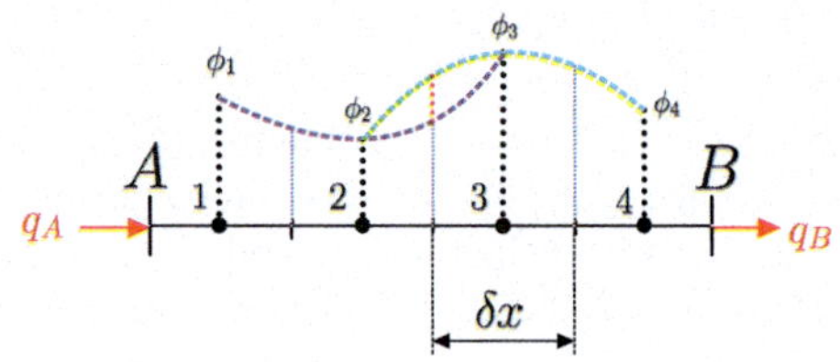

Fig. 3.8 Application of the scheme with quadratic interpolation curve for the calculation of the fluxes at the interface

The term *consistency* used here expresses a concept different from that of *consistency of a numerical method*, according to which, as the level of refinement increases, the truncation error must decrease, i.e., the difference between the solution of the analytical equation and the solution of the discretised equation decreases.

3.1.2.2 Boundedness

Considering Sect. 4.3, the confinement criterion is recalled here, according to which, in the absence of sources, the value of the quantity ϕ in the generic cell must lie between those of the two cells adjacent to it. In the case of the numerical example illustrated in Sect. 3.1.1, according to this criterion, the temperature within the computational domain must be between the values set at the ends of the domain as boundary conditions. If the considered scheme does not satisfy this criterion, it may cause the iterative solution process to fail to converge or present unrealistic oscillations—*wiggles*—as illustrated in Fig. 3.6.

3.1.2.3 Transportiveness

It is necessary here to define the *Péclet number* as

$$Pe = \frac{F}{D} = \frac{\rho u}{\Gamma/\delta x}.$$

This dimensionless number expresses a measure of the relative strength of convection and diffusion.

Figure 3.9 shows the transport of the quantity ϕ in the absence of convection: in this case, the fluid is at rest, $Pe = 0$, and the iso-level curves of ϕ are concentric circles with the centre located at the cell centres W and E. The value of ϕ at the cell

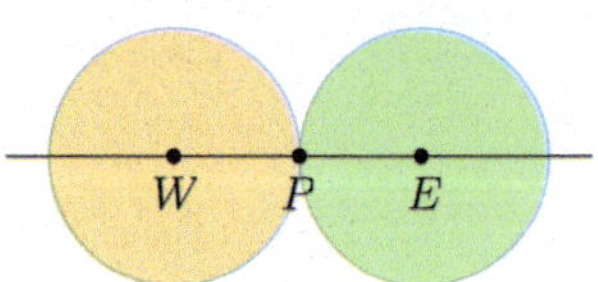

Fig. 3.9 Purely diffusive transport

Fig. 3.10 Convective-diffusive transport

Fig. 3.11 Purely convective transport

centre P depends both on the contributions from W and E, since the diffusive process proceeds in all directions indiscriminately.

Figure 3.10 shows the transport of the quantity ϕ in the case of the simultaneous presence of convection and diffusion: in this case, the fluid moves with velocity u, $Pe \neq 0$, and the iso-level curves of ϕ are translated ellipses in the direction of the fluid velocity. The value of ϕ at the cell centre P depends to a greater extent on the contribution from W.

Figure 3.11 shows the transport of the quantity ϕ in the absence of diffusion: in this case the fluid moves with velocity u, $Pe \to \infty$ and the iso-level curves of ϕ collapse into a half-line originated in W. The value of ϕ in the centre of cell P depends only on the contribution of W, while the centre of cell E does not influence the value in the centre of cell P at all. In conclusion, *transportiveness* describes the mutual influence of the nodes, depending on the Péclet number and the direction of the advection velocity.

3.1.3 *Assessment of the Central Scheme for Convection-Diffusion Cases*

With reference to the concepts just illustrated, the behaviour of the scheme with linear interpolation is now analysed.

Conservativeness: Sect. 3.1.2.1 showed that this scheme presents a consistent expression for the fluxes.

Boundedness: As demonstrated in Sect. 3.1.1, for the cell centre P, Eq. 3.17 applies with the expression of a_P shown in Eq. 3.18. Considering that a stationary one-dimensional flow must satisfy the continuity equation, it follows that $F_e - F_w = 0$, as shown in Eq. 3.11. Consequently, we can write that $a_P = a_W + a_E$. This expression implies that the coefficients of the central scheme satisfy the Scarborough criterion (see Sect. 4.3).

Particular attention should be given to the coefficient $a_E = D_e - \frac{F_e}{2}$, because in the case of strongly convective flows, the value of F_e could make the coefficient a_E negative. The limit condition is therefore expressed as:

$$\frac{F_e}{D_e} = Pe_e < 2. \tag{3.20}$$

Unlike cases 1 and 3, in case 2 of the numerical example seen before, this condition is violated, resulting in the presence of non-physical oscillations, as shown in Fig. 3.6.

It is interesting to note that the Péclet number is a combination of flow properties (i.e., the velocity u), fluid properties (i.e., the density ρ and the diffusion coefficient Γ), and properties of the computational grid (i.e., the spacing δx). Therefore, given the values of ρ and Γ for a specific fluid, the condition (3.20) can be satisfied either for low velocity values (i.e., flows with a low Reynolds number, where diffusion is dominant) or for small grid spacings (i.e., very fine grids).

Transportiveness: In this scheme, the computation of convective and diffusive fluxes does not take into account the direction of flow or the relative strength of convection versus diffusion. As a result, the scheme does not possess the property of transportiveness when applied to flows with a high Péclet number.

3.1.4 Upwind Scheme or Upwind Differencing (UD)

In this approach, the direction of the flow is considered in order to determine the value of ϕ_f. Specifically, ϕ_f is defined as the value of ϕ at the cell centre from which point f perceives the arriving flow (see Figs. 3.12 and 3.13). Using a nautical term, this is referred to as the *upwind* value of ϕ. The considered value is upwind relative to point f, which represents the intersection of the line connecting the centres of the two cells to which the face belongs, and the face itself, whose normal vector is denoted by $\mathbf{n}_f$. In formulas,

$$\begin{cases} \phi_f = \phi_P & \text{for} \quad (\mathbf{u} \cdot \mathbf{n})_f \geq 0, \\ \phi_f = \phi_N & \text{for} \quad (\mathbf{u} \cdot \mathbf{n})_f < 0. \end{cases}$$

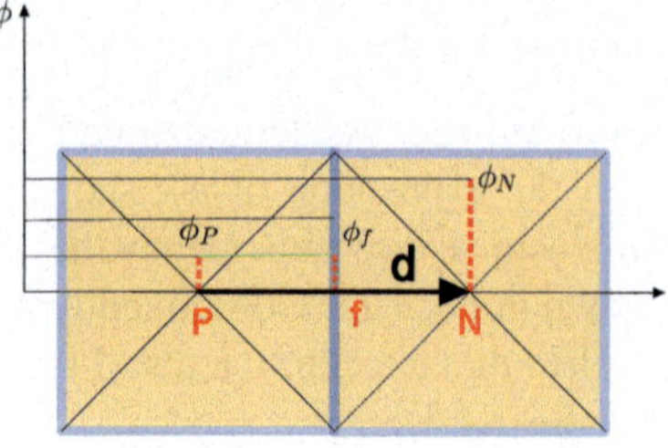

Fig. 3.12 Value of ϕ_f in the case $(\mathbf{u} \cdot \mathbf{n})_f \geq 0$

Fig. 3.13 Value of ϕ_f in the case $(\mathbf{u} \cdot \mathbf{n})_f < 0$

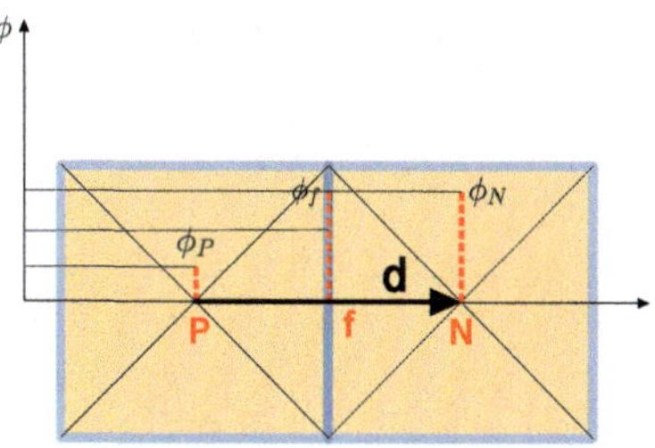

This scheme is accurate to the first order. Using the notation in Fig. 3.4, if the direction of the flow is positive, then $u_w > 0$, $u_e > 0$, and consequently $\phi_w = \phi_W$, $\phi_e = \phi_P$. The discretised equation for point P can then be written as

$$F_e\phi_P - F_w\phi_W = D_e\left(\phi_E - \phi_P\right) - D_w\left(\phi_P - \phi_W\right)$$

This can be rearranged as follows to highlight the coefficients of ϕ_P, ϕ_W, ϕ_E

$$\left[(D_w + F_w) + D_e + (F_e - F_w)\right]\phi_P = (D_w + F_w)\,\phi_W + D_e\phi_E. \tag{3.21}$$

When the flow direction is negative, $u_w < 0, u_e < 0$ and therefore $\phi_w = \phi_P, \phi_e = \phi_E$ and the discretised equation for point P is

$$\left[D_w + (D_e - F_e) + (F_e - F_w)\right]\phi_P = D_w\phi_W + (D_e - F_e)\,\phi_E. \tag{3.22}$$

Equations 3.21 and 3.22 can be rewritten in the form

$$a_P\phi_P = a_W\phi_W + a_E\phi_E$$

with

$$a_P = a_W + a_E + (F_e - F_w)\,, \quad a_W = D_W + \max(F_w, 0), \quad a_E = D_E + \max(0, -F_e).$$

To better understand the properties of this numerical scheme, it will be applied to the numerical example presented in Sect. 3.1.1. Once again, all cells will have the same values for the following quantities: $F_e = F_w = F = \rho u$, $D_e = D_w = D = D = \Gamma/\delta x$.

Applying the upwind scheme to cell 1, it can be written as

$$F_e\phi_P + F_A\phi_A = D_e\left(\phi_E - \phi_P\right) - D_A\left(\phi_P - \phi_A\right) \tag{3.23}$$

and for cell 5

$$F_B\phi_P - F_w\phi_W = D_B\left(\phi_B - \phi_P\right) - D_w\left(\phi_P - \phi_W\right). \tag{3.24}$$

Considering that $D_A = D_B = 2\Gamma/\delta x = 2D$ and that $F_A = F_B = F$, Eqs. 3.23 and 3.24 can be written in compact form as

$$a_P\phi_P = a_W\phi_W + a_E\phi_E + S_u \tag{3.25}$$

with

$$a_P = a_W + a_E + (F_E - F_W) - S_P \tag{3.26}$$

This includes the contribution of the boundary conditions as a source term. Table 3.5 summarises the values of the coefficients for the cells.

Considering the problem data and Eq. 3.14, the analytical solution for case 1 is given by

$$\phi(x) = \frac{2.7183 - \exp(x)}{1.7183}.$$

Table 3.6 and Fig. 3.14 summarise the comparison between the analytical and numerical solutions.

Regarding case 2, where $u = 2.5$ m/s, $F = \rho u = 2.5$, and $D = \Gamma/\delta x = 0.1/0.2 = 0.5$, the analytical solution is given by

$$\phi(x) = 1 + \frac{1 - \exp(25x)}{7.20 \cdot 10^{10}}.$$

Table 3.7 and Fig. 3.15 summarise the comparison between the analytical solution and the numerical solution obtained by applying the upwind scheme. It is evident

Table 3.5 Expression of the coefficients for the cells in the case of the upwind scheme

Node	a_W	a_E	S_P	S_u
1	0	D	$-(2D + F)$	$(2D + F)\phi_A$
2, 3, 4	$D + F$	D	0	0
5	$D + F$	0	$-2D$	$2D\phi_B$

Table 3.6 Comparison between numerical solution and analytical solution for case 1 solved using the upwind scheme

Node	Distance	Numerical solution	Analytical solution	Percentage error
1	0.1	0.9337	0.9387	0.53
2	0.3	0.7879	0.7963	1.05
3	0.5	0.6130	0.6224	1.51
4	0.7	0.4031	0.4100	1.68
5	0.9	0.1512	0.1505	−0.02

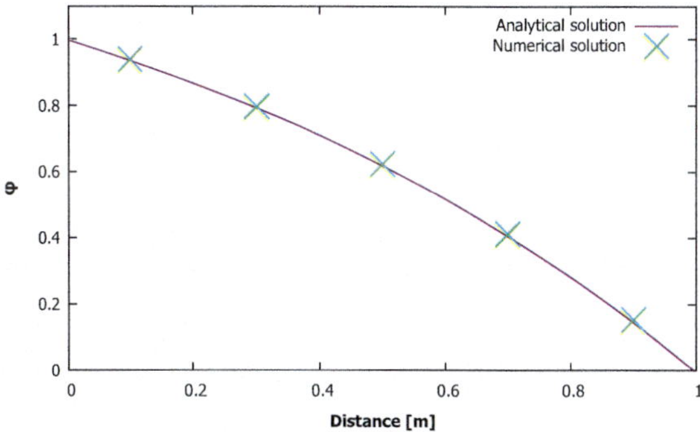

Fig. 3.14 Comparison between numerical solution and analytical solution for case 1 solved using the upwind scheme

Table 3.7 Comparison between numerical solution and analytical solution for case 2 solved using the upwind scheme

Node	Distance	Numerical solution	Analytical solution	Percentage error
1	0.1	0.9998	1	−3.56
2	0.3	0.9987	0.9999	13.05
3	0.5	0.9921	0.9999	−25.74
4	0.7	0.9524	0.9994	64.70
5	0.9	0.7143	0.9179	−168.48

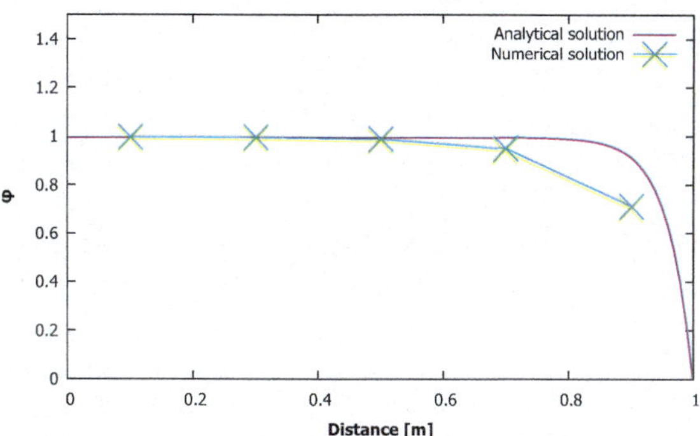

Fig. 3.15 Comparison between numerical solution and analytical solution for case 2 solved using the upwind scheme

that there is an improvement compared to the centred scheme, although a discrepancy remains between the analytical and numerical solutions at the nodes near the boundary B of the computational domain.

In relation to the properties describing the behaviour of a numerical scheme, it can be stated that the upwind scheme possesses the property of conservativeness because the expression for the flux at the interface is consistent. The upwind scheme also has the property of boundedness, as the coefficients of the discretised equation are always

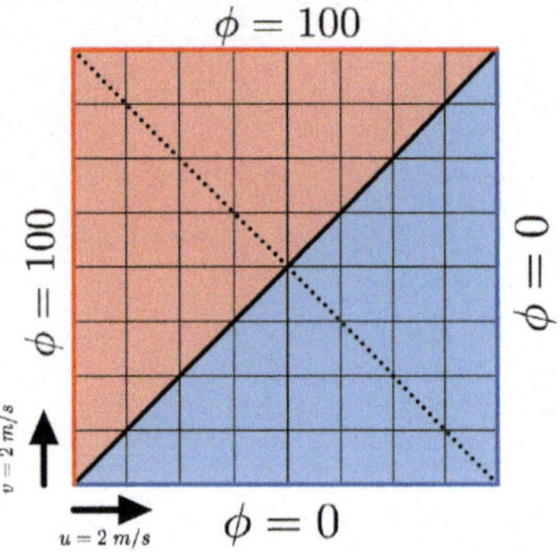

Fig. 3.16 Computational domain, initial and boundary conditions used to show the phenomenon of numerical diffusion

positive and the coefficient matrix is always diagonally dominant. Furthermore, the upwind scheme exhibits the property of transportiveness because it accounts for the direction of the flow. One of the most well-known problems associated with this scheme is *false diffusion*, also referred to as *numerical diffusion*. When the flow is not aligned with the grid, numerical diffusion causes the transported quantity ϕ to be redistributed over more than one cell (*smearing*), in a manner similar to the effect of physical diffusivity (see Fig. 2.21). This effect can be illustrated by analysing the transport of a scalar ϕ in a fluid that does not have a diffusion coefficient ($\Gamma = 0$) and in the absence of sources, applying the upwind scheme on a computational domain whose grid is inclined at a certain angle with respect to the direction of the fluid motion in which ϕ is transported. Figure 3.16 illustrates the computational domain, boundary conditions, and initial conditions used to demonstrate the phenomenon of numerical diffusion.

In Fig. 3.16, a dashed segment is shown, over which the values of the transported quantity are displayed at different levels of grid refinement in Fig. 3.17. The analytical solution for the case in Fig. 3.16 represents a flow with a direction parallel to the continuous line diagonal. Moving along the dashed diagonal from left to right, the value of the transported quantity remains constant (100) until it intersects with the continuous diagonal, where a step change occurs. After the intersection, the value of the transported quantity becomes constant and equal to 0 until the boundary of the domain. Figure 3.17 illustrates the variation in the value of the transported quantity, corresponding to the analytical solution. In the same figure, it can be observed that the smearing phenomenon is reduced as the level of grid refinement increases. In many practical cases, the level of refinement required may be unacceptable due to the computational load involved. Furthermore, it has been shown that even with high levels of grid refinement, numerical diffusion can still lead to unacceptable results for flows with a high Reynolds number.

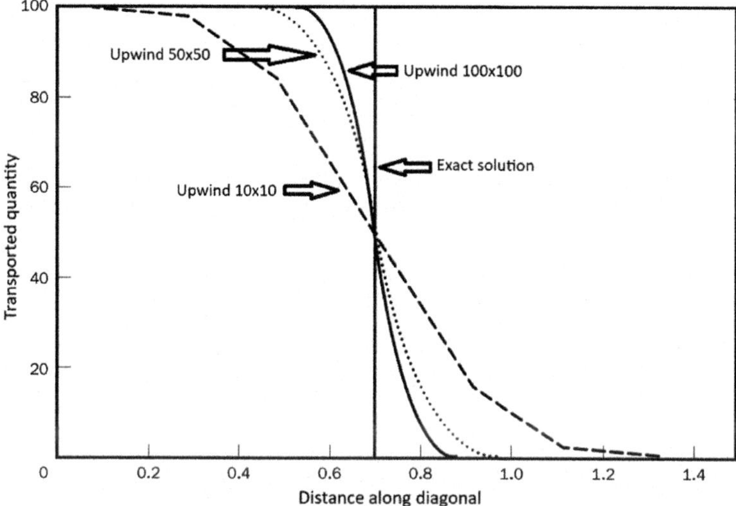

Fig. 3.17 Variation of the value of the transported quantity along the dashed segment shown in Fig. 3.16 in the case of the application of the upwind scheme

3.1.5 Linear Upwind Scheme

In central differencing, an *interpolation* is performed, whereas in the case of the linear upwind scheme—or *linear upwind differencing* (LUD)—a *linear extrapolation* is performed based on the value of ϕ and its gradient at the centre of the upwind cell. Essentially, this scheme can be thought of as an upwind scheme to which a corrective extrapolation term is added, derived from the use of the gradient value at the centre of the upwind cell (in Fig. 3.18, the slope of the segment connecting ϕ_W, ϕ_P, and ϕ_e), and the distance between the centre of the upwind cell and the centre of the face (in Fig. 3.18, $\delta x/2$). Figure 3.18 illustrates the calculation of the value of the transported quantity ϕ at the face e in a one-dimensional case with a uniform grid. The advection velocity u is positive at the considered face, and the gradient at the cell centre is calculated based on the values at the centres of the two upwind cells. In formulas,

$$\phi_e = \begin{cases} \phi_P + \frac{1}{2}(\phi_P - \phi_W) & for \quad F_e \geq 0, \\ \phi_P + \frac{1}{2}(\phi_P - \phi_E) & for \quad F_e < 0. \end{cases}$$

In constructing the matrix equation representing the conservation equations for each of the cells, the term related to the "upwind part" of this scheme will contribute to the coefficient matrix and, therefore, will be calculated implicitly through the inversion of the coefficient matrix itself. The extrapolation correction term will contribute to determining the vector of known terms and, hence, will be calculated explicitly—using the values of the necessary quantities obtained from the last iteration or from the initial conditions. For this scheme, errors in the gradient calculation due to skewness (see Sect. 3.4) are typically not taken into account, because the extrapolation correction term is computed considering the line joining the centre of the upwind cell and the centre of the face, rather than the line joining the centres of the two cells to which the face belongs. Although this second-order accurate scheme can produce acceptable solutions, it does not possess the property of boundedness. Therefore, in cases of very high gradients, it may cause non-physical oscillations in the value of the transported quantity. With the aim of eliminating these undesirable behaviours, Total Variation Diminishing (TVD) schemes have been introduced, as will be shown later.

3.1.6 QUICK Scheme (Quadratic Upwind Interpolation for Convective Kinetics)

This is one of the first so-called *higher-order* schemes, i.e., schemes with an order of accuracy greater than second order. For the calculation of the flow at the interface, this scheme uses a quadratic function, whose value at the two centres near the face under consideration, as well as at the nearest cell centre in the upwind direction, is equal to the value of the considered quantity at these centres. In the case where, as shown in Fig. 3.19, $u_w > 0$ and $u_e > 0$, a quadratic curve passing through the points ϕ_P, ϕ_W, and ϕ_{WW} is used to calculate ϕ_w, while a quadratic curve passing through the points ϕ_E, ϕ_P, and ϕ_W is used to calculate ϕ_e. Referring to Figs. 3.19 and 3.20, we aim to find the expression of the quadratic function whose profile is chosen to approximate the trend of the quantity ϕ within the cell centred at P:

$$\phi = a_0 + a_1 x + a_2 x^2.$$

In the case where $u_w > 0$, the coefficients a_0, a_1, and a_2 of such a function can be determined by assigning the passage through the points $i-2$, $i-1$, and i with

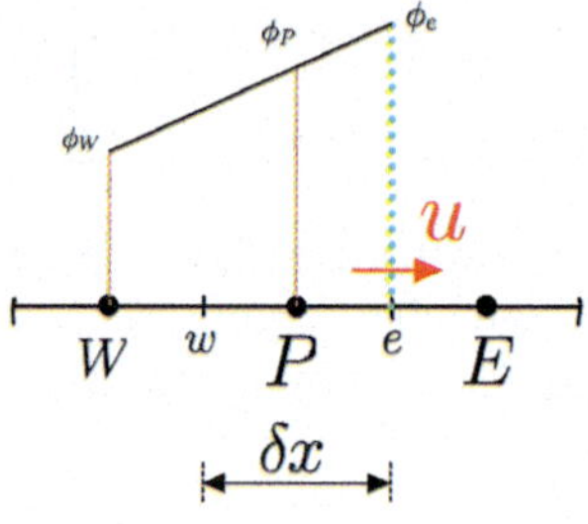

Fig. 3.18 Linear upwind scheme

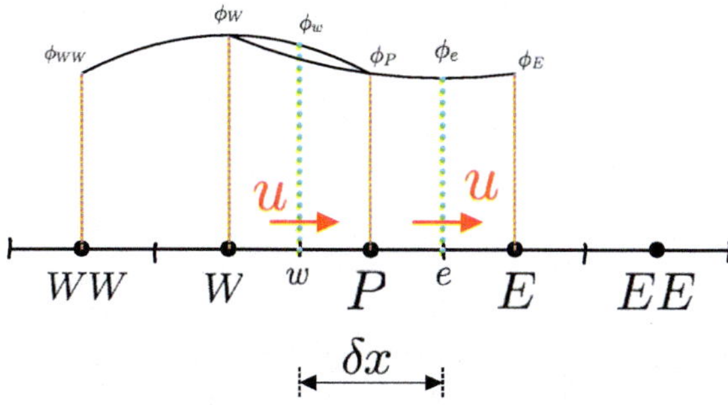

Fig. 3.19 Profile of quadratic function used for the calculation of the flow at the interface

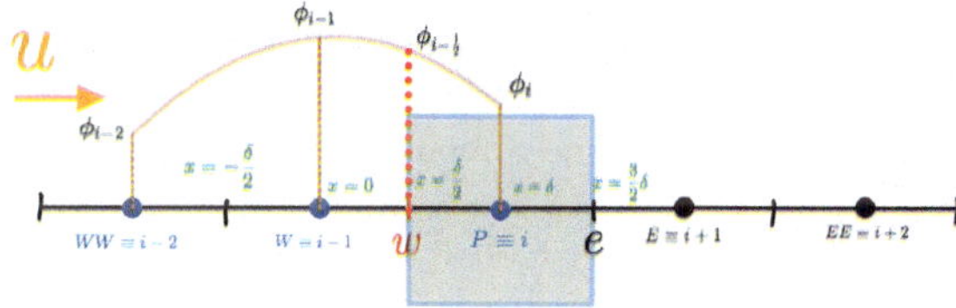

Fig. 3.20 Discretised computational domain

coordinates $(-\delta, \phi_{i-2})$, $(0, \phi_{i-1})$, and (δ, ϕ_i), respectively. By imposing the passage through the point $i-1$, we obtain:

$$\phi_{i-1} = a_0. \tag{3.27}$$

Imposing the passage through the point i we get

$$\phi_i = a_0 + a_1\delta + a_2\delta^2. \tag{3.28}$$

Imposing the passage through the point $i-2$ we get

$$\phi_{i-2} = a_0 - a_1\delta + a_2(-\delta)^2. \tag{3.29}$$

From Eqs. 3.27, 3.28, 3.29 we get

$$a_0 = \phi_{i-1}, \qquad a_1 = \frac{\phi_i - \phi_{i-2}}{2\delta}, \qquad a_2 = \frac{\phi_i + \phi_{i-2} - 2\phi i - 1}{2\delta^2}.$$

Knowing the values of the coefficients a_0, a_1, a_2 it is possible to calculate the value of ϕ_w at $x = \delta/2$:

$$\phi_w = a_0 + a_1\frac{\delta}{2} + a_2(\frac{\delta}{2})^2$$

from which

$$\phi_w = \frac{6}{8}\phi_{i-1} + \frac{3}{8}\phi_i - \frac{1}{8}\phi_{i-2}$$

or, equivalently

$$\phi_w = \frac{6}{8}\phi_W + \frac{3}{8}\phi_P - \frac{1}{8}\phi_{WW}.$$

Similarly, for $u_e > 0$ we get

$$\phi_e = \frac{6}{8}\phi_P + \frac{3}{8}\phi_E - \frac{1}{8}\phi_W.$$

For the calculation of diffusive fluxes on uniform grids, the application of the centred scheme is equivalent to the application of the QUICK scheme. This is because the slope of the tangent at the midpoint of a parabolic arc is the same as the segment connecting the two ends of the arc considered. In the case where $F_w > 0$ and $F_e > 0$, applying the QUICK scheme for convective terms and the centred scheme for diffusive terms, the one-dimensional convective-diffusive transport equation can be written as:

$$\left[F_e\left(\frac{6}{8}\phi_P + \frac{3}{8}\phi_E - \frac{1}{8}\phi_W\right) - F_w\left(\frac{6}{8}\phi_W + \frac{3}{8}\phi_P - \frac{1}{8}\phi_{WW}\right)\right] =$$
$$D_e\left(\phi_E - \phi_P\right) - D_w\left(\phi_P - \phi_W\right)$$

when rearranged to highlight the coefficients of the fluxes at the cell centres, this becomes:

$$\left(D_w - \frac{3}{8}F_w + D_e + \frac{6}{8}F_e\right)\phi_P =$$
$$\left(D_w + \frac{6}{8}F_w + D_e + \frac{1}{8}F_e\right)\phi_W + \left(D_e - \frac{3}{8}F_e\right)\phi_E - \frac{1}{8}F_w\phi_{WW}.$$

In general, for the QUICK scheme, it can be written

$$a_P\phi_P = a_W\phi_W + a_E\phi_E + a_{WW}\phi_{WW} + a_{EE}\phi_{EE}$$

with

$$a_P = a_W + a_E + a_{WW} + a_{EE} + (F_e - F_w)$$

and

$$a_W = D_w + \frac{6}{8}\alpha_w F_w + D_e + \frac{1}{8}\alpha_e F_e,$$
$$a_{WW} = -\frac{1}{8}\alpha_w F_w + \frac{3}{8}(1-\alpha_w)F_w,$$
$$a_E = D_e - \frac{3}{8}\alpha_e F_e - \frac{6}{8}\left(1-\alpha_e\right)F_e - \frac{1}{8}(1-\alpha_w)F_w$$

while

$$\alpha_w = 1\ for\ F_w > 0\ \ ;\ \ \alpha_e = 1\ for\ F_e > 0$$
$$\alpha_w = 0\ for\ F_w < 0\ \ ;\ \ \alpha_e = 0\ for\ F_e < 0.$$

The use of two nodes adjacent to the considered one, along with the upwind node, ensures that the scheme is conservative and endowed with the property of transportiveness. By using a quadratic curve, the QUICK scheme achieves third-order accuracy in the case of a uniform grid. On the other hand, this scheme exhibits conditionally stable behaviour due to the possibility that the coefficients a_{WW} and a_{EE}, even for small values of the Péclet number ($Pe > 8/3$), can become negative. A modified form of this scheme that is conservative, bounded, and transportive is known as *Hayase's QUICK scheme*.[2] This scheme uses an appropriate source term that prevents the coefficients from becoming negative, and can be illustrated as follows: the value of the transported quantity is calculated as:

$$\begin{aligned}
\phi_w &= \phi_W + \frac{1}{8}(3\phi_P - 2\phi_W - \phi_{WW})\ \ for\ \ F_w > 0,\\
\phi_e &= \phi_P + \frac{1}{8}(3\phi_E - 2\phi_P - \phi_W)\ \ for\ \ F_e > 0,\\
\phi_w &= \phi_P + \frac{1}{8}(3\phi_W - 2\phi_P - \phi_E)\ \ for\ \ F_w < 0,\\
\phi_e &= \phi_E + \frac{1}{8}(3\phi_P - 2\phi_E - \phi_{EE})\ \ for\ \ F_e < 0.
\end{aligned}$$

The discretisation equation is written in the form

$$a_P\phi_P = a_W\phi_W + a_E\phi_E + \overline{S}$$

with

$$\begin{aligned}
a_P &= a_W + a_E + (F_e - F_w),\\
a_W &= D_w + \alpha_w F_w,\\
a_E &= D_e - (1-\alpha_e)\,F_e,\\
\overline{S} &= \frac{1}{8}(3\phi_P - 2\phi_W - \phi_{WW})\alpha_w F_w + \frac{1}{8}(\phi_W - 2\phi_P - 3\phi_E)\alpha_e F_e\\
&\quad + \frac{1}{8}(3\phi_W - 2\phi_P - \phi_E)(1-\alpha_w)F_w + \frac{1}{8}(2\phi_E - 2\phi_{EE} - 3\phi_P)(1-\alpha_e)F_e
\end{aligned}$$

and

$$\alpha_w = 1\ for\ F_w > 0\ \ ;\ \ \alpha_e = 1\ for\ F_e > 0$$

---

[2] T. Hayase, J. A. C. Humphrey, R. Greif, *A consistently formulated QUICK scheme for fast and stable convergence using finite-volume iterative calculation procedures*, "Journal of Computational Physics", January 1992, 98(1), pp. 108–118.

$$\alpha_w = 0 \; for \; F_w < 0 \quad ; \quad \alpha_e = 0 \; for \; F_e < 0.$$

The computation of the source term in the parts of the discretisation that contain negative coefficients is referred to as *deferred correction*, because the value of the terms necessary for the calculation of $\overline{S}$ at the n-th iteration is taken from the $(n-1)$-th iteration. In other words, the correction is deferred by one iteration. Although the QUICK scheme has many positive aspects (such as reduced numerical diffusivity and a higher degree of accuracy), it, like the centred scheme, can produce non-physical oscillations. In some cases, this can lead to results that are not acceptable, such as in turbulence modelling, where negative values of quantities that can only be physically positive may occur.

3.1.7 Total Variation Diminishing (TVD) Schemes

This technique aims to mitigate the accuracy issues of the upwind scheme, as well as the stability and boundedness issues of the centred scheme. For simplicity, consider the calculation of the transported quantity ϕ at the face e in a one-dimensional case with a uniform grid, where the advection velocity u is positive at the considered face (see Fig. 3.18). Initially, considering the LUD scheme, it is:

$$\phi_e = \phi_P + \frac{\phi_P - \phi_W}{\delta x}\frac{\delta x}{2} = \phi_P + \frac{1}{2}\left(\phi_P - \phi_W\right). \tag{3.30}$$

Therefore, the LUD scheme can be considered an upwind first-order accurate scheme, which, when modified with additional terms, becomes second-order accurate. The additional term is always constructed in accordance with the upwind strategy that takes into account the direction of the flow and is an estimate of the gradient $\left(\frac{\phi_P - \phi_W}{\delta x}\right)$ of the transported quantity, multiplied by the distance between the cell centre and the considered face. In terms of the total flux $F_e \phi_e$ at the interface, the LUD scheme can be viewed as the convective flux $F_e \phi_P$ calculated using the UD scheme, to which a corrective term $F_e \left(\frac{\phi_P - \phi_W}{2}\right)$ is added to increase the order of accuracy. According to this logic, the QUICK scheme can also be considered an upwind scheme to which a corrective term is added to further increase the order of accuracy:

$$\phi_e = \phi_P + \frac{1}{8}\left[3\phi_E - 2\phi_P - \phi_W\right].$$

The same applies to the centred scheme:

$$\phi_e = \phi_P + \frac{1}{2}\left(\phi_E - \phi_P\right).$$

The three aforementioned schemes can, therefore, be summarised in a single formula that makes use of an appropriate function ψ.

$$\phi_e = \phi_P + \frac{1}{2}(\phi_E - \phi_P)\psi$$

from which, it can once again be observed that the total flux $F_e\phi_e$ at the interface can be considered as the convective flux $F_e\phi_P$ calculated using the UD scheme, to which a corrective term $F_e\psi(\phi_E - \phi_P)/2$ is added. Through the function ψ, this corrective term is proportional to the corrective term of the centred scheme $F_e(\phi_E - \phi_P)/2$, which represents the variation of ϕ as the transition occurs from cell centre P to cell centre E. It is immediately apparent that for $\psi = 0$, the upwind scheme is obtained, and for $\psi = 1$, the centred scheme is recovered. Furthermore, to determine the value of ψ in the LUD case, it is necessary to rewrite Eq. 3.30 as:

$$\phi_e = \phi_P + \frac{1}{2}\frac{\phi_P - \phi_W}{\phi_E - \phi_P}(\phi_E - \phi_P)$$

From this, it is evident that for the LUD scheme, $\psi = \frac{\phi_P - \phi_W}{\phi_E - \phi_P}$. Following the same strategy, in the case of the QUICK scheme, it is:

$$\phi_e = \phi_P + \frac{1}{2}\left[\left(3 + \frac{\phi_P - \phi_W}{\phi_E - \phi_P}\right)\frac{1}{4}\right](\phi_E - \phi_P)$$

with

$$\psi = \left(3 + \frac{\phi_P - \phi_W}{\phi_E - \phi_P}\right)\frac{1}{4}.$$

The coefficient ψ can be thought of as a function of the ratio r between the variation (gradient) $\phi_P - \phi_W$ of the quantity ϕ from the upwind side and the variation (gradient) $\phi_E - \phi_P$ of the quantity ϕ from the downwind side with respect to point P. In formulas:

$$\psi = \psi(r) \qquad \text{with} \qquad r = \frac{\phi_P - \phi_W}{\phi_E - \phi_P}$$

and so

$$\phi_e = \phi_P + \frac{1}{2}(\phi_E - \phi_P)\psi(r)$$

With $\psi(r) = 0$ for the upwind scheme, $\psi(r) = 1$ for the centred scheme, $\psi(r) = r$ for the LUD scheme, and $\psi(r) = \frac{3+r}{4}$ for the QUICK scheme. Figure 3.21 shows the $r - \psi$ diagram, graphically describing the function $\psi(r)$. The function $\psi(r)$, called the *flux limiter*, determines the weight of the additional term, compared to that derived from the simple upwind scheme, in calculating the value of the quantity at the face. Its aim is to maximise accuracy, stability, and boundedness. The value of

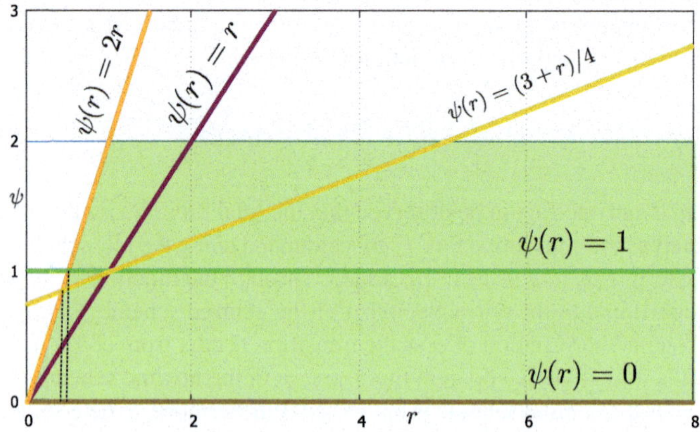

Fig. 3.21 $r - \psi$ diagram; in the shaded area the values of ψ that make the scheme TVD

$\psi(r)$ changes depending on the cell considered, even though the expression used to calculate it remains the same for the entire computational domain.

3.1.7.1 Total Variation

Schemes such as upwind have a low order of accuracy but do not give rise to oscillations. Conversely, higher-order schemes such as the centred scheme, linear upwind, and QUICK exhibit oscillations, although they have a higher order of accuracy. It has been shown that a higher-order scheme that is stable and does not present oscillations possesses the property of *monotonicity-preserving*. For a scheme to be monotonicity-preserving, it must *not*...

- create local maxima or minima;
- accentuate the value of any maxima or minima present in the solution.

It is now possible to introduce the concept of *total variation*, which, with reference to Fig. 3.22, can be defined as...

$$TV(\phi) = |\phi_2 - \phi_1| + |\phi_3 - \phi_2| + |\phi_4 - \phi_3| + |\phi_5 - \phi_4| = |\phi_3 - \phi_1| + |\phi_5 - \phi_3|$$

In the literature,[3] total variation was initially analysed for the case of non-stationary

Fig. 3.22 Variation of the quantity ϕ

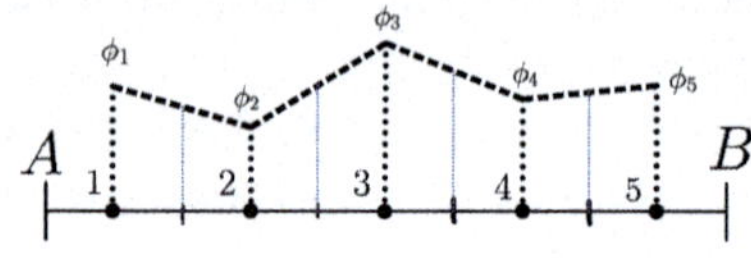

one-dimensional transport equations. For this reason, we refer to *TVD (Total Variation Diminishing)* schemes when the total variation of the solution, $TV(\phi^n)$, at a certain time step n is greater than the total variation of the solution, $TV(\phi^{n+1})$, at the next time step $n+1$, i.e., $TV(\phi^{n+1}) < TV(\phi^n)$. A monotonicity-preserving scheme is also TVD.

3.1.7.2 TVD Schemes

A necessary and sufficient condition for a scheme to possess the TVD property is that

- for $0 < r < 1$ it must be $\psi(r) \leq 2r$;
- for $r \geq 1$ it must be $\psi(r) \leq 2$.

Referring to Fig. 3.21, this condition is equivalent to having values of ψ contained within the shaded area of the $r - \psi$ diagram. From the same figure, it can be noted that:

- the upwind scheme is TVD;
- the LUD scheme is not TVD for $r \geq 2$;
- the centred scheme is not TVD for $r < 0.5$;
- the QUICK scheme is not TVD for $r < 3/7$ and for $r > 5$.

The goal is to find a particular function $\psi(r)$ such that its values satisfy the necessary and sufficient condition mentioned above for every value of r. In other words, the objective is to determine a function $\psi(r)$ that limits the flux $F_e\psi(r)(\phi_E - \phi_P)/2$, which, when added to the flux $F_e\phi_P$, makes the scheme of higher order. For this reason, $\psi(r)$ is called the *limiter function*.

For a limiter function to make the scheme second-order accurate, it must satisfy the following condition:

1. $\psi(r = 1) = 1$ that is, it must pass through the point with coordinates $(1, 1)$ of the $r - \psi$ diagram,
2. for $0 < r < 1$ it must be $r \leq \psi(r) \leq 1$,
3. for $r \geq 1$ it must be $1 \leq \psi(r) \leq r$.

Considering the first condition, it becomes clear that the upwind scheme is not second-order accurate, while both the centred and QUICK schemes are. From the second and third conditions, it is evident that the values of the limiter must be constrained between those assumed for the centred scheme and those assumed for the LUD scheme. These three conditions correspond to the shaded areas on the $r - \psi$ diagram shown in Fig. 3.23. Some of the most well-known limiting functions (or *flux limiters*) are now listed.

---

[3] A. Harten, *On a class of high-resolution total-variation stable finite-difference schemes*, SIAM Journal on Numerical Analysis, 21(1), 1–23, 1984.

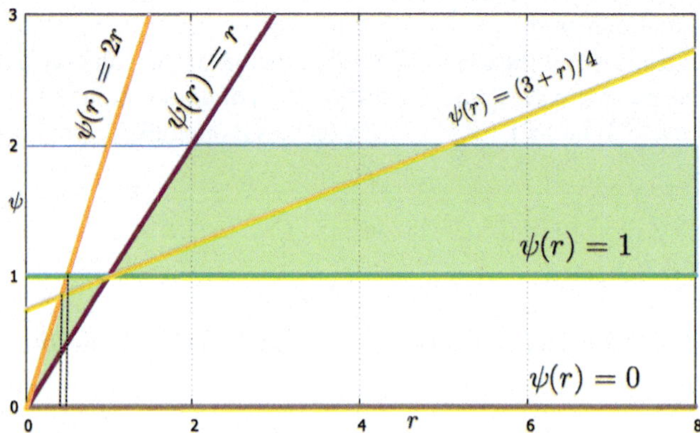

Fig. 3.23 Areas of the $r - \psi$ diagram for which the limiting function $\psi(r)$ makes the scheme accurate to the second order

- Van Leer: $\psi(r) = \dfrac{r + |r|}{1 + r}$;
- Van Albada: $\psi(r) = \dfrac{r + r^2}{1 + r^2}$;
- Min-Mod: $\psi(r) = \begin{cases} \min(r, 1) & for \quad r > 0 \\ 0 \quad for & r \leq 0 \end{cases}$;
- SUPERBEE: $\psi(r) = \max[0, \min(2r, 1), \min(r, 2)]$;
- Sweby: $\psi(r) = \max[0, \min(\beta r, 1), \min(r, \beta)]$;
- QUICK: $\psi(r) = \max[0, \min(2r, (3 + r)/4, 2]$;
- UMIST: $\psi(r) = \max[0, \min(2r, (1 + 3r)/4, (3 + r)/4), 2]$.

All the listed limiters have values in the TVD part of the $r - \psi$ graph and pass through the point (1,1) on the same graph, making them all TVD second-order accurate schemes. It is noted that Van Leer and Van Albada are continuous functions, while the others are piecewise linear functions. The Min-Mod function represents the lower edge of the TVD part of the $r - \psi$ graph, while the SUPERBEE function captures its upper edge. The Sweby limiter combines Min-Mod and SUPERBEE using the parameter β: for $\beta = 1$, it yields the Min-Mod, and for $\beta = 2$, it gives the SUPERBEE. Typically, the Sweby limiter is used with $\beta = 1.5$.

3.1.7.3 Implementation of TVD Schemes

In this regard, the one-dimensional convection-diffusion equation is considered:

$$\frac{d}{dx}(\rho u \phi) = \frac{d}{dx}\left(\Gamma \frac{d\phi}{dx}\right).$$

Referring to Fig. 3.1, the discretised form of the equation is given by:

$$F_e\phi_e - F_w\phi_w = D_e(\phi_E - \phi_P) - D_w(\phi_P - \phi_W). \tag{3.31}$$

In the case of velocity $u > 0$, using the TVD approach, the coefficients ϕ_e and ϕ_w can be expressed as

$$\phi_e = \phi_P + \frac{1}{2}\psi(r_e)(\phi_E - \phi_P),$$

$$\phi_w = \phi_W + \frac{1}{2}\psi(r_w)(\phi_E - \phi_W),$$

$$\text{with } r_e = \frac{\phi_P - \phi_W}{\phi_E - \phi_P} \quad \text{and} \quad r_w = \frac{\phi_W - \phi_{WW}}{\phi_P - \phi_W}.$$

Recalling that r represents the ratio between the upwind gradient of ϕ and the downwind gradient of ϕ, $\psi(r_e)$ and $\psi(r_w)$ can take the form of one of the flux-limiting functions listed above. By substituting the two expressions for $\psi(r_e)$ and $\psi(r_w)$ into Eq. 3.31, it becomes:

$$(D_e + F_e + D_w)\phi_P =$$

$$(D_w + F_w)\phi_W + D_e\phi_E - F_e\left[\frac{1}{2}\psi(r_e)(\phi_E - \phi_P)\right] + F_w\left[\frac{1}{2}\psi(r_w)(\phi_P - \phi_W)\right].$$

To highlight the coefficients of ϕ_P, ϕ_W, and ϕ_E, as well as the source term, the equation can be rewritten as:

$$a_P\phi_P = a_W\phi_W + a_E\phi_E + S_u^{DC}$$

with

$$a_P = a_W + a_E + (F_e - F_w)$$

$$a_W = D_w + F_w$$

$$a_E = D_e$$

$$S_u^{DC} = -F_e\left[\frac{1}{2}\psi(r_e)(\phi_E - \phi_P)\right] + F_w\left[\frac{1}{2}\psi(r_w)(\phi_P - \phi_W)\right].$$

It can be noted that the coefficients a_P, a_W, and a_E are identical to those of the upwind scheme, which ensures the stability of the TVD scheme. The additional flux term that makes the scheme second-order accurate, and which contains the limiting function, is expressed in the form of source terms with deferred correction—hence the DC superscript. As discussed in Sect. 3.1.6, this avoids the possibility of negative values appearing for the coefficients, which could destabilise the solution. In this manner, the behaviour of the final solution remains TVD. To indicate the case with velocity $u > 0$, the source term can be rewritten using the superscript $^+$ for the term r:

$$S_u^{DC} = -F_e\left[\frac{1}{2}\psi\left(r_e^+\right)(\phi_E - \phi_P)\right] + F_w\left[\frac{1}{2}\psi\left(r_w^+\right)(\phi_P - \phi_W)\right].$$

In the case of velocity $u < 0$, using the TVD approach, the coefficients ϕ_e and ϕ_w can be expressed as:

$$\phi_e = \phi_E + \frac{1}{2}\psi(r_e^-)(\phi_P - \phi_E),$$
$$\phi_w = \phi_P + \frac{1}{2}\psi(r_w^-)(\phi_W - \phi_P),$$
$$\text{with } r_e^- = \frac{\phi_{EE} - \phi_E}{\phi_E - \phi_P} \quad \text{and} \quad r_w^- = \frac{\phi_E - \phi_P}{\phi_P - \phi_W}$$

In this case, the superscript $^-$ is used to indicate the negative value of the velocity. By substituting the two expressions now found for $\psi(r_e^-)$ and $\psi(r_w^-)$ into Eq. 3.31, and highlighting the coefficients of ϕ_P, ϕ_W, and ϕ_E as well as the source term, it becomes:

$$a_P\phi_P = a_W\phi_W + a_E\phi_E + S_u^{DC}$$

with

$$\begin{aligned} a_P &= a_W + a_E + (F_e - F_w), \\ a_W &= D_w, \\ a_E &= D_e - F_e, \\ S_u^{DC} &= -F_e\left[\frac{1}{2}\psi\left(r_e^-\right)(\phi_E - \phi_P)\right] - F_w\left[\frac{1}{2}\psi\left(r_w^-\right)(\phi_P - \phi_W)\right]. \end{aligned}$$

For implementation purposes, a single expression is used to encompass both cases, $u > 0$ and $u < 0$:

$$\begin{aligned} a_P &= a_W + a_E + (F_e - F_w), \\ a_W &= D_w + \max(F_w, 0), \\ a_E &= D_e + \max(-F_e, 0), \\ S_u^{DC} &= \frac{1}{2}F_e\left[(1-\alpha_e)\psi\left(r_e^-\right) - \alpha_e\psi\left(r_e^+\right)\right](\phi_E - \phi_P) \\ &+ \frac{1}{2}F_w\left[\alpha_w\psi\left(r_w^+\right) - (1-\alpha_w)\psi\left(r_w^-\right)\right](\phi_P - \phi_W) \end{aligned}$$

with

$$\alpha_w = 1 \; for \; F_w > 0 \;\; ; \;\; \alpha_e = 1 \; for \; F_e > 0$$
$$\alpha_w = 0 \; for \; F_w < 0 \;\; ; \;\; \alpha_e = 0 \; for \; F_e < 0.$$

3.1.8 *The Case of Unstructured Grids*

The interpolation schemes discussed so far were initially developed for orthogonal structured grids. In the case of unstructured grids, determining the cell centres to be used in the application of the scheme becomes more complicated, as they do not all lie on the same line. One approach to address this problem involves reformulating the schemes of interest in terms of the gradient $\nabla\phi_P$ of the transported quantity ϕ at the cell centre and the gradient at the considered face, $\nabla\phi_f$. For the upwind scheme, it will be $\phi_f = \phi_P$. For linear interpolation, it will be $\phi_f = \phi_P + \nabla\phi_f \cdot \mathbf{d}_{Pf}$, where $\mathbf{d}_{Pf}$ is the vector connecting the cell centre P with the face centre f. For the linear upwind scheme, $\phi_f = \phi_P + \left(2\nabla\phi_P - \phi_f\right) \cdot \mathbf{d}_{Pf}$. It is evident that, in this case, accurately calculating the value of the gradient of ϕ both at the cell centre (see Sect. 3.4) and at the centroid of the considered face (see Sect. 3.4.1) is fundamental. An alternative approach is described below and is implemented in most commercial solvers. Known as Barth and Jespersen's method, this scheme involves the use of a limiter. Specifically, it will be:

$$\phi_f = \begin{cases} \phi_P + \psi_f \nabla\phi_P \cdot \mathbf{d}_{Pf} & for \quad \mathbf{F} \geq 0, \\ \phi_N + \psi_f \nabla\phi_N \cdot \mathbf{d}_{Nf} & for \quad \mathbf{F} < 0. \end{cases}$$

The symbol ψ_f represents the limiter, which is necessary to avoid overestimates or underestimates resulting from the calculation of the gradient at the cell centre.

3.2 Reconstruction

The application of the finite volume method results in a set of cells within which the value of the considered quantity is assumed to be constant, leading to a piecewise constant behaviour. In practice, this results in the loss of information regarding the spatial distribution within each cell. Given that the values are only known at the corresponding cell centres, the objective of the reconstruction process is to recover the spatial distribution of the considered quantity inside the cell using continuous functions, which are typically represented by polynomials of varying degrees. A fundamental constraint that these polynomials must satisfy is that they must be *conservative*, meaning that their average value within the cell must equal the value at the cell centre. Each cell has a distinct polynomial, determined based on the values assumed by the quantity at the centres of adjacent cells. The selection of these cell centres for polynomial construction defines what is referred to in the scientific literature as the *stencil*. The number of cell centres considered in this process determines the degree of the polynomial. When the stencil remains the same for every cell, the approach is referred to as *linear reconstruction*. This is the case for the central differencing, linear upwind, and QUICK schemes previously discussed. The use of numerical schemes with an order of accuracy higher than the first becomes necessary when discretising equations that give rise to strong gradients or discontinuities, even when the initial solutions do not exhibit such features. In general, the error in computing the numerical solution increases significantly as the variation of the considered quantity approaches the minimum resolvable by the computational grid. For equations that always ensure smooth solutions, even when starting from continuous initial conditions, the numerical error can be reduced by refining both the spatial resolution and the time integration step. However, for equations—such as the Burgers equation (Eq. 1.10)–that can develop discontinuous solutions even from continuous initial conditions, such an approach does not yield satisfactory results. In these cases, numerical schemes with higher-order accuracy are required. Higher-order schemes inevitably introduce unwanted spurious numerical oscillations, particularly near strong gradients or discontinuities. To minimise these oscillations as much as possible, specialised techniques have been developed, among which the *non-linear reconstruction* approach is notable. In non-linear reconstruction, different stencils are used depending on the considered cell. Examples of this approach include the TVD schemes, the Essentially Non-Oscillatory (ENO) schemes, and the Weighted Essentially Non-Oscillatory (WENO) schemes, which are discussed below.

3.2.1 Essentially Non Oscillatory (ENO) Schemes

This interpolation method is particularly suitable for cases where the considered quantity exhibits discontinuities or strong gradients. As observed in the QUICK scheme, classical interpolation methods employ an interpolating polynomial of degree n that passes through each of the $n + 1$ centres of the considered stencil. In contrast, the present approach utilises a single *global interpolation polynomial*.

In the ENO approach, multiple polynomials of degree less than n are considered, each characterised by the following properties:

- *local definition*: each polynomial is defined within the interval determined by the considered stencil;
- *conservativeness*: dividing the integral of the considered polynomial, limited to the extent of the given cell, by the cell's extent yields the value of the quantity at the cell centre, which was used to construct the polynomial itself. Considering the one-dimensional case of cell i, let ϕ_i be the value of the considered quantity at the cell centre. The cell i has an extent Δx, bounded by the points $x_{i-1/2}$ and $x_{i+1/2}$. The conservativeness property of the interpolating polynomial $p_i(x)$ is expressed as

$$\frac{1}{\Delta x}\int_{x_{i-1/2}}^{x_{i+1/2}} p_i(x)dx = \phi_i;$$

- *adaptability*: the selection of the polynomial to be used must depend on the values assumed by the quantity at the centres belonging to the considered stencil.

Fig. 3.24 Cells of one-dimensional discretised computational domain

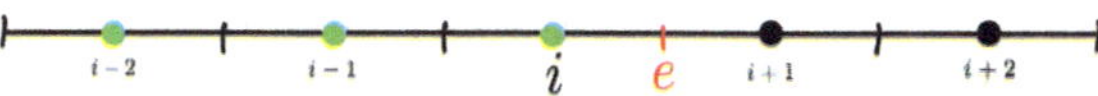

Fig. 3.25 Cell centres (in gray) used for the construction of the polynomial p_0^i

To better understand this approach, and with reference to Fig. 3.24, consider the case of a one-dimensional domain with a stencil consisting of five cell centres used to determine the value of the quantity at the interface between two cells.

The classical approach in this case prescribes the determination of a single global interpolating polynomial of fourth degree, passing through the five cell centres that define the considered stencil. The ENO approach, on the other hand, establishes that if second-degree polynomials (thus achieving third-order accuracy) are preferred, three different polynomials must be determined, each constructed on three different centres of the stencil shown in Fig. 3.24. To determine the value of the quantity ϕ at the interface e, recalling the concepts introduced for the QUICK scheme in Sect. 3.1.6, the first second-degree polynomial that can be considered, denoted as $p_i^{(0)}$, is the one whose value coincides with that of ϕ at the centres $i-2$, $i-1$, and i (see Fig. 3.25). This polynomial leads to the following expression:

$$\phi_e^{(0)} = \frac{3}{8}\phi_{i-2} - \frac{10}{8}\phi_{i-1} + \frac{15}{8}\phi_i. \tag{3.32}$$

The second second-degree polynomial, denoted as $p_i^{(1)}$, that can be considered is the one whose value coincides with that of ϕ at the centres $i-1$, i, and $i+1$ (Fig. 3.26). This polynomial leads to the following expression:

$$\phi_e^{(1)} = -\frac{1}{8}\phi_{i-1} + \frac{6}{8}\phi_i + \frac{3}{8}\phi_{i+1}. \tag{3.33}$$

The third second-degree polynomial, denoted as $p_i^{(2)}$, that can be considered is the one whose value coincides with that of ϕ at the centres i, $i+1$, and $i+2$. This polynomial leads to the following expression (Fig. 3.27):

$$\phi_e^{(2)} = \frac{3}{8}\phi_i + \frac{6}{8}\phi_{i+1} - \frac{1}{8}\phi_{i+2}. \tag{3.34}$$

Notice that by substituting i with the value $i-1$ in the subscripts of Eqs. 3.32, 3.33, and 3.34, the corresponding values of $\phi_w^{(0)}$, $\phi_w^{(1)}$, and $\phi_w^{(2)}$ are obtained. For the selection of the single value among the three obtained to be used, the ENO method

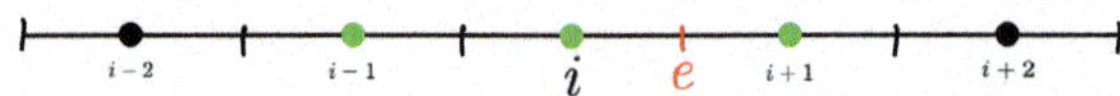

Fig. 3.26 Cell centres (in gray) used for the construction of the polynomial p_1^i

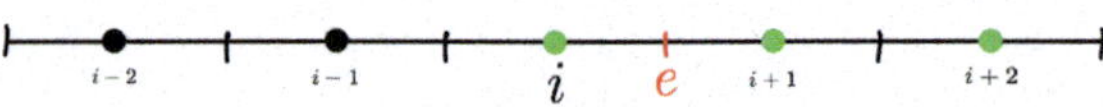

Fig. 3.27 Cell centres (in grey) used for the construction of the polynomial p_2^i

defines the *smoothness indicator* $\beta^{(k)}$, which is defined for each of the polynomials $p_i^{(0)}$, $p_i^{(1)}$, and $p_i^{(2)}$ as

$$\beta^k = \sum_{j=1}^{n} \Delta x^{2j-1} \int_{i-1/2}^{i+1/2} \left(\frac{d^j}{dx^j} p_i^{(k)}(x) \right)^2 dx \tag{3.35}$$

and calculated at the considered interface. In Eq. 3.35, $p_i^{(k)}$ is the polynomial of degree n ($n = 2$ in this case) with the value at i equal to that of ϕ at the cell centre i, associated with the sub-stencil k (where $k = 0, 1, 2$ in this case). The smoothness factor is, in fact, the sum of the squares of the derivatives of the polynomials $p_i^{(k)}$: the polynomial characterised by lower gradients will have a lower smoothness factor. For the second-degree polynomials considered in this example, at the interface e, it will be:

$$\begin{aligned}
\beta_e^0 &= \frac{1}{3}\left(4\phi_{i-2}^2 - 19\phi_{i-2}\phi_{i-1} + \phi_{i-1}^2 + 11\phi_{i-2}\phi_i - 31\phi_{i-1}\phi_i + 10\phi_i^2\right), \\
\beta_e^1 &= \frac{1}{3}\left(4\phi_{i-1}^2 - 13\phi_{i-1}\phi_i + 13\phi_i^2 + 5\phi_{i-1}\phi_{i+1} - 13\phi_i\phi_{i+1} + 4\phi_{i+1}^2\right), \\
\beta_e^2 &= \frac{1}{3}\left(10\phi_i^2 - 31\phi_i\phi_{i+1} + 25\phi_{i+1}^2 + 11\phi_i\phi_{i+2} - 19\phi_{i+1}\phi_{i+2} + 4\phi_{i+2}^2\right).
\end{aligned}$$

Once again, by substituting i with the value $i - 1$ in the subscripts of the expressions for β_e^k, the expression for β_w^k is obtained. The polynomial characterised by the lowest value of the smoothness factor at the considered interface will determine the choice of the value of ϕ to be attributed to the interface e. Notice that in this way, only one of the three values, $\phi_e^{(0)}$, $\phi_e^{(1)}$, $\phi_e^{(2)}$, will be used, and the remaining two will be discarded.

3.2.2 *Weighted Essentially Non Oscillatory (WENO) Schemes*

These types of schemes are derived from the ENO schemes. The main difference between ENO and WENO schemes is that the latter take into account the information produced by the interpolating polynomials, which, having a high smoothness factor, are discarded in the ENO schemes. To better understand this type of scheme, consider the example presented in the previous section related to the ENO schemes. In particular, it is observed that the total number of cell centres used is five. With this number of points, a global interpolation polynomial of fourth degree (fifth order of accuracy) could be constructed. The basis of the WENO method is that the global fourth-degree polynomial can be written as a linear combination of the three second-degree polynomials, $p_i^{(0)}$, $p_i^{(1)}$, and $p_i^{(2)}$:

$$\phi_e = \gamma_0 \phi_e^{(0)} + \gamma_1 \phi_e^{(1)} + \gamma_2 \phi_e^{(2)}$$

In this case, the coefficients γ_0, γ_1, and γ_2 must satisfy the condition $\gamma_0 + \gamma_1 + \gamma_2 = 1$ and are known as *linear weights*. To achieve a fifth-order accuracy, the values are $\gamma_0 = \frac{1}{16}$, $\gamma_1 = \frac{5}{8}$, and $\gamma_2 = \frac{5}{16}$. In the absence of discontinuities, or strong gradients, of ϕ in the interval defined by the (in this example, five) cells identified for the definition of the polynomials, considering the polynomials $p_i^{(0)}$, $p_i^{(1)}$, and $p_i^{(2)}$ leads to equivalent results as employing a single fifth-degree polynomial. However, in the presence of a discontinuity, this is no longer the case. As previously seen, the ENO method uses the smoothness coefficient to choose the polynomial capable of ensuring the maximum accuracy order (in this example, the third). In contrast, the WENO method determines the best approximation as a convex combination of the three values $\phi_e^{(0)}$, $\phi_e^{(1)}$, and $\phi_e^{(2)}$. It is worth recalling that a convex combination is a linear combination of elements made with non-negative coefficients summing to one:

$$\phi_e = \omega_0 \phi_e^{(0)} + \omega_1 \phi_e^{(1)} + \omega_2 \phi_e^{(2)}$$

with $\omega_0 + \omega_1 + \omega_2 = 1$ and $\omega_k \geqslant 0$ for $k = 0, 1, 2$. The coefficients ω_k are called *non-linear weights*. In calculating the value of the nonlinear weights, it is necessary to keep in mind the following constraints:

- $\omega_k \approx \gamma_k$ for $k = 0, 1, 2$ in the case where ϕ does not have a discontinuity in the stencil consisting of all five cell centres;
- $\omega_k \approx 0$ for $k = 0, 1, 2$ in the case where ϕ has a discontinuity in the stencils consisting of the three cell centres used to construct the polynomials $p_i^{(k)}$, except for one (i.e., in at least one of the three stencils there must not be a discontinuity).

Given these constraints, the non-linear weights are defined as

$$\omega_k = \frac{\tilde{\omega}_k}{\tilde{\omega}_0 + \tilde{\omega}_1 + \tilde{\omega}_2} \qquad \text{with} \qquad \tilde{\omega}_k = \frac{\gamma_j}{(\epsilon + \beta_k)^2}$$

in which ϵ is a constant necessary to avoid the case of a zero denominator: the value typically associated with it is 10^{-6}.

3.3 Interpolation of Diffusive Fluxes

In this case, it is necessary to calculate, at the centroid of the face, the component of the gradient of the transported quantity along the direction normal to the face itself. In fact, referring to Eq. 3.3, it is possible to write

$$\int_{V_P} \nabla \cdot (\rho \Gamma_\phi \nabla \phi)\, dV \approx \sum_f \mathbf{S}_f \cdot \overline{(\rho \Gamma_\phi \nabla \phi)}_f =$$
$$\sum_f \mathbf{S}_f \cdot (\rho \Gamma_\phi \nabla \phi)_f = \sum_f \rho_f \left(\Gamma_\phi\right)_f |\mathbf{S}_f| \left(\frac{\partial \phi}{\partial n}\right)_f \tag{3.36}$$

In this case, the term $\left(\Gamma_\phi\right)_f$ represents the diffusivity coefficient, which, depending on ϕ, could be either viscosity or thermal diffusivity, calculated at the centroid of the face. The value of $\left(\Gamma_\phi\right)_f$ can be obtained using one of the interpolation schemes described above (e.g. central differencing, upwind, etc.).

Given the values of ϕ at the centres of the two cells to which the face belongs, it is possible to calculate the component $(\partial \phi / \partial n)_f$ of the gradient $\nabla \phi_f$ on the face, along the direction defined by the two centres. In the case where this direction coincides with the normal to the face (see Fig. 3.28), it will not be necessary to calculate any additional components. However, if the directions of the normal to the face and the line joining the two centres do not coincide (see Fig. 3.29), the angle between these two directions is referred to as *non-orthogonality*. In the presence of non-orthogonality, the vector normal to the face can be decomposed into two components. The first, known as the *normal component*, is along the direction joining the centres of the two cells to which the face belongs. The second component, determined by the chosen decomposition method, is referred to as *transverse diffusion* or "*cross diffusion*".

In the case of two cells of an orthogonal grid, as shown in Fig. 3.28, the diffusive flux can be calculated with a second-order approximation using a centred difference.

$$\mathbf{S} \cdot \nabla \phi_f = |\mathbf{S}| \frac{\phi_N - \phi_P}{|\mathbf{d}|} \tag{3.37}$$

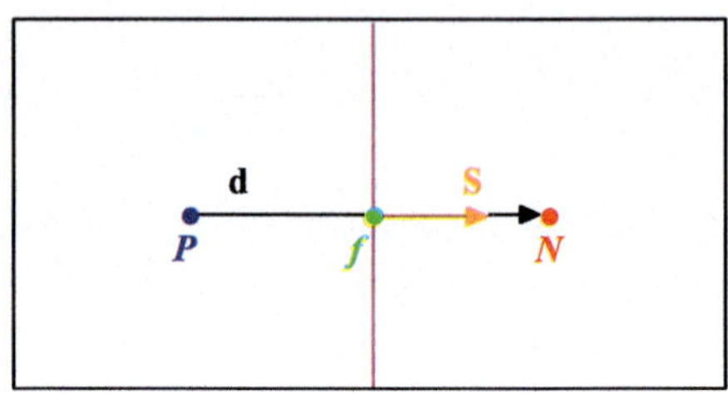

Fig. 3.28 Orthogonal grid

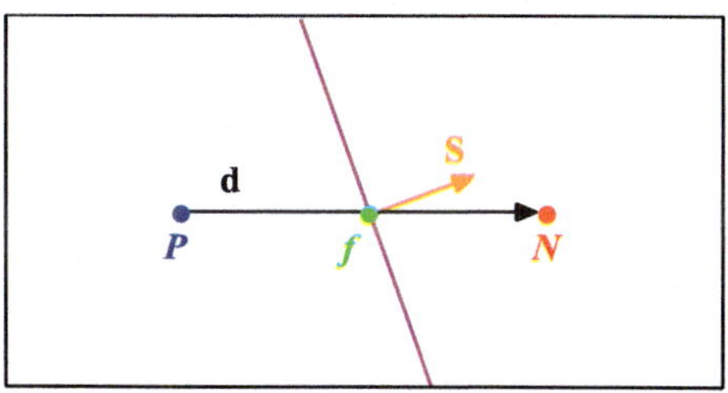

Fig. 3.29 Non-orthogonal grid

where **S** is the vector normal to the face, ϕ_N is the value of ϕ in the centre of cell N, ϕ_P is the value of ϕ in the centre of cell P, and **d** is the vector connecting the two cell centres.

In the case of a non-orthogonal grid (see Fig. 3.29), the vector **S** can be decomposed into its components along two directions:

- the direction connecting the two cell centres identified by the unit vector Δ;
- the direction parallel to that of the face and identified by the unit vector **k**.

Indicating with $\hat{\mathbf{n}}$ the unit vector normal to the face it is $\hat{\mathbf{n}} = \Delta + \mathbf{k}$ (see Fig. 3.30). At this point, it is possible to write Eq. 3.36 as

$$\sum_f \mathbf{S}_f \cdot (\rho\Gamma_\phi \nabla\phi)_f = \sum_f |\mathbf{S}_f| \left[(\rho\Gamma_\phi \nabla\phi)_f \cdot \hat{\mathbf{n}}_f\right] = \sum_f S_f \left[(\rho\Gamma_\phi \nabla\phi)_f \cdot \hat{\mathbf{n}}_f\right]$$

in which $\hat{\mathbf{n}}$ is the unit vector of the vector **S**. That is

$$\begin{aligned}&\sum_f S_f \left[(\rho\Gamma_\phi \nabla\phi)_f \cdot \hat{\mathbf{n}}_f\right] = \sum_f S \left[(\rho\Gamma_\phi \nabla\phi)_f \cdot (\Delta + \mathbf{k})_f\right] = \\ &\sum_f S_f \left[(\rho\Gamma_\phi \nabla\phi)_f \cdot \Delta_f\right] + \sum_f S_f \left[(\rho\Gamma_\phi \nabla\phi)_f \cdot \mathbf{k}_f\right]. \end{aligned} \tag{3.38}$$

In Eq. 3.38, the orthogonal term is clearly visible

$$\sum_f S_f \left[(\rho\Gamma_\phi)_f \nabla\phi_f \cdot \Delta_f\right]$$

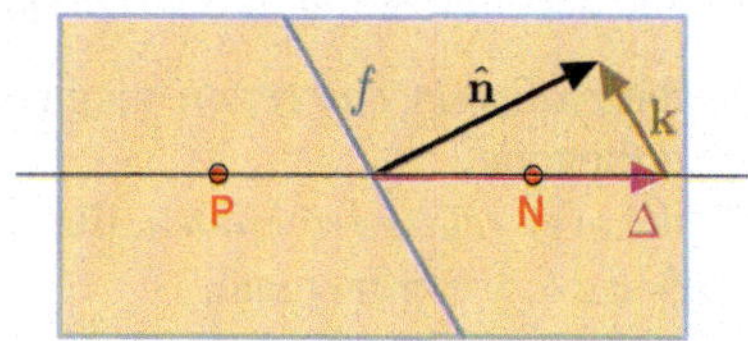

Fig. 3.30 Over-relaxed decomposition method

and the non-orthogonal term, also called *non-orthogonal correction*,

$$\sum_f S_f \left[(\rho \Gamma_\phi)_f \nabla \phi_f \cdot \mathbf{k}_f \right].$$

The orthogonal term can be calculated considering that

$$\nabla \phi_f \cdot \Delta = |\Delta| \frac{\phi_N - \phi_P}{|\mathbf{d}|}.$$

As demonstrated in Sect. 3.1, the discretisation process leads to the formulation of a discretised conservation equation for each grid element. The equations for all the elements are then reorganised to form a system, where each row contains the elements of the discretised conservation equation calculated at the centre of the generic cell. The elements on the main diagonal correspond to quantities related to the centre of the cell, while the off-diagonal elements represent quantities related to the centres of adjacent cells. The contribution of the orthogonal term appears in the coefficient matrix of this system of equations and can therefore be solved implicitly, which enhances the numerical stability of the solution process. Conversely, the contribution of the non-orthogonal term ($\nabla \phi_f \cdot \mathbf{k}$) acts as a source term, appearing in the column vector of known terms. As a result, it must be solved explicitly, considering the values of ϕ calculated at the previous iteration or determined from the initial conditions. This explicit calculation, however, contributes to a reduction in the stability of the numerical solution process. As the level of non-orthogonality between the cells increases, the relative magnitude of the non-orthogonal terms grows, which in turn decreases the stability of the numerical solution. To mitigate the effects of the non-orthogonal contribution, two approaches can be considered:

1. generate a computational grid with a low degree of non-orthogonality.
2. artificially limit the value of the non-orthogonal contribution so that it does not exceed the value of the orthogonal contribution or a fraction thereof (see Sect. 6.1.4).

Regarding point 2, it is important to specify that a greater artificial reduction of the non-orthogonal contribution corresponds to greater stability, but also to reduced accuracy, which is acceptable if the number of cells with high non-orthogonality remains relatively small. It should be noted that, given the vector **S** and, therefore, its unit vector $\hat{\mathbf{n}}$, as well as the direction of the vector Δ—determined by the line connecting the two cell centres to which the considered face belongs—the possibilities for the decomposition of **S** are infinite. Among these, we highlight two approaches:

1. the *minimum correction* method, which ensures that the vectors Δ and **k** are orthogonal,
2. the *over-relaxed approach*, illustrated in Fig. 3.30, which ensures that the vectors **S** and **k** are orthogonal.

The *over-relaxed* approach offers greater numerical stability because, as the non-orthogonality level increases, both the non-orthogonal and orthogonal components increase. In contrast, in the minimum correction approach, as the level of non-orthogonality increases, the non-orthogonal component increases while the orthogonal component decreases. In summary,

$$(\nabla\phi)_f\cdot\mathbf{k}_f = (\nabla\phi)_f\cdot\left(\mathbf{S}_f-\mathbf{d}_f\right) = \begin{cases}(\nabla\phi)_f\cdot\left(\hat{\mathbf{n}}-\hat{\mathbf{d}}\cos\theta\right)S_f & \text{minimum correction,}\\ (\nabla\phi)_f\cdot\left(\hat{\mathbf{n}}-\hat{\mathbf{d}}\right)S_f & \text{normal correction,}\\ (\nabla\phi)_f\cdot\left(\hat{\mathbf{n}}-\hat{\mathbf{d}}\dfrac{1}{\cos\theta}\right)S_f & \text{over-relaxed}\end{cases}$$

where $\mathbf{S}=\hat{\mathbf{n}}S$ is the vector normal to the face, $\mathbf{d}$ is the vector connecting the two cell centres, and $\hat{\mathbf{d}}$ is its unit vector. Additionally, $\mathbf{k}$ represents the vector of transverse diffusion, and θ is the angle formed between the line connecting the two cell centres and the direction normal to the face. It is clear that the transverse diffusion term cannot be calculated using the values of ϕ at the cell centres. Therefore, the calculation proceeds by first determining the gradient $\nabla\phi$ at the cell centres and then interpolating to obtain its value on the face, $(\nabla\phi)_f$ (see Fig. 3.2).

$$(\nabla\phi)_f = f_x\,(\nabla\phi)_P + (1-f_x)\,(\nabla\phi)_N$$

where the gradients at the two cell centres are calculated using Eq. 3.40. This method of calculating the gradient on the face is also referred to as the *Green-Gauss method* (*Green-Gauss cell-based gradient*), and it is second-order accurate. Once the value of the transverse diffusion term is obtained, it is added to the algebraic equation of the cell as a source term. In summary, for orthogonal grids (see Fig. 3.28), the diffusive flux can be calculated using a second-order accurate approximation via Eq. 3.37. In the case of a non-orthogonal grid (see Fig. 3.29), the vector $\mathbf{S}$ can be calculated as the sum of two vectors (see Fig. 3.30). In the over-relaxed method, it turns out to be

$$\Delta = \frac{\mathbf{d}}{\mathbf{d}\cdot\mathbf{S}}|\mathbf{S}|^2.$$

From which

$$\mathbf{S}\cdot(\nabla\phi)_f = \underbrace{|\Delta|\frac{\phi_N-\phi_P}{|\mathbf{d}|}}_{orthogonal\ contribution} + \overbrace{\mathbf{k}\cdot(\nabla\phi)_f}^{contribution\ non\ orthogonal} \tag{3.39}$$

3.4 Calculation of the Gradient at the Cell Centre

The computation of the gradient at the cell centre is essential for both the application of discretisation schemes for the convective terms of the general transport equation (see Sect. 6.1.2) and for the discretisation of specific terms (e.g., the gradient of the pressure) that appear in the discretised momentum conservation equation (see Eq. 2.42). Moreover, the calculation of the velocity gradient at the cell centre is necessary for determining the production terms of turbulent kinetic energy in turbulence modelling or for the determination of shear stress in flows with non-Newtonian fluids. A widely used approach for calculating the gradient is based on the Gauss theorem (see Sect. 1.1.6), which states that, for each closed volume V bounded by a surface ∂V, the integral of the gradient of a generic quantity (scalar or vector) ϕ over the volume V is equal to the integral of ϕ over the surface ∂V. In mathematical terms:

$$\int_V \nabla\phi \, dV = \oint_{\partial V} \phi \, d\mathbf{S}$$

in which $d\mathbf{S}$ is the vector normal to the surface, representing the generic elementary surface element. To discretise this equation, the mean value theorem can be applied (see Sect. 1.5.2) to the left-hand side.

$$\int_V \nabla\phi \, dV \approx \overline{\nabla\phi} V$$

where $\overline{\nabla\phi}$ is the average value of the gradient within the volume V. Therefore, it can be expressed as

$$\overline{\nabla\phi} \approx \frac{1}{V} \oint_{\partial V} \phi \, d\mathbf{S}$$

and for the generic cell with centroid P

$$\overline{\nabla\phi_P} \approx \frac{1}{V_P} \oint_{\partial V_P} \phi \, d\mathbf{S}.$$

Then, by applying the mean value theorem to the faces that bound the cell, it becomes

$$\nabla\phi_P \approx \frac{1}{V_P} \sum_{f \sim nb(P)} \phi_f \mathbf{S}_f \tag{3.40}$$

where we have set $\overline{\nabla\phi_P} = \nabla\phi_P$. $nb(P)$ is the number of faces that delimit the cell with centroid P, ϕ_f is the average value of ϕ on the generic face f of cell P, and $\mathbf{S}_f$ is the vector exiting from cell P, normal to face f, with the area of f as its magnitude.

The calculation of ϕ_f can be performed using two possible approaches: the first, known as *cell-based*, and the second, known as *node-based*.

Referring to Fig. 3.31, in the *cell-based* approach, it is

$$\phi_f = g_P \phi_P + (1 - g_P)\phi_N \tag{3.41}$$

where ϕ_P is the value of ϕ at the centroid of the considered cell, ϕ_N is the value of ϕ at the centroid of the cell that shares face f with the considered cell, and g_P is the weight factor that depends on the geometric characteristics of the two cells sharing the considered face:

$$g_P = \frac{d_{Pf}}{d}$$

where d_{Pf} is the distance between the centroid P of the considered cell and the centroid of face f, while d is the distance between P and the centroid N of the cell that shares face f with the cell of centroid P. Equation 3.41 is a second-order accurate approximation only in the case where the intersection between face f and the line connecting the two centroids P and N coincides with the centroid of the face. This condition is not satisfied in most non-orthogonal structured grid cases, as well as in most unstructured grid cases. Referring to Fig. 3.31, where f denotes both the face considered and its centroid, and f' denotes the intersection between face f and the line connecting the two centroids P and N, the "*skewness error*" is defined as the distance between f and f'. Therefore, in the case of non-zero skewness error, Eq. 3.41 will provide the value of $\phi_{f'}$. In the case of grids characterised by high skewness error values, one can use the *node-based* approach, in which the value of ϕ_f is obtained as the average of the ϕ values at the vertices of the face considered. The value at the vertices is in turn determined by calculating the weighted average of ϕ at the centres of the cells sharing the considered vertex.

The value of ϕ at the vertex n of the face considered can be obtained using the following formula

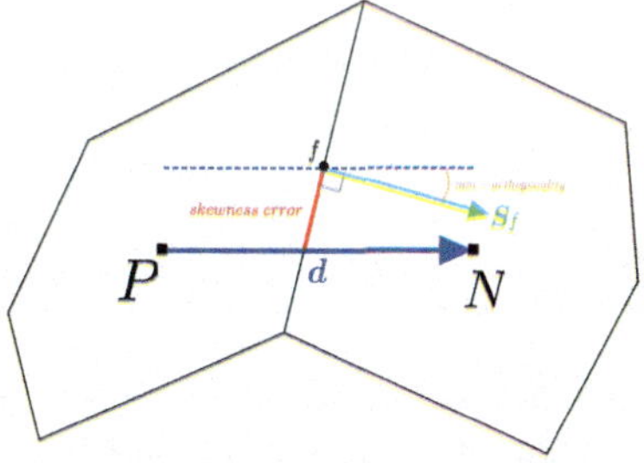

Fig. 3.31 Distortion and non-orthogonality

$$\phi_n = \frac{\sum_{k=1}^{NB(n)} \frac{\phi_{F_k}}{\|\mathbf{r}_n - \mathbf{r}_{F_k}\|}}{\sum_{k=1}^{NB(n)} \frac{1}{\|\mathbf{r}_n - \mathbf{r}_{F_k}\|}}$$

where ϕ_{F_k} denotes the value of ϕ on the k-th face to which the considered vertex belongs, calculated by simple linear interpolation between the values of ϕ at the centres of the two cells to which the k-th face belongs. $NB(n)$ is the total number of cells to which the considered vertex belongs, and $\|\mathbf{r}_n - \mathbf{r}_{F_k}\|$ is the distance between the vertex n and the centroid of the considered cell. Once the values at all the vertices of the considered face are known, the face is divided into triangles, each of which has for vertices two vertices of the face and the intersection point between the face and the line connecting the two centres of the cells to which the face belongs. In the presence of skewness error, this last point does not coincide with the centroid of the face. For each triangle, the average value ϕ_T of the values at the vertices is calculated. Subsequently, the value ϕ_f of the quantity on the face is obtained as a weighted average, as follows:

$$\phi_f = \frac{\sum S_T \phi_T}{S_f}$$

where S_T is the area of each triangle and S_f is the area of the considered face. This approach entails a computational burden due to the necessity of managing the information associated with the vertices of the faces. However, it provides greater accuracy in the presence of distorted cells, given its independence from the distortion error. Another method for computing the gradient at the cell centre is known as the *Least-Squares Fit* (LSF). Consistent with the constraint of second-order accuracy, this method assumes a linear variation of the quantity ϕ and defines an error function for each neighbouring cell N surrounding the considered cell P:

$$\epsilon_N = \phi_N - \left[\phi_P + \mathbf{d} \cdot (\nabla\phi)_P\right]$$

where the vector $\mathbf{d}$ connects the two cell centres P and N. It then proceeds to minimise the mean squared error (*least-squares error*) defined as

$$\epsilon_P^2 = \sum_N w_N^2 \epsilon_N^2$$

in which the weight function w is defined as $w_N = \frac{1}{|\mathbf{d}|}$. The expression used for the calculation of the gradient at the centre of cell P is

$$(\nabla\phi)_P = \sum_N w_N^2 \mathbf{G}^{-1} \cdot \mathbf{d}\,(\phi_N - \phi_P)$$

with

$$\mathbf{G} = \sum_N w_N^2 \mathbf{dd}.$$

This calculation method, like the vertex-based method, is unaffected by the distortion error, which, in contrast, influences the results obtained using the cell-based approach.

3.4.1 *Calculation of the Gradient on the Centroid of the Faces*

As observed, the need to perform this calculation arises from the discretisation process of the diffusive terms of the general transport equation, particularly when the computational grid contains non-orthogonal cells. In such cases, it is necessary to calculate the value of the non-orthogonal correction term. One possible approach for calculating the gradient at the centroid of the faces involves correcting the average gradient value computed at the centres of the two cells to which the face belongs. Specifically, with reference to Fig. 3.31, the gradient of ϕ at the centroid of the face, denoted as $\nabla\phi_f$, will be

$$\nabla\phi_f = \overline{\nabla\phi_f} + \left[\frac{\phi_N - \phi_P}{d} - \left(\overline{\nabla\phi_f} \cdot \mathbf{e}\right)\right]\mathbf{e}$$

in which

$$\overline{\nabla\phi_f} = g_P \nabla\phi_P + g_N \nabla\phi_N, \quad g_P = \frac{d_{Pf}}{d}, \quad g_N = \frac{d_{Nf}}{d} \quad \mathbf{e} = \frac{\mathbf{d}}{d}, \quad \mathbf{d} = \mathbf{r}_N - \mathbf{r}_p$$

with $\mathbf{e}$ being the unit vector of the vector $\mathbf{d}$, $\mathbf{r}_P$ the position vector of the cell centre P, and $\mathbf{r}_N$ the position vector of the cell centre N.

3.5 Calculation of the Time Derivative or Transient Term

Typically, the evolution of a quantity ϕ is described by an equation of the type

$$\frac{\partial(\rho\phi)}{\partial t} + L(\phi) = 0$$

in which the term $L(\phi)$ is an operator that includes all terms (convection, diffusion, sources, etc.) that are not dependent on time. Integrating over the cell with centroid C, we obtain

$$\int_{V_C} \frac{\partial(\rho\phi)}{\partial t}\, dV + \int_{V_C} L(\phi)\, dV = 0$$

and discretising in space, we obtain

$$\frac{\partial\left(\rho_C\phi_C\right)}{\partial t}V_C + L(\phi_C^t) = 0$$

where V_C is the cell volume, while $L(\phi_C^t)$ is the spatial discretisation operator at time t. Traditionally, the approach used for time discretisation involves the application of finite differences, where the Taylor series expansion of the term $\dfrac{\partial\left(\rho\phi\right)}{\partial t}$ is employed to express the derivative in terms of the values at the cell centres. The finite volume method, on the other hand, involves applying strategies similar to those used in the spatial discretisation of the convective term, but instead of integrating in space, the integration is performed in time. To better understand this approach, consider a two-dimensional computational domain in which only one direction of integration in space is considered. This integration in space will correspond to the time evolution. Let Δt denote the time integration interval.

- The value at the centre of the spatial discretisation cell corresponds to the value of ϕ at a specific time instant t;
- The value on the faces of the spatial discretisation cell corresponds to the value of ϕ at the instant $t \pm \Delta t/2$.

The time discretisation cell will therefore have its centre positioned at the time coordinate t, and its two faces at the time coordinates $t - \Delta t/2$ and $t + \Delta t/2$. Integrating then over the time interval $[t - \Delta t/2,\ t + \Delta t/2]$ it is (Fig. 3.32)

$$\int_{t-\Delta t/2}^{t+\Delta t/2} \frac{\partial\left(\rho_C\phi_C\right)}{\partial t}V_C\,dt + \int_{t-\Delta t/2}^{t+\Delta t/2} L(\phi_C)dt = 0.$$

Considering that V_C is constant over time, the first integral results in a simple difference, while, by applying the rule of the average value, the second integral yields only the value of the integrand at time t.

$$V_C\left(\rho_C\phi_C\right)^{t+\Delta t/2} - V_C\left(\rho_C\phi_C\right)^{t-\Delta t/2} + L(\phi_C^t)\Delta t = 0.$$

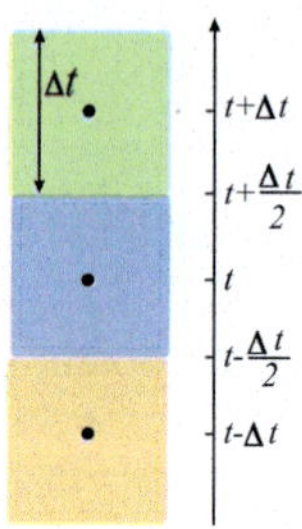

Fig. 3.32 Temporal discretisation

Dividing by Δt, we obtain

$$\frac{(\rho_C \phi_C)^{t+\Delta t/2} - (\rho_C \phi_C)^{t-\Delta t/2}}{\Delta t} V_C + L(\phi_C^t) = 0 \tag{3.42}$$

From this, it is clear that interpolation between the values at times t, $t - \Delta t$, etc., is required to obtain the values at the "intermediate" times $t - \Delta t/2$ and $t + \Delta t/2$. Just as in the case of spatial discretisation of convective terms, the choice of interpolation method will influence the accuracy of the results.

3.5.1 *Implicit Euler Scheme*

This time integration scheme is derived from the upwind spatial integration scheme. It is, in fact, assumed that

$$(\rho_C \phi_C)^{t+\Delta t/2} = (\rho_C \phi_C)^t \qquad \textit{and} \qquad (\rho_C \phi_C)^{t-\Delta t/2} = (\rho_C \phi_C)^{t-\Delta t}.$$

Substituting into Eq. 3.42, we obtain

$$\frac{(\rho_C \phi_C)^{t+\Delta t} - (\rho_C \phi_C)^{t}}{\Delta t} V_C + L(\phi_C^t) = 0.$$

3.5.2 *Crank-Nicolson Scheme or Central Difference Profile*

This time integration scheme performs a linear interpolation between the value of the quantity at time $t - \Delta t$ and the value at time $t + \Delta t$. It is, therefore,

$$(\rho_C \phi_C)^{t+\Delta t/2} = \frac{1}{2} (\rho_C \phi_C)^{t+\Delta t} + \frac{1}{2} (\rho_C \phi_C)^{t},$$
$$(\rho_C \phi_C)^{t-\Delta t/2} = \frac{1}{2} (\rho_C \phi_C)^{t} + \frac{1}{2} (\rho_C \phi_C)^{t-\Delta t}.$$

Substituting into Eq. 3.42, we obtain

$$\frac{(\rho_C \phi_C)^{t+\Delta t} - (\rho_C \phi_C)^{t-\Delta t}}{2\Delta t} V_C + L(\phi_C^t) = 0.$$

3.5.3 Backward Scheme or Second Order Upwind Euler

This time integration scheme is derived from the linear upwind spatial integration scheme. In fact, we set

$$(\rho_C \phi_C)^{t+\Delta t/2} = \frac{3}{2}(\rho_C \phi_C)^t - \frac{1}{2}(\rho_C \phi_C)^{t-\Delta t},$$

$$(\rho_C \phi_C)^{t-\Delta t/2} = \frac{3}{2}(\rho_C \phi_C)^{t-\Delta t} - \frac{1}{2}(\rho_C \phi_C)^{t-2\Delta t}.$$

Substituting into Eq. 3.42, we obtain

$$\frac{3(\rho_C \phi_C)^t - 4(\rho_C \phi_C)^{t-\Delta t} + (\rho_C \phi_C)^{t-2\Delta t}}{2\Delta t} V_C + L(\phi_C^t) = 0.$$

Chapter 4
Linear Systems and Their Solution

In the previous chapter, we observed that the discretisation process results in a discretised conservation equation for each of the N grid elements. It is possible (see Sect. 3.1.1) to assemble the equations corresponding to all N elements into a system of equations, which, in its compact form, can be written as

$$\mathbf{A}\phi = \mathbf{b}. \tag{4.1}$$

In the following representation, the elements of the main diagonal correspond to the contribution of the generic cell and are represented by black squares. White squares denote the off-diagonal elements of the matrix, which account for the contributions of adjacent elements that share at least one face with the considered cell. The first row corresponds to the first cell of the discretised domain, while the last row represents the last cell. Since each cell shares a limited number of faces with its neighbouring cells, most elements of the matrix $\mathbf{A}$ will be zero. In the case of structured grids, the non-zero elements will be arranged along the main diagonal and the secondary diagonals.

$$\begin{bmatrix} a_1 & \square & & \square & & & & & \\ \square & \blacksquare & \square & & \square & & & & \\ & \ddots & \ddots & \ddots & & \ddots & & & \\ \ddots & & \ddots & \ddots & \ddots & & \ddots & & \\ & \square & & \square & a_C & \square & & \square & \\ & & \ddots & & \ddots & \ddots & \ddots & & \\ & & & \square & & \square & \blacksquare & \square & \\ & & & & \square & & \square & a_N & \end{bmatrix} \begin{bmatrix} \phi_1 \\ \phi_2 \\ \vdots \\ \\ \phi_C \\ \vdots \\ \\ \phi_N \end{bmatrix} = \begin{bmatrix} b_1 \\ b_2 \\ \vdots \\ \\ b_C \\ \vdots \\ \\ b_N \end{bmatrix}.$$

G. Caramia and E. Distaso, *A Practical Approach to Computational Fluid Dynamics Using OpenFOAM®*, https://doi.org/10.1007/978-3-031-88957-8_4

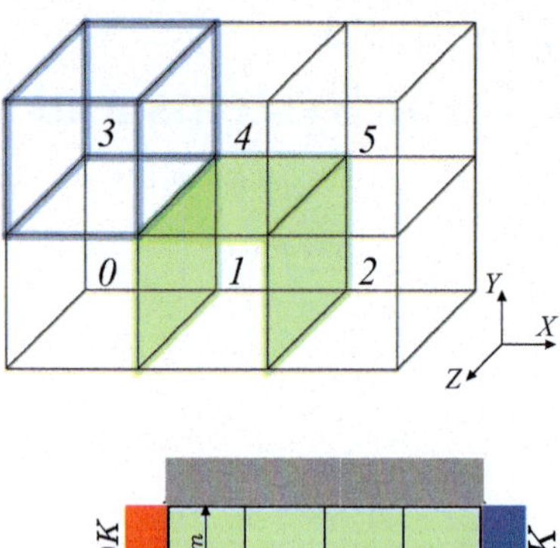

Fig. 4.1 Example of a structured grid

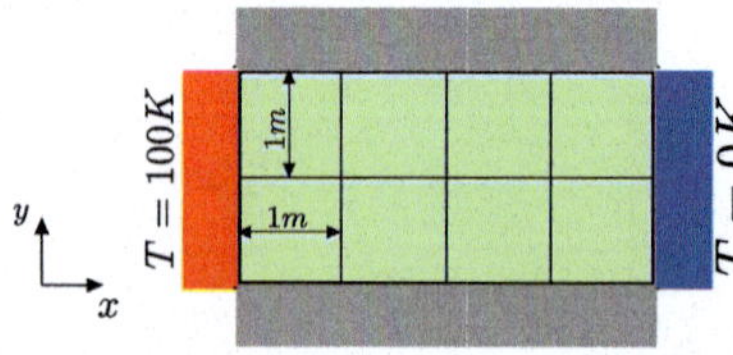

Fig. 4.2 Computational domain

An example of a structured grid comprising six hexahedral cells is shown in Fig. 4.1. The edges of cell 3 are highlighted with a thicker line, while the faces of cell 1 that are shared with other cells are shaded. For this grid, the coefficient matrix can be assembled as shown in Fig. 4.2.

The coefficients resulting from the application of the generic conservation equation to a single cell are all placed in a single row. By ordering the rows consistently with the numbering assigned to the cells, the coefficients associated with cell 0 will be placed in row 0 of the matrix, those related to cell 1 in row 1, and so forth.

Cell 0 shares one face with cell 1 and another with cell 3. Therefore, the computation of fluxes through these two faces is required for the discretisation of the considered conservation equation applied to this cell. Consequently, the terms a_{01}—corresponding to the fluxes through the face shared between cell 0 and cell 1—and a_{03}—corresponding to the fluxes through the face shared between cell 0 and cell 3—will be non-zero. Faces that belong exclusively to a single cell are referred to as "boundary" faces. Subsequently, specific boundary conditions applied to these faces will be considered.

$$\begin{bmatrix} a_{00} & a_{01} & 0 & a_{03} & 0 & 0 \\ a_{10} & a_{11} & a_{12} & 0 & a_{14} & 0 \\ 0 & a_{21} & a_{22} & 0 & 0 & a_{25} \\ a_{30} & 0 & 0 & a_{33} & a_{34} & 0 \\ 0 & a_{41} & 0 & a_{43} & a_{44} & a_{45} \\ 0 & 0 & a_{52} & 0 & a_{54} & a_{55} \end{bmatrix}. \tag{4.2}$$

It can be observed that the coefficient matrix is significantly influenced by the properties of the grid:

- the number of cells in the grid corresponds to the size of the matrix and thus to the number of elements on the main diagonal.

- the number of non-zero coefficients above the main diagonal is equal to the number of non-zero coefficients below it, and both are equal to the number of internal faces of the grid.
- each coefficient on the main diagonal corresponds to the associated cell (identified by the row number).
- the coefficients outside the main diagonal are associated with the cells (identified by the column number) adjacent to the given cell.

Finding the values of the unknowns ϕ_i in Eq. 4.1 requires inverting the matrix $\mathbf{A}$, yielding $\boldsymbol{\phi} = \mathbf{A}^{-1}\mathbf{b}$. Among the characteristics that the coefficient matrix $\mathbf{A}$ must satisfy, the most important being the values of the coefficients themselves, as they are strongly influenced by the geometric properties of the cells, such as orthogonality and skewness. An individual cell with unfavourable values can lead to the failure of the inversion process (*divergence*). The methods for inverting a matrix can initially be classified as *direct* and *indirect*. For reasons related to excessive computer memory and processing requirements, the former are impractical. Indirect methods iteratively apply a solution algorithm until the predefined level of convergence is reached, thus eliminating the requirement to compute the final solution in a single iteration.

4.1 The Jacobi Method

This method is the simplest of the iterative methods for solving linear systems. The solution process begins by assigning an initial guess to each element of the unknowns column vector $\boldsymbol{\phi}$ (*initialisation*). Assuming that the elements of the main diagonal are non-zero and using the initial guess values of ϕ, the first equation is solved to obtain a new estimate of ϕ_1, the second to obtain a new estimate of ϕ_2, and so on until ϕ_N is reached. Once the new estimate of ϕ_N has been obtained, the first iteration is concluded and the new values of $\boldsymbol{\phi}$ can then be used to begin a new iteration. The iterative process continues until the difference between successive iterations is negligible or a stopping criterion is met. The expression for the new estimate of $\boldsymbol{\phi}$ is

$$\phi_j^{(n)} = \frac{1}{a_{ii}} \left(b_j - \sum_{j=1, j\neq i}^{N} a_{ij} \phi_j^{(n-1)} \right) \qquad i, j = 1, 2, 3, \ldots, N$$

which, in matrix form, can be expressed using the following decomposition of the coefficient matrix

$$\begin{bmatrix} a_{11} & 0 & \ldots & 0 & 0 \\ 0 & a_{22} & \ldots & 0 & 0 \\ \vdots & \vdots & \ldots & \vdots & \vdots \\ 0 & 0 & \ldots & 0 & a_{NN} \end{bmatrix} \begin{bmatrix} \phi_1 \\ \phi_2 \\ \vdots \\ \vdots \\ \phi_N \end{bmatrix} + \begin{bmatrix} 0 & a_{12} & \ldots & a_{1N-1} & a_{1N} \\ a_{21} & 0 & \ldots & a_{2N-1} & a_{2N} \\ \vdots & \vdots & \ldots & \vdots & \vdots \\ a_{N1} & a_{N2} & \ldots & a_{NN-1} & 0 \end{bmatrix} \begin{bmatrix} \phi_1 \\ \phi_2 \\ \vdots \\ \vdots \\ \phi_N \end{bmatrix} = \begin{bmatrix} b_1 \\ b_2 \\ \vdots \\ \vdots \\ b_N \end{bmatrix}.$$

The updated value $\phi^{(n)}$ can be obtained as

$$\begin{bmatrix} \phi_1^{(n)} \\ \phi_2^{(n)} \\ \vdots^{(n)} \\ \vdots \\ \phi_N^{(n)} \end{bmatrix} = \begin{bmatrix} a_{11} & 0 & \ldots & 0 & 0 \\ 0 & a_{22} & \ldots & 0 & 0 \\ \vdots & \vdots & \ldots & \vdots & \vdots \\ 0 & 0 & \ldots & 0 & a_{NN} \end{bmatrix}^{-1} \cdot$$
$$\left(\begin{bmatrix} b_1 \\ b_2 \\ \vdots \\ \vdots \\ b_N \end{bmatrix} - \begin{bmatrix} 0 & a_{12} & \ldots & a_{1N-1} & a_{1N} \\ a_{21} & 0 & \ldots & a_{2N-1} & a_{2N} \\ \vdots & \vdots & \ldots & \vdots & \vdots \\ a_{N1} & a_{N2} & \ldots & a_{NN-1} & 0 \end{bmatrix} \begin{bmatrix} \phi_1^{(n-1)} \\ \phi_2^{(n-1)} \\ \vdots \\ \vdots \\ \phi_N^{(n-1)} \end{bmatrix} \right). \tag{4.3}$$

The coefficient matrix $\mathbf{A}$ can be decomposed into

$$\mathbf{A} = \mathbf{D} + \mathbf{L} + \mathbf{U}$$

where $\mathbf{D}$ is a diagonal matrix of order N whose diagonal elements are the diagonal elements of $\mathbf{A}$, $\mathbf{L}$ is a matrix of order N whose subdiagonal elements are the subdiagonal elements of $\mathbf{A}$ and all other elements are zero, and $\mathbf{U}$ is a matrix of order N whose superdiagonal elements are the superdiagonal elements of $\mathbf{A}$ and all other elements are zero. Considering this decomposition, Eq. 4.3 can be written as

$$\phi^{(n)} = \mathbf{D}^{-1}\mathbf{b} \; - \; \mathbf{D}^{-1}\,(\mathbf{L} + \mathbf{U})\,\phi^{(n-1)}.$$

4.2 The Gauss-Seidel Method

This method is similar to that of the Jacobi method, although it is typically preferred due to its superior convergence properties and its lower memory storage requirements. Unlike the Jacobi method, this method always uses the most updated value of each of the unknowns considered. As previously described, the resolution process begins with the initialisation phase. Once the value of ϕ_1 is obtained by solving the first equation, the value of ϕ_2 is obtained by solving the second equation using the value of ϕ_1 just calculated, rather than the initial guess, as in the Jacobi method. Once the value of ϕ_2 is obtained from the second equation, the value of ϕ_3 is obtained by solving the third equation using the value of ϕ_2 just calculated, and so on, until the value of ϕ_N is calculated. The expression for the new estimate of ϕ is

$$\phi_j^{(n)} = \frac{1}{a_{ii}} \left(b_j - \sum_{j=1}^{i-1} a_{ij}\phi_j^{(n)} - \sum_{j=i+1}^{N} a_{ij}\phi_j^{(n-1)} \right) \qquad i, j = 1, 2, 3, \ldots, N \quad (4.4)$$

which, in matrix form, becomes

$$\boldsymbol{\phi}^{(n)} = -\,(\mathbf{D}+\mathbf{L})^{-1}\,\mathbf{U}\boldsymbol{\phi}^{(n-1)} \;+\; (\mathbf{D}+\mathbf{L})^{-1}\,\mathbf{b}.$$

The set of operations that results in obtaining the new value of ϕ is often referred to as a *sweep*. This method requires less memory capacity since the new value of each unknown overwrites the old value, thus eliminating the need to store both the old and new values separately.

4.2.1 Numerical Example

To better illustrate the characteristics of the Gauss-Seidel method, a numerical example is provided. This numerical example demonstrates the solution to a linear system resulting from the application of the finite volume method to a case of steady heat conduction on a two-dimensional domain. In Fig. 4.2, the computational domain consists of the shaded part. The boundary conditions are as follows: the left border with a constant temperature of 100 K, the right border with a constant temperature of 0 K, and the upper and lower borders, which are adiabatic. The computational domain has been discretised into eight cells, whose numbering is shown in Fig. 4.3. Recalling Sect. 3.1, the equation describing the phenomenon of steady heat conduction is

$$0 = \nabla \cdot (k \nabla T) + S \quad (4.5)$$

where k denotes the thermal conductivity of the considered material, set to 1 W/m · K. T is the temperature, and S is the source term, including the contribution from the

Fig. 4.3 Cell numbering

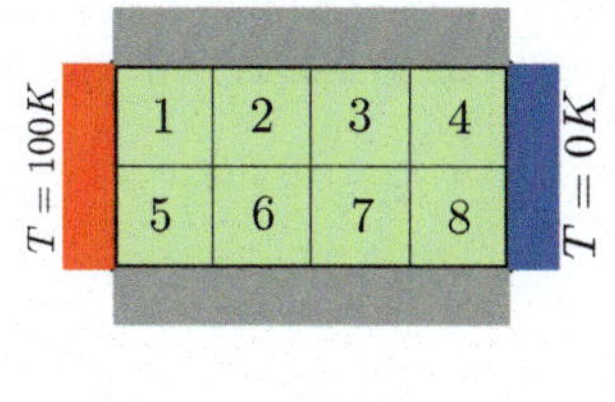

boundary conditions. According to the finite volume method, the discretised form of Eq. 4.5 applied to the cell with centre P and to the neighbouring cell with centre N is

$$T_P \sum_f \frac{A_f k_f}{|\mathbf{x}_N - \mathbf{x}_P|} + \sum_f T_N \left(- \frac{A_f k_f}{|\mathbf{x}_N - \mathbf{x}_P|} \right) = S_P V_P, \tag{4.6}$$

where A_f is the area of the generic face (in the two-dimensional case, a length), the distance $|\mathbf{x}_N - \mathbf{x}_P|$ is the distance between the centres of the two considered cells, and V_P is the volume of the cell (in the two-dimensional case, an area). For simplicity, the expression of Eq. 4.6 is given in the case of cell number 2 of Fig. 4.3:

$$T_2 \left(\frac{1 \cdot A_{f12}}{x_2 - x_1} + \frac{1 \cdot A_{f23}}{x_3 - x_2} + \frac{1 \cdot A_{f26}}{y_2 - y_6} \right) - T_1 \frac{1 \cdot A_{f12}}{x_2 - x_1} - T_3 \frac{1 \cdot A_{f23}}{x_3 - x_2} - T_6 \frac{1 \cdot A_{f26}}{y_2 - y_6} = 0 \tag{4.7}$$

The symbol A_{ij} denotes the face shared between the cell centred at i and the cell centred at j. Applying Eq. 4.6 to all eight cells in the computational domain results in the following system of equations, expressed in matrix form:

$$\begin{bmatrix} 2 & -1 & 0 & 0 & -1 & 0 & 0 & 0 \\ -1 & 3 & -1 & 0 & -1 & -1 & 0 & 0 \\ 0 & -1 & 3 & -1 & -1 & 0 & -1 & 0 \\ 0 & 0 & -1 & 2 & 0 & 0 & 0 & -1 \\ -1 & 0 & 0 & 0 & 2 & -1 & 0 & 0 \\ 0 & -1 & 0 & 0 & -1 & 3 & -1 & 0 \\ 0 & 0 & -1 & 0 & 0 & -1 & 3 & -1 \\ 0 & 0 & 0 & -1 & 0 & 0 & -1 & 2 \end{bmatrix} \begin{bmatrix} T_1 \\ T_2 \\ T_3 \\ T_4 \\ T_5 \\ T_6 \\ T_7 \\ T_8 \end{bmatrix} = \begin{bmatrix} b_1 \\ b_2 \\ b_3 \\ b_4 \\ b_5 \\ b_6 \\ b_7 \\ b_8 \end{bmatrix}$$

in which the terms b_j represent the known terms, derived solely from the application of the boundary conditions, since there are no sources within the computational domain. Applying the Gauss-Seidel method, the first equation of the system (4.2) has the form

$$2T_1 - T_2 - T_5 = b_1.$$

The first estimate of T_1 is obtained as

$$T_1 = \frac{1}{2}(T_2 + T_5 + b_1) \tag{4.8}$$

in which the initial guess values for both T_2 and T_5 are used. The algorithm then proceeds to the second equation

$$-T_1 + 3T_2 - T_3 - T_6 = b_2$$

to obtain the first estimate of T_2 using the previously calculated value for T_1 and the initial guesses for T_3 and T_6:

$$T_2 = \frac{1}{3}(T_1 + T_3 + T_6 + b_2) \tag{4.9}$$

Having completed the calculation of T_2, the values of the remaining temperatures up to T_6 can be calculated, thereby concluding the first iteration (sweep). The process is repeated until the preselected convergence criterion is reached. Note that the Gauss-Seidel algorithm determines the updated value of the unknown—in this case, the temperature—according to the numbering of the cells: the value of T_1 is first calculated, then T_2, then T_3, and so on. The finite volume method then determines the value of the unknown as the weighted average of the same unknown in the neighbouring cells, along with the contribution of the source term, which includes the contribution from the boundary conditions (see Eqs. 4.8 and 4.9) in accordance with the general formula (4.4). Assuming an initial condition of a temperature of 0 K for all cells, at the first iteration, the following value of temperature is obtained for cell 1:

$$T_1 = \frac{1}{2}(0+0) + \frac{1}{2}\frac{100}{1/2} = 100\,\text{K}.$$

The contribution to the determination of T_1 is evident from the values of cells 2 and 5, as well as the boundary condition. In the case of cell 2, only cells 1, 3, and 6 contribute to determining the corresponding temperature value, and, of these, only cell 1 includes the contribution derived from the boundary condition:

$$T_2 = \frac{1}{3}(100+0+0) + \frac{1}{3}0 = 33.3\,\text{K}.$$

In the case of cell 3, only cells 2, 4, and 7 contribute to determining the corresponding temperature value, and, of these, only cell 2 includes the contribution derived from the boundary condition:

$$T_3 = \frac{1}{3}(33.3+0+0) + \frac{1}{3}0 = 11.1\,\text{K}.$$

At this point, it is clear that in the Gauss-Seidel algorithm, the information originating from the boundary condition is propagated within the computational domain according to the order initially assigned to the cells. The numbering of the cells thus assumes fundamental importance in determining the convergence of the entire iterative process. By changing the cell numbering, a *slowdown* in the convergence occurs, as can be seen by comparing the temperature values obtained above with those derived from the alternative numbering shown in Fig. 4.4:

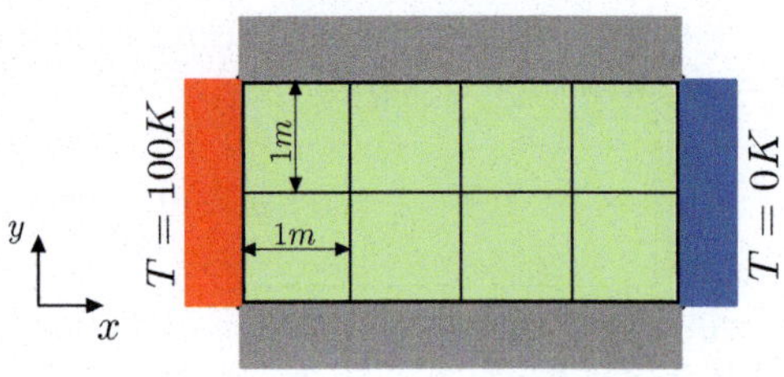

Fig. 4.4 Alternative cell numbering

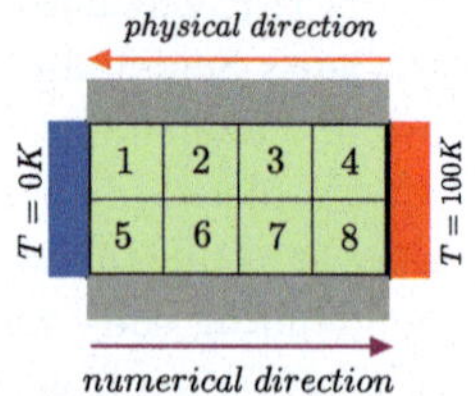

Fig. 4.5 Example of a case where the numerical and physical directions of information propagation do not coincide

$$
\begin{aligned}
T_1 &= \frac{1}{2}(0+0) + \frac{1}{2}\frac{100}{1/2} = 100\,\text{K}, \\
T_2 &= \frac{1}{2}(0+0) + \frac{1}{2}0 = 0\,\text{K}, \\
T_3 &= \frac{1}{3}(0+0+0) + \frac{1}{3}0 = 0\,\text{K}.
\end{aligned}
$$

The preceding discussion highlights the need for algorithms such as Cuthill and inverse Cuthill to achieve the correct cell numbering, as grid generation software typically assigns a numbering based on requirements different from those needed for the efficient execution of iterative methods for matrix inversion. The (numerical) direction of information propagation resulting from the execution of the Gauss-Seidel algorithm may not align with the (physical) direction dictated by the boundary conditions. Referring to Fig. 4.5, consider the case where the two boundary conditions are swapped: convergence will slow due to the numerical direction of information propagation proceeding from left to right (in accordance with the cell numbering), while the physical direction propagates from right to left, given the initial condition of zero temperature across the entire domain and a temperature of 100 K on the right edge of the computational domain. To overcome this issue, when the physical direction of information propagation is not known a priori, the *symmetric Gauss-Seidel* algorithm is used, which involves reversing the numerical direction of information propagation between one iteration and the next.

4.3 Diagonal Dominance and Scarborough Criterion

A *diagonally dominant matrix* is a square matrix of order n in which the absolute value of each diagonal element is greater than or equal to the sum of the absolute values of all the remaining elements in the same row. If a_{ij} denotes the generic element of the matrix **A**, the following condition must hold:

$$|a_{ii}| \geqslant \sum_{j=1, j \neq i}^{n} |a_{ij}|.$$

A diagonally dominant matrix is always singular (that is, it has a determinant different from zero and, therefore, is invertible). The Scarborough criterion provides a sufficient condition for convergence in the inversion of the coefficient matrix using iterative methods. Specifically, adherence to this condition ensures the existence of at least one iterative method that results in convergence. However, as a sufficient condition, convergence could also be achieved even if the criterion is not met. The Scarborough criterion can be expressed in terms of the coefficients of the generic discretised equation present in the matrix. **A** as

$$\frac{\sum |a_{nb}|}{|a_P|} \begin{cases} \leqslant 1 & \text{for all equations} \\ < 1 & \text{for at least one equation} \end{cases} \tag{4.10}$$

where a_P is the coefficient linked to the cell with centroid P, while a_{nb} are the coefficients linked to the cells that share at least one face with the cell with centroid P, or those cells involved based on the chosen discretisation scheme. A coefficient matrix that respects this criterion is certainly diagonally dominant. This ensures that the *boundedness criterion* is also satisfied. According to the boundedness criterion, the absolute value of a generic transported quantity ϕ in a cell is never greater than the same quantity in the adjacent cells, in the absence of source terms. To achieve diagonal dominance, it is therefore necessary to have high values for the coefficients on the main diagonal and low values for the off-diagonal terms. This goal can be achieved through:

- constructing a computational grid with favourable geometric characteristics (orthogonality and skewness),
- ensuring that the source terms, moved to the right-hand side of the equation, are negative,
- reducing the time integration step size,
- resorting to the under-relaxation technique.

4.4 Residue and Correction/Error

As mentioned above, iterative methods for solving systems of equations involve executing numerous iterations to obtain updated values of the unknown quantity ϕ in each cell. The value ϕ^n, obtained at the nth iteration, does not necessarily satisfy Eq. 4.1. At this point, it is possible to define the *residual error*, or simply the *residual*, as

$$\mathbf{R}^n = \mathbf{b} - \mathbf{A}\phi^n. \tag{4.11}$$

It is useful to remember that both $\mathbf{R}^n$ and ϕ^n are column matrices, the number of elements of which coincides with the number of cells with which the computational domain has been discretised. Indicating by ϕ^{n+1} the value of ϕ obtained at the $(n+1)$-th iteration and assuming that this is the exact value of ϕ (the one that satisfies Eq. 4.1), it can be written

$$\mathbf{A}\phi^{n+1} = \mathbf{b}.$$

Now, indicating by ϕ' the difference in the value of ϕ between the n-th iteration and the $(n+1)$-th iteration, it can be written

$$\phi^{n+1} = \phi^n + \phi'. \tag{4.12}$$

The term ϕ' is called *correction* or *error* and is defined as the difference between the exact and the approximate value (the one obtained at the n-th iteration) of ϕ. From Eq. 4.12, it follows that

$$\mathbf{A}\left(\phi^n + \phi'\right) = \mathbf{b}.$$

Keeping Eq. 4.11 in mind, it will be

$$\mathbf{A}\phi' = \mathbf{R}^n$$

known as the *correction form* of Eq. 4.1, from which it differs in that, at convergence, both terms on the left and on the right-hand side become zero.

Once the residual is defined, its L_1 and L_2 norms can also be defined as

$$L1\ norm : R1 = \sum_{k=1}^{N} |R_k|, \qquad L2\ norm : R2 = \sqrt{\sum_{k=1}^{N} (R_k)^2}.$$

The L_2 norm is also known as the *Euclidean norm* and is often calculated as

$$R2 = \sqrt{\mathbf{R}^T\mathbf{R}}.$$

4.5 Stopping Criteria

In the case where an iterative method is used, it is necessary to define a criterion upon which to stop the iterative process. Many of these criteria are based on the concept of the residual. One such criterion is the one that stops the iterations if the maximum value of the residual falls below a certain threshold value, ϵ.

$$\max_{i=1}^{N}\left|b_i - \sum_{j=1}^{N} a_{ij}\phi_j^n\right| \leqslant \epsilon.$$

Another criterion requires that the mean squared error be less than a certain value, ϵ.

$$\frac{\sum_{i=1}^{N}\left(b_i - \sum_{j=1}^{N} a_{ij}\phi_j^n\right)^2}{N} \leqslant \epsilon.$$

A further criterion involves stopping the calculation if the normalised difference between two consecutive values falls below a certain threshold.

$$\max_{i=1}^{N}\left|\frac{\phi_i^n - \phi_i^{n-1}}{\phi_i^n}\right| \times 100 \leqslant \epsilon.$$

A final criterion involves stopping the calculation once the maximum number of iterations is reached.

4.6 LU Factorisation Method

Given a matrix $A \in \mathbb{R}^{N\times N}$, and assuming that there exist a lower triangular matrix L and an upper triangular matrix U such that

$$A = LU, \tag{4.13}$$

Equation 4.13 is called the LU *factorisation* (or *decomposition*). In particular, it will be

$$U = \begin{bmatrix} u_{11} & u_{12} & \dots & u_{1N-1} & u_{1N} \\ 0 & u_{22} & \dots & u_{2N-1} & u_{2N} \\ \vdots & \vdots & \dots & \vdots & \vdots \\ 0 & 0 & \dots & 0 & a_{NN} \end{bmatrix} \quad \text{and} \quad L = \begin{bmatrix} 1 & 0 & \dots & 0 & 0 \\ l_{21} & 1 & \dots & 0 & 0 \\ \vdots & \vdots & \dots & \vdots & \vdots \\ l_{N1} & l_{N2} & \dots & l_{N\,N-1} & 1 \end{bmatrix}$$

The elements of the main diagonal of the matrix L are set equal to 1 to make the factorisation unique. Solving the matrix equation $A\mathbf{x} = \mathbf{b}$ is equivalent to solving the two simpler triangular systems

$$L\mathbf{y} = \mathbf{b} \quad \text{and} \quad U\mathbf{x} = \mathbf{y}.$$

Since L is lower triangular, the first row of the system $L\mathbf{y} = \mathbf{b}$ will have the form $l_{11}y_1 = b_1$, from which we derive the value of y_1 assuming $l_{11} \neq 0$. Substituting the value found for y_1 into the subsequent $N-1$ equations, we obtain a system whose unknowns are $y_2, \ldots, y_N$, for which we can proceed in the same manner. Proceeding forward, equation by equation, we calculate all the unknowns using the following algorithm, called *forward substitution*. In a completely analogous manner, the system $U\mathbf{x} = \mathbf{y}$ can be solved: in this case, the first unknown to be calculated will be x_N, and then, in reverse, all the remaining unknowns x_i for i ranging from $N-1$ to 1. Given a matrix $A \in \mathbb{R}^{n\times n}$, its LU factorisation exists and is unique if and only if the principal submatrices A_i of A of order $i = 1, \ldots, N-1$ (i.e., those obtained by limiting A to the first i rows and columns) are non-singular. Some classes of matrices satisfy the condition just stated, and among these, the following can be mentioned:

- strictly diagonally dominant matrices: For simplicity, we recall here that a matrix is said to be *row diagonally dominant* if

$$|a_{ii}| \geqslant \sum_{j=1,\ j\neq i}^{N} |a_{ij}| \qquad i = 1, \ldots, N.$$

It is said to be *column diagonally dominant* if

$$|a_{ii}| \geqslant \sum_{j=1,\ j\neq i}^{N} |a_{ji}| \qquad i = 1, \ldots, N$$

f the sign $>$ can replace the sign $\geqslant$, there will be *strict diagonal dominance* (for rows or columns, respectively).
Real symmetric and positive definite matrices. A matrix is said to be *positive definite* if

$$\forall \mathbf{x} \in \Re^N \quad \text{with} \quad \mathbf{x} \neq 0, \qquad \mathbf{x}^T A\mathbf{x} > 0;$$

a matrix is said to be *semi positive defined* if

$$\forall \mathbf{x} \in \Re^N \quad \text{with} \quad \mathbf{x} \neq 0, \qquad \mathbf{x}^T A\mathbf{x} \geqslant 0.$$

If $A \in \mathbb{R}^{N\times N}$ is symmetric and positive definite, there exists a special factorisation called the *Cholesky factorisation*, expressed by

$$A = R^T R$$

where R is an upper triangular matrix, obtained using a suitable algorithm, with positive elements on the main diagonal. The Cholesky factorisation generates the filling of the band, a phenomenon called *fill-in*, whereby the factorisation process tends to fill the L and U matrices, modifying the structure of the corresponding triangle of the initial A matrix. For example, if the A matrix is coarse, the L and U matrices may have non-zero elements where in the A matrix they were null. A square matrix of size N is called *coarse* if it has a number of non-null elements of the order of N. Furthermore, the *pattern of a coarse matrix* is the set of its non-null elements. To overcome the fill-in of a matrix, reordering techniques can be adopted, which permute rows and columns of the matrix before performing the factorisation.

4.6.1 Preconditioning

Given a symmetric and positive-definite linear system $\mathbf{A}\phi = \mathbf{b}$, preconditioning the system means conditioning the matrix of coefficients before applying any iterative method; conditioning the matrix aims to improve its condition number. The *condition number* of a symmetric matrix is defined as:

$$K(\mathbf{A}) = \frac{\lambda_M}{\lambda_m}$$

where λ_M is the maximum eigenvalue and λ_m is the minimum eigenvalue of the matrix $\mathbf{A}$. To be *well-conditioned*, a matrix must have a condition number $K(\mathbf{A}) \approx 1$. The larger the condition number of $\mathbf{A}$, the slower the convergence will be. The preconditioning technique is used to improve this number, and therefore the robustness and computational efficiency of the iterative methods used to solve the linear system. In practice, preconditioning consists of defining a matrix $\mathbf{M}$ of the same dimensions as $\mathbf{A}$, called the *preconditioning matrix* or *preconditioner*, with the following characteristics:

- non-singular, i.e., invertible;
- symmetric, that is, $\mathbf{M}^T = \mathbf{M}$;
- positive-definite.

and transform the previous system into the equivalent preconditioned system:

$$\mathbf{M}^{-1}\mathbf{A}\phi = \mathbf{M}^{-1}\mathbf{b}$$

which can be solved faster than the initial one, provided that $K\left(\mathbf{M}^{-1}\mathbf{A}\right) \ll K(\mathbf{A})$. An example is the *diagonal or Jacobi preconditioner*, in which the matrix $\mathbf{M}$ is constructed from the matrix $\mathbf{A}$, considering only the elements of the main diagonal:

$$\mathbf{M} = diag(\mathbf{A}).$$

In this case, the preconditioner performs a simple 'scaling' of the initial matrix **A**. In the case where **A** is also coarse, in addition to being symmetric and positive-definite, the preconditioning matrix can be constructed through an *incomplete Cholesky factorisation* of the matrix **A**. This is referred to as *ILU decomposition* (Incomplete Lower-Upper) of a symmetric matrix. With this strategy, a lower triangular matrix $\mathbf{R}_P$ is constructed. $\mathbf{R}_P$ approximates the factor $\mathbf{R}^T$ of the Cholesky factorisation.

$$\mathbf{M} = \mathbf{R}_P \mathbf{R}_P^T.$$

A widely used technique to construct $\mathbf{R}_P$ is known as ICT (Incomplete Cholesky with Threshold Dropping). To implement it, the following steps are performed:

1. choose a drop tolerance $\epsilon_d > 0$;
2. perform the Cholesky factorisation algorithm (appropriately modified to generate a lower triangular matrix) to construct the elements of $\mathbf{R}_P$. Off-diagonal elements that are less than $c_j \epsilon_d$ are ignored (c_j is the norm of the j-th column vector of the lower triangle of **A**).

The *nofill* technique, on the other hand, ignores all elements of $\mathbf{R}_P$ where the corresponding positions in **A** contain null elements. In fact, it is a factorisation without fill-in. This is referred to as *ILU(0) decomposition*. A simplified version of the ILU decomposition is known as *diagonal ILU* (DILU), in which only the elements of the main diagonal are modified.

4.6.2 The Gradient and Conjugate Gradient Methods

Given the square linear system $\mathbf{Ax} = \mathbf{b}$ of dimension N, it is possible to define the function $\Phi : \mathbb{R}^N \to \mathbb{R}$

$$\Phi(\mathbf{x}) = \frac{1}{2}\mathbf{x}^T \mathbf{A}\mathbf{x}\mathbf{x}^T \mathbf{b}.$$

If **A** is symmetric and positive definite, Φ is a *convex* function, i.e., $\forall\, \mathbf{x}, \mathbf{y} \in \mathbb{R}^N$. If Φ is *convex*, then $\forall \alpha \in [0, 1]$, it holds

$$\Phi\left(\alpha \mathbf{x} + (1-\alpha)\,\mathbf{y}\right) \leqslant \alpha\Phi(\mathbf{x}) + (1-\alpha)\Phi(\mathbf{y})$$

and Φ admits a unique stationary point, $\mathbf{x}^*$, which is also a point of local and absolute minimum. From this, it follows that

$$\mathbf{x}^* = \underset{\mathbf{x} \in \Re^N}{\operatorname{argmin}}\, \Phi(\mathbf{x}) \tag{4.14}$$

is the only solution of the equation

$$\nabla\Phi(\mathbf{x}) = \mathbf{Ax} - \mathbf{b} = 0.$$

Solving the minimum problem (4.14) is equivalent to solving the linear least squares system $\mathbf{Ax} = \mathbf{b}$ of dimension N, in which $\mathbf{A}$ is symmetric and positive definite. Starting from a generic point $\mathbf{x}^{(0)} \in \mathbb{R}^N$, the gradient and conjugate gradient methods construct a sequence of vectors $\mathbf{x}^{(k)}$ converging to $\mathbf{x}^*$, exploiting the information provided by the gradient vector of Φ. In fact, for a generic $\overline{\mathbf{x}} \in \mathbb{R}^N$ different from $\mathbf{x}^*$, $\nabla\Phi(\overline{\mathbf{x}})$ is a non-zero vector in $\mathbb{R}^N$ that identifies the direction along which the maximum growth of Φ occurs. Consequently, $-\nabla\Phi(\overline{\mathbf{x}})$ identifies the direction of maximum decrease of Φ starting from $\overline{\mathbf{x}}$. Recall that the residual vector at the point $\overline{\mathbf{x}}$ is defined as $\overline{\mathbf{r}} = \mathbf{b} - \mathbf{A}\overline{\mathbf{x}}$. It can be written

$$\overline{\mathbf{r}} = -\nabla\Phi(\overline{\mathbf{x}})$$

From this, it is noted that the residual identifies a possible direction in which to move in order to approach the minimum point $\mathbf{x}^*$. More generally, if the following conditions are met

$$\begin{cases} \mathbf{d}^T\nabla\Phi(\overline{\mathbf{x}}) < 0 \quad if & \nabla\Phi(\overline{\mathbf{x}}) \neq 0, \\ \mathbf{d} = 0 \quad if & \nabla\Phi(\overline{\mathbf{x}}) = 0, \end{cases}$$

the vector $\mathbf{d}$ represents a *direction of descent* for Φ at the point $\overline{\mathbf{x}}$.
The *descent methods* are thus defined:
given a vector $\mathbf{x}^{(0)} \in \Re^N$,

- a direction of descent $\mathbf{d}^{(k)} \in \Re^N$ is determined;
- a step $\alpha_k \in \Re$ is determined;
- the update rule is set as $\mathbf{x}^{(k+1)} = \mathbf{x}^{(k)} + \alpha_k\mathbf{d}^{(k)}$.

for $k = 0, 1, \ldots$ until convergence.

The gradient and conjugate gradient methods are both descent methods, differing in the choice of descent directions. The determination of the steps α_k is common to both methods and is performed using the following formula:

$$\alpha_k = \frac{\left(\mathbf{d}^{(k)}\right)^T \mathbf{r}^{(k)}}{\left(\mathbf{d}^{(k)}\right)^T \mathbf{A}\mathbf{d}^{(k)}}. \tag{4.15}$$

The *gradient method* is characterised by the choice

$$\mathbf{d}^{(k)} = \mathbf{r}^{(k)} = -\nabla\Phi(\mathbf{x}^{(k)}), \qquad k = 0, 1, \ldots$$

That is, the direction of descent at each step is opposite to the direction of the gradient of the function Φ (hence the name of the method). The *conjugate gradient method* constructs a system of descent directions $\mathbf{d}^{(k)}{}_{k=0}^{N-1}$ in $\mathbb{R}^N$ that are all linearly independent and therefore constitute a basis for $\mathbb{R}^N$. Moreover, the descent directions are such that the values $\alpha_k{}_{k=0}^{N-1}$, calculated with the formula (4.15), are precisely the coefficients of the decomposition of $\left(\mathbf{x}^* - \mathbf{x}^{(0)}\right)$ with respect to the basis $\mathbf{d}^{(k)}{}_{k=0}^{N-1}$.

This implies that the term $\mathbf{x}^{(N)}$ obtained at the N-th iteration coincides with the exact solution $\mathbf{x}^*$.

4.7 Multigrid Methods

A multigrid algorithm improves the performance of an iterative method for solving systems of equations by using a hierarchy of grids generated from an initial grid.

4.7.1 *The Smoothing Property of Iterative Methods*

Many iterative schemes used to solve a linear system obtained by appropriately discretising a generic PDE have the property of eliminating, in a few iterations, the high-frequency errors, while leaving the low-frequency errors almost unchanged. The system $A\phi = 0$, obtained by discretising the one-dimensional problem, is considered.

$$\begin{cases} \phi'' = 0 \quad \text{with} \quad x \in (0, 1) \\ \phi(0) = \phi(1) = 0. \end{cases}$$

Defined the error as the difference between the exact and approximate values. For this problem, the exact solution is $\phi = 0$, therefore, given an approximate solution $\mathbf{v}$, the error is trivially $-\mathbf{v}$. As the initial solution, the vector $\mathbf{v}^k$ can be considered, defined as

$$v_j^k = \sin\left(\frac{jk\pi}{n}\right).$$

In this context, the index j represents the j-th component of the vector $\mathbf{v}$, while k is called the *wave number* or *frequency* of the signal. A signal is defined as *low frequency*, or *smooth*, if $1 \leqslant k < \frac{n}{2}$, where n is the number of discretisation intervals. A signal is defined as *high frequency* if $\frac{n}{2} \leqslant k < n$. Taking, for example, four initial vectors with $k = 1, 6, 16, 32$, the trend of the norm of the error is shown in Fig. 4.6. It is observed that the error decreases with each iteration regardless of the initial data, but the rate at which it decreases is very different: it is higher in the case of $k = 32$, or for high-frequency signals. This simple example shows that the high-frequency components of the error are significantly reduced in a few iterations, while the low-frequency components require a greater number of iterations. In the case where $\mathbf{v} = \mathbf{v}^1 + \mathbf{v}^6 + \mathbf{v}^{16} + \mathbf{v}^{32}$, the method manages to eliminate the low-frequency component of the error, and therefore, in Fig. 4.7, it is observed that, in the first iterations, the error decreases considerably and then stabilises. This property of eliminating the high frequencies of the error (*smoothing*) is characteristic of many iterative algorithms (Jacobi, Gauss-Seidel, red-block Gauss-Seidel, etc.) and

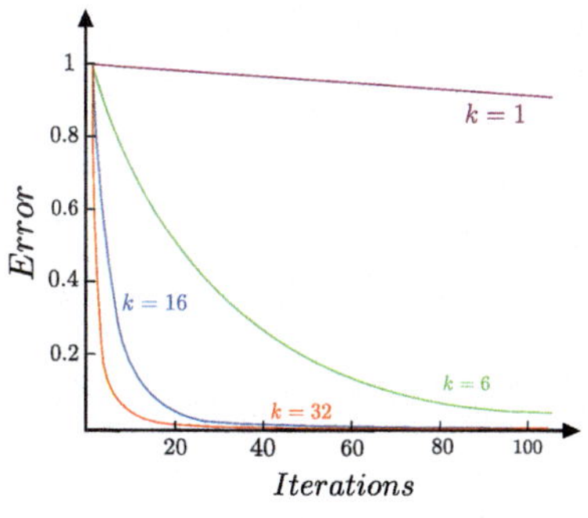

Fig. 4.6 Error as a function of the number of iterations for the system $A\phi = 0$, using 4 vectors initial at different frequency

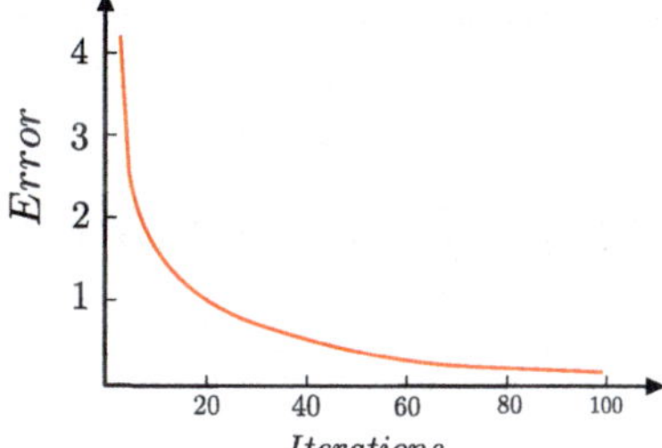

Fig. 4.7 Error trend for the system $A\phi = 0$, using the initial vector $\mathbf{v} = \mathbf{v}^1 + \mathbf{v}^6 + \mathbf{v}^{16} + \mathbf{v}^{32}$

represents the starting point for multigrid methods. From the definition of high/low-frequency signals, it is understood that as the number of discretisation intervals n varies, the same signal, identified by the wave number k, can be considered high or low frequency: specifically, a low-frequency signal for a high number of intervals becomes high frequency for a low number of intervals.

4.7.2 *Geometric Multigrid*

To better understand this method, also known as FAS (Full-Approximation Storage) multigrid, initially consider the case where only two grids are used: a coarse one and a fine one. The example referred to is that of the Poisson equation solved on a two-dimensional rectangular domain with a structured grid and Dirichlet boundary conditions on all boundaries. It is important to keep in mind that the accuracy of the final solution must still be that of the fine grid. This method is implemented in an algorithm consisting of various steps, a brief description of which is given below.

4.7.2.1 Generation of Grids (Agglomeration)

Starting from the fine grid with which the computational domain has been discretised, the first step consists of generating the coarse grid. As shown in Fig. 4.8, each node of the coarse grid will share its position with a node of the fine grid. Indicating by (I, J) the identifying indices of the coarse grid and by (i, j) the indices of the fine grid, we can write with reference to Fig. 4.8:

$$(i, j) = (2I - 1, 2J - 1).$$

Being N_F and N_C the total number of points for the fine grid and the coarse grid, respectively, along the direction I/i, and M_F and M_C the total number of points for the fine grid and the coarse grid, respectively, along the direction J/j, it will be

$$N_F = 2N_C - 1, \qquad M_F = 2M_C - 1.$$

As an example, from a fine grid of dimensions 21×21 we would obtain a coarse grid of dimensions 11×11.

4.7.2.2 Initialisation

In this phase, the initial value $\phi_{i,j}^{F(0)}$ of the dependent variable at each point of the fine grid is set.

4.7.2.3 Smoothing on the Fine Grid

In this phase, the system of equations $\mathbf{A}_F \phi_F = \mathbf{b}_F$, resulting from the application of the Poisson equation to each cell of the fine grid, is solved. Keeping in mind what was discussed in Sect. 4.7.1 and with the aim of eliminating only the high-frequency components of the error, only a reduced number of iterations is performed to solve this system (*smoothing*). The iterative method chosen—also called the *solver* or *smoother*—to solve this system must be selected from those computationally less expensive, as the reduction of errors in the multigrid method is mainly due

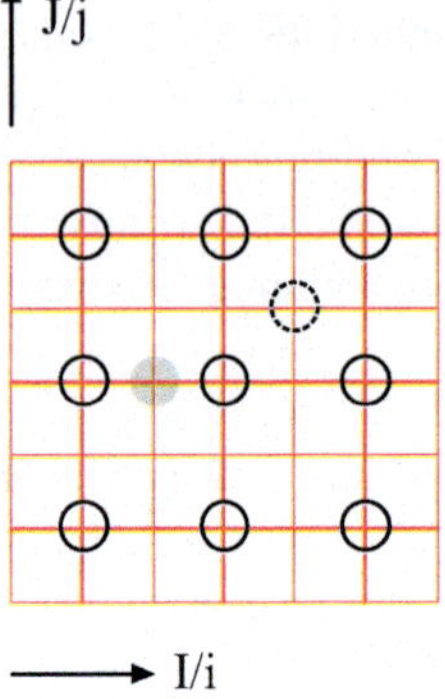

Fig. 4.8 Generation of the coarse grid (continuous line circles) from the fine grid. Grey circle: point of the fine grid positioned on a line of coarse grid. Dashed circle: point of the fine grid not positioned on a line of coarse grid

to the use of grids with different levels of cell density. The Gauss-Seidel solver is normally preferred over more complex solvers, such as those based on gradient analysis. The solution obtained in this phase is affected by an error characterised by long wavelengths for the fine grid, as the shorter wavelength components of the error have been eliminated by the smoothing process. The subsequent steps of the multigrid method aim to transfer this error to the coarse grid.

4.7.2.4 Calculation of the Residual on the Fine Grid

Here, the residual on the fine grid is calculated as

$$\mathbf{R}_F = \mathbf{b}_F - \mathbf{A}_F \phi_F$$

and then its L2 norm as

$$R2_F = \sqrt{\mathbf{R}_F^T \mathbf{R}_F}.$$

4.7.2.5 Transfer of the Fine Grid Residuals to the Coarse Grid (Restriction)

This phase, known as *restriction*, ensures that the residuals of the nodes of the fine grid that have a counterpart in the coarse grid are copied onto the same nodes of the coarse grid. These residuals will be indicated with the symbol $\mathbf{R}^{C \leftarrow F}$. In the event that there is no correspondence between the nodes of the fine grid and the coarse grid, it will be necessary to use an interpolation method. Note that the component

4.7.2.6 Smoothing on the Coarse Grid

Considering the equation in correction form for the coarse grid

$$\mathbf{A}_C \phi'_C = \mathbf{R}^{C \leftarrow F}$$

a reduced number of iterations is performed to solve this system with the aim of eliminating only the high-frequency components (for the coarse grid) of the correction ϕ'_C. Here too, as done for smoothing on the fine grid, the method to solve this system is chosen from those computationally less expensive.

4.7.2.7 Transfer of the Correction for the Coarse Grid to the Fine Grid (Prolongation)

The correction ϕ'_C for the coarse grid, obtained at the end of the smoothing phase on the coarse grid, is transferred to the fine grid with a process known as *prolongation*. The prolongation phase involves the use of interpolation, as there are points on the fine grid that are not present on the coarse grid. In this regard, and with reference to Fig. 4.8, three different cases can occur.

1. The point of the fine grid coincides with the point of the coarse grid: in this case, interpolation is not necessary;
2. the point of the fine grid is positioned on a grid line of the coarse grid: in this case, interpolation between the two points of the coarse grid adjacent to the considered fine grid point can be performed;
3. the point of the fine grid is not positioned on a grid line of the coarse grid: in this case, interpolation must be performed considering the four points of the coarse grid adjacent to the considered fine grid point.

The symbol $\phi'^{F \leftarrow C}$ will denote the correction on the fine grid obtained from the correction on the coarse grid.

4.7.2.8 Updating the Solution on the Fine Grid

In this phase, the initial solution previously obtained for the fine grid is updated by considering the correction obtained from the coarse grid:

$$\phi_F = \phi_F + \phi'^{F \leftarrow C}.$$

4.7.2.9 Checking the Level of Convergence

The L2 norm of the residual, calculated after the smoothing phase on the fine grid, is compared with the threshold value ϵ_{tol}, which is defined to determine whether the solution obtained is acceptable. This value of the residual, rather than the one corresponding to the last update of the solution for the fine grid, is used to avoid calculating the residual twice in the same multigrid cycle. If the convergence criterion $(R2_F < \epsilon_{tol})$ is not met, the process is repeated, starting from the smoothing phase on the fine grid.

4.7.3 V-Cycle

In general, the multigrid algorithm involves the use of more than two grids. Specifically, we refer to a hierarchy of grids to denote the set of grids employed, each with a different level of refinement. To better understand the use of the multigrid algorithm with multiple grids, we will now consider the algorithm known as *V-cycle multigrid*. Figure 4.9 shows a V-cycle multigrid with three grid levels. The shaded boxes represent the smoothing operation, the box with a thick continuous line represents the correction operation, and the dashed box represents the operation of updating the correction and the final solution. The arrows pointing down represent the restriction operation, while the arrows pointing up represent the prolongation operation. Through the restriction operation, the residual obtained from the partial solution of the original system is transferred to the intermediate grid. At this point, the equation in correction form, $\mathbf{A}_2\phi_2' = \mathbf{R}_{2\leftarrow 1}$, for the intermediate grid is partially solved, and then the corresponding residual $\mathbf{R}_2 = \mathbf{R}_{2\leftarrow 1} - \mathbf{A}_2\phi_2'$ is calculated. The residual corresponding to the intermediate grid is then transferred to the coarse grid. In the ideal case, the coarse grid allows for a direct solution from the corresponding system resulting from the application of the equation in correction form, $\mathbf{A}_3\phi_3' = \mathbf{R}_{3\leftarrow 2}$. In practice, a direct solution of this system is not possible, so we proceed with a partial solution using an iterative method, typically the one used for the smoothing phase performed in the previous steps. Subsequently, the coarse grid correction is transferred to the intermediate grid. The intermediate grid correction is calculated as $\phi_2' = \phi_2' + \phi_{2\leftarrow 3}'$. The transfer to the fine grid and the calculation of the corresponding correction is done via $\phi_1' = \phi_1' + \phi_{1\leftarrow 1}'$. The process ends with the updating of the solution on the fine grid. The total number of grids used is determined by the computational costs associated with interpolation operations and data storage for each grid level. In addition to the V-cycle algorithm, the W-cycle and full multigrid algorithms are also widely used. These are based on the principles outlined here and will not be discussed further.

4.7.4 Algebraic Multigrid

As seen earlier, one of the factors that most influences the efficiency of the iterative process for solving systems of equations is a coefficient matrix characterised by elements whose ratio between the maximum and minimum value is very high. In this case, we speak of *anisotropy of the coefficients*, which, in some cases, can lead to poorly conditioned matrices. Keeping in mind what was discussed in Chap. 3 and referring to Eq. 4.7, the dependence of the value of the coefficients on the geometric characteristics of the cells is evident. In the case of rectangular cells with high ratios between the lengths of the two sides, an anisotropic matrix of coefficients is obtained. The solution would advance at different speeds depending on the considered direction, slowing down the entire resolution process. The presence of physical phenomena with direction-dependent characteristics also contributes to the anisotropy of the coefficient matrix, even in the presence of a regular computational grid.

Stating the dependence of the coefficient matrix on the geometry of the grid, it is clear that, for geometric multigrid, the grid geometry determines the agglomeration

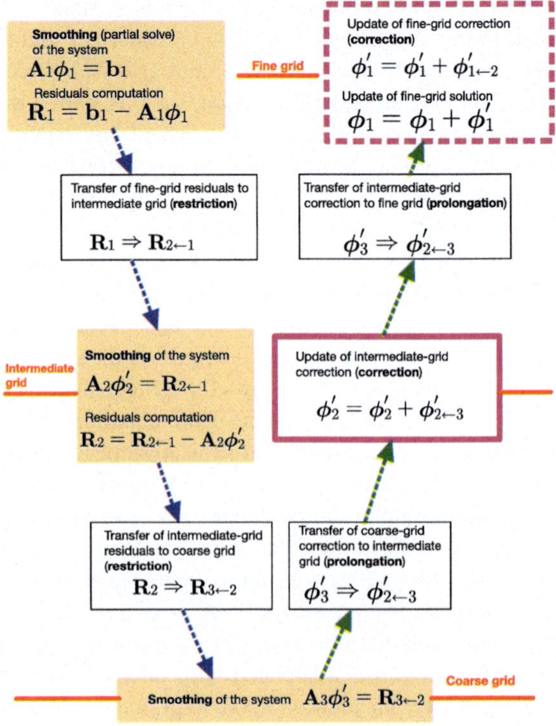

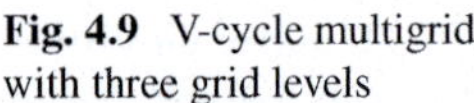
Fig. 4.9 V-cycle multigrid with three grid levels

process. Conversely, in algebraic multigrid, the values of the elements of the fine grid coefficient matrix are used to construct a coarse grid whose coefficient matrix has better isotropy characteristics. In algebraic multigrid, both the influence of geometry and physical phenomena on the isotropy of the coefficient matrix are taken into account, leading to a general improvement in the resolution process. In algebraic multigrid, the basic strategy of the geometric multigrid method is maintained, as grids with different levels of refinement continue to be considered. Specifically, the transition from fine grids to coarse grids involves the phases of restriction, setting up/updating the system of equations for the coarser grid, and smoothing on the coarser grid. The transition from coarse grids to fine grids involves the phases of prolongation, correction of the solution on the finer grid, and smoothing on the finer grid.

4.7.4.1 Generation of Grids (Agglomeration/Coarsening)

Various approaches are possible for the creation of grids at different levels of refinement that will be used in the multigrid procedure. For example, one can start from a coarse grid and then gradually refine it. This approach implies an excessive dependence of the finer grid on the starting coarse grid. In general, it is therefore preferred to start from the finer grid in order to obtain the coarser grid through the union (agglomeration) of cells of the fine grid. The agglomeration process can be based on both geometric criteria and criteria related to the values assumed by the coefficients of the algebraic equations in the various cells of the fine grid. Note: what is referred to in OpenFOAM® when discussing Geometric-Algebraic Multi-Grid (GAMG) is a solver for linear systems of the algebraic multi-grid type, with an agglomeration process that, depending on the settings, can be based on both geometric criteria and the values of the elements of the coefficients matrix.

4.7.4.2 Initialisation and Smoothing on the Fine Grid

Once an initial value for the unknown in each cell of the computational domain is set, and using the chosen solver as an iterative method for the solution of linear systems, a limited number of iterations is performed to obtain a first approximate solution on the fine grid. Based on this approximate solution, the vector consisting of the values of the residual in each of the cells of the computational domain is calculated.

4.7.4.3 Calculation of Residuals on the Fine Grid

The general linear conservation equation for the cell centred at C can be written in the form

$$a_C \phi_C + \sum_{F=NB(C)} a_F \phi_F = b_C \tag{4.16}$$

where $NB(C)$ is the number of faces that bounds the cell. Applying Eq. 4.16 to all the cells of the computational domain, the system $\mathbf{A}\phi = \mathbf{b}$ is obtained. For each cell, it can be written

$$a_i \phi_i + \sum_{j=NB(i)} a_{ij} \phi_j = b_i$$

and, using the index k to refer to the fine grid,

$$a_i^{(k)} \phi_i^{(k)} + \sum_{j=NB(i)} a_{ij}^{(k)} \phi_j^{(k)} = b_i^{(k)}.$$

By definition, the residual on the generic cell i of the fine grid will be

$$r_i^{(k)} = b_i^{(k)} - \left(a_i^{(k)} \phi_i^{(k)} + \sum_{j=NB(i)} a_{ij}^{(k)} \phi_j^{(k)} \right).$$

Recalling the definition described in Sect. 4.4, the same residual can be written in terms of correction ϕ':

$$\tilde{r}_i^{(k)} = b_i^{(k)} - \left[a_i^{(k)} \left(\phi_i^{(k)} + \phi_i^{\prime(k)} \right) + \sum_{j=NB(i)} a_{ij}^{(k)} \left(\phi_j^{(k)} + \phi_j^{\prime(k)} \right) \right]$$

that is

$$\tilde{r}_i^{(k)} = r_i^{(k)} - \left(a_i^{(k)} \phi_i^{\prime(k)} + \sum_{j=NB(i)} a_{ij}^{(k)} \phi_j^{\prime(k)} \right). \tag{4.17}$$

4.7.4.4 Transfer of Residuals and Coefficients from the Fine Grid to the Coarse Grid (Restriction)

Using the index $k+1$ to refer to the coarse grid, the residuals on the coarse grid can be calculated in terms of the residuals on the fine grid as

$$\mathbf{r}^{k+1} = \mathbf{I}_k^{k+1} \mathbf{r}^k$$

where $\mathbf{I}_k^{k+1}$ is the restriction operator in the transition from the fine grid to the coarse grid resulting from the process of agglomeration. In algebraic multigrid, this operator (the interpolation matrix) is defined in order to obtain

$$r_I^{k+1} = \sum_{i \in I} r_i^k$$

where the subscript i refers to the cells of the grid at level k (the fine grid) which, in the agglomeration process, have been grouped to form the cell I of the $k+1$ level grid (the coarse grid).

The coefficients of the coarse grid are calculated from those of the fine grid using the following relationships:

$$a_I^{k+1} = \sum_{i \in I} a_i^k + \sum_{i \in I} \sum_{j \in I} a_{ij}^k, \qquad a_{IJ}^{k+1} = \sum_{i \in I} \sum_{\substack{j \notin I \\ j \in NB(I)}} a_{ij}^k.$$

4.7.4.5 Smoothing on the Coarse Grid

Imposing that for each cell I of the coarse grid the residual value is zero is equivalent to requiring that

$$\sum_{i \in I} \tilde{r}_i^{(k)} = 0$$

which, as referred to in Eq. 4.17, becomes

$$0 = \sum_{i \in I} r_i^{(k)} - \left(\sum_{i \in I} a_i^{(k)} \phi_i'^{(k)} + \sum_{i \in I} \sum_{j=NB(i)} a_{ij}^{(k)} \phi_j'^{(k)} \right) \tag{4.18}$$

which is the correction form of the equation for the coarse grid, written according to the numbering of the fine grid. Using the numbering of the coarse grid, Eq. 4.18 can be rewritten as

$$a_I^{k+1} \phi_I'^{(k+1)} + \sum_{J=NB(I)} a_{IJ}^{(k+1)} \phi_J'^{(k+1)} = r_I^{k+1}. \tag{4.19}$$

A reduced number of iterations of the chosen linear system solution algorithm is executed on the system resulting from the application of the correction form of Eq. 4.19 to each cell of the coarse grid. This results in the value $\phi'^{(k+1)}$ of the correction on the coarse grid.

4.7.4.6 Transfer of the Correction for the Coarse Grid to the Fine Grid (Prolongation)

This phase can be implemented according to different approaches. One possibility is to set the correction value for all the cells i of the fine grid, which together form the cell I, equal to the value obtained for the cell I of the coarse grid.

4.7.4.7 Smoothing of the Correction for the Fine Grid

In the case where the fine grid used is not the starting fine grid, a reduced number of iterations of the chosen linear system solution algorithm is executed on the system resulting from the application of the correction form of Eq. 4.19 to each cell of the fine grid. This results in an updated value $\phi'^{(k+1)}$ of the correction on the fine grid.

4.7.4.8 Updating the Solution on the Fine Grid

In this phase, the solution previously obtained on the fine grid is updated by considering the correction from the coarse grid.

4.7.4.9 Checking the Level of Convergence

If the convergence criterion ($R2_F < \epsilon_{tol}$) is not reached, the process is repeated, starting from the smoothing phase on the fine grid.

4.7.5 Application Example

As an application example, a stationary diffusion problem is considered in a thermally insulated metal bar of length $L = 1m$, with a constant cross-sectional area of $A = 0.1\text{m}^2$, with its ends kept at constant temperatures of 100 °C and 500 °C, respectively. Inside the bar, heat is produced at a constant volumetric power density of $q = 2000\frac{\text{kW}}{\text{m}^3}$, and the material of the bar has a constant thermal conductivity of $k = 5\frac{\text{W}}{\text{m K}}$. Assuming the coordinate x is associated with the length of the bar, the equation that describes this phenomenon is

$$\frac{d}{dx}\left(k\frac{dT}{dx}\right) + q = 0. \tag{4.20}$$

The equation that generally describes the stationary one-dimensional diffusion phenomenon of a quantity ϕ is

$$\frac{d}{dx}\left(\Gamma\frac{d\phi}{dx}\right) + S = 0 \tag{4.21}$$

where Γ is the diffusion coefficient and S is the source term. Referring to Fig. 4.10, by integrating and discretising, it becomes

$$\int_{\Delta V}\frac{d}{dx}\left(\Gamma\frac{d\phi}{dx}\right)dV + \int_{\Delta V} q dV = \left(\Gamma A\frac{d\phi}{dx}\right)_e - \left(\Gamma A\frac{d\phi}{dx}\right)_w + \overline{S}\Delta V = 0 \tag{4.22}$$

where $\overline{S}$ is the average value of the source term within the control volume ΔV. Assuming a linear approximation to calculate the value of Γ at the interfaces e and w, it becomes

$$\Gamma_w = \frac{\Gamma_W + \Gamma_P}{2} \quad \text{and} \quad \Gamma_e = \frac{\Gamma_P + \Gamma_E}{2}.$$

The diffusive flows are expressed as

$$\left(\Gamma A\frac{\phi}{dx}\right)_e = \Gamma_e A_e\left(\frac{\phi_E - \phi_P}{dx_{PE}}\right) \quad \text{and} \quad \left(\Gamma A\frac{\phi}{dx}\right)_w = \Gamma_w A_w\left(\frac{\phi_P - \phi_W}{dx_{WP}}\right).$$

To account for cases where the source term S is a function of the dependent variable, it is expressed in linear form:

$$\overline{S}\Delta V = S_u + S_P\phi_P.$$

It is now possible to rewrite Eq. 4.22 as

$$\Gamma_e A_e\left(\frac{\phi_E - \phi_P}{dx_{PE}}\right) - \Gamma_w A_w\left(\frac{\phi_P - \phi_W}{dx_{WP}}\right) + (S_u + S_P\phi_P)$$

which, rearranged, becomes

$$\left(\frac{\Gamma_E}{dx_{PE}}A_e + \frac{\Gamma_W}{dx_{WP}}A_w - S_P\right)\phi_P = \left(\frac{\Gamma_w}{dx_{WP}}A_w\right)\phi_W + \left(\frac{\Gamma_e}{dx_{WP}}A_w\right)\phi_E + S_u$$

or, in compact form,

$$a_P\phi_P = a_W\phi_W + a_E\phi_E + S_u \tag{4.23}$$

having set

$$a_W = \left(\frac{\Gamma_w}{dx_{WP}}A_w\right), \quad a_E = \left(\frac{\Gamma_e}{dx_{WP}}A_w\right), \quad a_P = a_W + a_E - S_P.$$

The analysis of the one-dimensional computational domain, constituted by the thickness of the bar, is now performed. Such a domain can be divided into intervals, each constituting the control volume in which the governing Eq. 4.20 is integrated. The integration of Eq. 4.20 over the control volume highlighted in Fig. 4.10 leads to the formulation.

$$\int_{\Delta V}\frac{d}{dx}\left(k\frac{dT}{dx}\right)dV + \int_{\Delta V} q dV = \left[\left(kA\frac{dT}{dx}\right)_e - \left(kA\frac{dT}{dx}\right)_w\right] + q\Delta V = 0 \tag{4.24}$$

that is

$$\left[k_e A\left(\frac{T_E - T_P}{\delta x}\right) - k_w A\left(\frac{T_P - T_W}{\delta x}\right)\right] + qA\delta x = 0. \tag{4.25}$$

Setting $k_e = k_w = k$ and rearranging, we get

$$\left(\frac{kA}{\delta x} + \frac{kA}{\delta x}\right)T_P = \left(\frac{kA}{\delta x}\right)T_W + \left(\frac{kA}{\delta x}\right)T_E + qA\delta x.$$

Using the compact notation (4.23), it is

$$a_W = \frac{k}{\delta x}A, \quad a_E = \frac{k}{\delta x}A, \quad a_P = a_W + a_E - S_P, \quad S_P = 0, \quad S_u = qA\delta x.$$

The approach for boundary nodes 1 and 5 is slightly different. In the case of node 1, point P coincides with point 1, and the temperature on face w—which coincides with point A—is known as it is prescribed by the boundary condition. The integration of

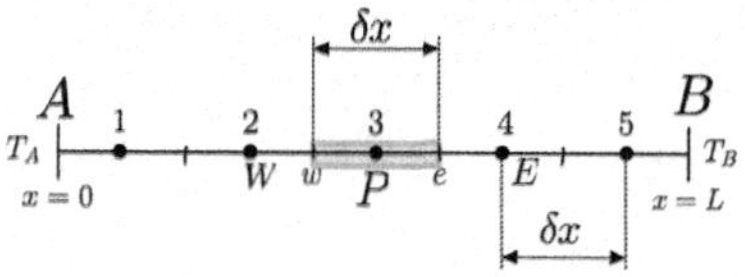

Fig. 4.10 Discretisation of the computational domain

the governing equation still leads to Eq. 4.24. Assuming a linear temperature trend between point A and point 1, for the control volume centred at point 1, Eq. 4.25 can be expressed as

$$\left[k_e A\left(\frac{T_E - T_P}{\delta x}\right) - k_w A\left(\frac{T_P - T_A}{\delta x/2}\right)\right] + qA\delta x = 0.$$

Using the compact notation of Eq. 4.23, it is

$$a_W = 0,\quad a_E = \frac{k}{\delta x}A,\quad a_P = a_W + a_E - S_P,\quad S_P = -\frac{2k}{\delta x}A,\quad S_u = qA\delta x + \frac{2k}{\delta x}A\,T_A.$$

In the case of node 5, point P coincides with point 5, and the temperature on face *e*—which coincides with point B—is known as it is prescribed by the boundary condition. The integration of the governing equation again leads to Eq. 4.24. Assuming a linear variation of the temperature between point 5 and point B, for the control volume centred at point 5, Eq. 4.25 can

$$\left[k_e A\left(\frac{T_B - T_P}{\delta x/2}\right) - k_w A\left(\frac{T_P - T_W}{\delta x}\right)\right] + qA\delta x = 0.$$

Using the compact notation of Eq. 4.23, one obtains

$$a_W = \frac{k}{\delta x}A,\quad a_E = 0,\quad a_P = a_W + a_E - S_P,\quad S_P = -\frac{2k}{\delta x}A,\quad S_u = qA\delta x + \frac{2k}{\delta x}A\,T_B.$$

Table 4.1 summarises the expressions derived so far for the coefficients a_W, a_E, S_u, S_P, and a_P as functions of the considered node.

Given these premises and assuming the entire length of the bar is discretised into 20 intervals, we obtain $\delta x = 0.05$ m, while the values of the coefficients a_W, a_E, S_u, S_P, and a_P are those shown in Table 4.2.

Table 4.1 Summary table of coefficients

Node	a_W	a_E	S_u	S_P	a_P
First node	0	$\frac{kA}{\delta x}$	$qA\delta x + \frac{2k}{\delta x}A\,T_A$	$-\frac{2k}{\delta x}A$	$a_W + a_E - S_P$
Non-border nodes	$\frac{kA}{\delta x}$	$\frac{kA}{\delta x}$	$qA\delta x$	0	$a_W + a_E - S_P$
Last node	$\frac{kA}{\delta x}$	0	$qA\delta x + \frac{2k}{\delta x}A\,T_B$	$-\frac{2k}{\delta x}A$	$a_W + a_E - S_P$

Table 4.2 Numerical value of the coefficients a_W, a_E, S_u, S_P and a_P

Node	a_W	a_E	S_u	S_P	a_P
1	0	1	210	−2	3
2, ...,19	1	1	10	0	2
20	1	0	1010	−2	3

By constructing the matrix equation resulting from the application of the conservation equation, integrated over each of the 20 control volumes, one obtains

$$\begin{bmatrix} 3 & -1 & 0 & \dots & \dots & \dots & 0 \\ -1 & 2 & -1 & 0 & \dots & \dots & 0 \\ 0 & -1 & 2 & -1 & 0 & \dots & 0 \\ \vdots & \vdots & \dots & \vdots & \vdots & \dots & 0 \\ \dots & \dots & \dots & \dots & -1 & 2 & -1 \\ 0 & 0 & \dots & 0 & \dots & -1 & 3 \end{bmatrix} \begin{bmatrix} y_1 \\ y_2 \\ y_3 \\ \vdots \\ y_{19} \\ y_{20} \end{bmatrix} = \begin{bmatrix} 210 \\ 10 \\ 10 \\ \vdots \\ 10 \\ 1010 \end{bmatrix}. \tag{4.26}$$

4.7.5.1 Iterations on the Fine Grid

Once the matrix Eq. 4.26 is determined, representing the system of algebraic equations for the considered system, the multigrid algorithm calls for a limited number of iterations using any iterative method for solving linear systems. In the example considered here, after setting the initial guess solution as one assuming a constant temperature of 150 °C throughout the entire computational domain, five iterations of the Gauss-Seidel algorithm are performed. The obtained solution vector is

$$\mathbf{y}^h = \begin{bmatrix} y_1 \\ y_2 \\ y_3 \\ \vdots \\ y_{19} \\ y_{20} \end{bmatrix} = \begin{bmatrix} 116.755 \\ 141.994 \\ 160.427 \\ \vdots \\ 394.392 \\ 468.130 \end{bmatrix}.$$

The residual vector is defined as $\mathbf{r} = \mathbf{b} - \mathbf{A}\mathbf{y}$, where $\mathbf{b}$ is the column matrix of known terms and $\mathbf{A}$ is the matrix of coefficients. Referring to the case of the fine grid, this becomes $\mathbf{r}^h = \mathbf{b}^h - \mathbf{A}^h\mathbf{y}^h$, which numerically results in

$$\mathbf{r}^h = \begin{bmatrix} r_1^h \\ r_2^h \\ r_3^h \\ \vdots \\ r_{19}^h \\ r_{20}^h \end{bmatrix} = \begin{bmatrix} 210 \\ 10 \\ 10 \\ \vdots \\ 10 \\ 1010 \end{bmatrix} - \begin{bmatrix} 3 & -1 & 0 & \dots & \dots & \dots & 0 \\ -1 & 2 & -1 & 0 & \dots & \dots & 0 \\ 0 & -1 & 2 & -1 & 0 & \dots & 0 \\ \vdots & \vdots & \dots & \vdots & \vdots & \dots & 0 \\ \dots & \dots & \dots & \dots & -1 & 2 & -1 \\ 0 & 0 & \dots & 0 & \dots & -1 & 3 \end{bmatrix} \begin{bmatrix} y_1 \\ y_2 \\ y_3 \\ \vdots \\ y_{19} \\ y_{20} \end{bmatrix} = \begin{bmatrix} 1.728 \\ 3.193 \\ 4.658 \\ \vdots \\ 7.461 \\ 0 \end{bmatrix}.$$

4.7.5.2 Transfer of Residuals and Coefficients from the Fine Grid to the Coarse Grid (Restriction)

The multigrid algorithm involves the use of multiple grids, each with a different level of refinement. The simplest way to construct grids with a lower degree of refinement, starting from the initial grid, is to pair the cells of the initial grid, as shown in Fig. 4.11. In the present example, three grids will be used:

- the finest initial grid, shown at the top in Fig. 4.11, has intervals of width $\delta x = 0.05$. this grid will be associated with the superscript h;
- the intermediate grid, shown in the centre of Fig. 4.11, has intervals of width $\delta x = 0.1$. this grid will be associated with the superscript $2h$;
- the coarsest grid, shown at the bottom of Fig. 4.11, has intervals of width $\delta x = 0.2$. this grid will be associated with the superscript $4h$.

Note that the $2h$ grid has half the number of intervals compared to the h grid, and the $4h$ grid has half the number of intervals compared to the $2h$ grid. Once the grids are defined, a residual vector must be associated with the two intermediate and coarse grids, starting from the residual vector initially calculated for the fine grid (restriction). Since each centre of an interval on the intermediate grid is equidistant from the two centres of the fine grid intervals from which it originated, the residual associated with each interval on the intermediate grid will be the average of the residuals associated with the pair of fine grid intervals from which the intermediate grid interval originated. The same procedure is applied to calculate the residuals for

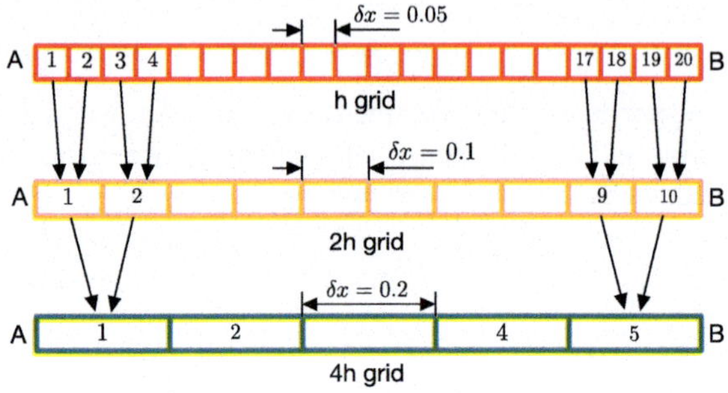

Fig. 4.11 Grids used in the example of the multigrid algorithm application

the coarse grid, starting from those of the intermediate grid. The residuals associated with the intermediate grid will therefore be

$$\mathbf{r}^{2h} = \begin{bmatrix} r_1^{2h} \\ r_2^{2h} \\ r_3^{2h} \\ \vdots \\ r_9^{2h} \\ r_{10}^{2h} \end{bmatrix} = \begin{bmatrix} 2.460 \\ 5.317 \\ 7.506 \\ \vdots \\ 28.173 \\ 3.730 \end{bmatrix}.$$

Once the vector of residuals on the intermediate grid is determined, it is necessary to determine the coefficient matrix $\mathbf{A}^{2h}$ for the matrix equation $\mathbf{A}^{2h}\mathbf{e}^{2h} = \mathbf{r}^{2h}$, which the residual vector $\mathbf{r}^{2h}$ satisfies. In this example, this matrix is not calculated by interpolating the corresponding matrix from the fine grid but by reapplying the procedure shown in the Table 4.1 to the intermediate grid. The result is

$$\begin{bmatrix} 1.5 & -0.5 & 0 & \dots & \dots & \dots & 0 \\ -0.5 & 1 & -0.5 & 0 & \dots & \dots & 0 \\ 0 & -1 & 2 & -1 & 0 & \dots & 0 \\ \vdots & \vdots & \dots & \vdots & \vdots & \dots & 0 \\ \dots & \dots & \dots & \dots & -0.5 & 1 & -0.5 \\ 0 & 0 & \dots & 0 & \dots & -0.5 & 1.5 \end{bmatrix} \begin{bmatrix} e_1^{2h} \\ e_2^{2h} \\ e_3^{2h} \\ \vdots \\ e_9^{2h} \\ e_{10}^{2h} \end{bmatrix} = \begin{bmatrix} 2.460 \\ 5.317 \\ 7.506 \\ \vdots \\ 28.173 \\ 3.730 \end{bmatrix}. \tag{4.27}$$

Even for the error vector on the intermediate grid, $\mathbf{e}^{2h}$, the same system as in Eq. 4.27 is solved using the Gauss-Seidel procedure. This time, more iterations are performed compared to the case of the fine grid, as there are fewer elements, and the computational cost of each iteration is reduced. After ten iterations, the following solution vector is obtained.

$$\mathbf{e}^{2h} = \begin{bmatrix} e_1^{2h} \\ e_2^{2h} \\ e_3^{2h} \\ \vdots \\ e_9^{2h} \\ e_{10}^{2h} \end{bmatrix} = \begin{bmatrix} 19.156 \\ 58.310 \\ 96.049 \\ \vdots \\ 158.591 \\ 55.351 \end{bmatrix}.$$

Given that the system in Eq. 4.27 was solved using an iterative method, a residual can be defined as

$$\widehat{\mathbf{r}}^{2h} = \mathbf{r}^{2h}_{initial} - \mathbf{A}^{2h}\mathbf{e}^{2h} = \begin{bmatrix} 2.881 \\ 4.609 \\ 5.929 \\ \vdots \\ 0.9192 \\ 0.000 \end{bmatrix}$$

whose numerical values correspond to the values of $\mathbf{e}^{2h}$ after ten iterations. As done previously for the intermediate grid, these residuals can be interpolated and transferred onto the coarse grid to obtain the residual vector $\mathbf{r}^{4h}$. Similarly, as done for the intermediate grid, it is possible to determine the matrix $\mathbf{A}^{4h}$. The matrix $\mathbf{A}^{4h}$ is then used to solve the system $\mathbf{A}^{4h}\mathbf{e}^{4h} = \mathbf{r}^{4h}$ using the same iterative method. After ten iterations, the vector $\mathbf{e}^{4h}$ will be

$$\mathbf{e}^{4h} = \begin{bmatrix} e_1^{4h} \\ e_2^{4h} \\ e_3^{4h} \\ e_4^{4h} \\ e_5^{4h} \end{bmatrix} = \begin{bmatrix} 23.408 \\ 55.831 \\ 63.731 \\ 47.205 \\ 16.348 \end{bmatrix}.$$

4.7.5.3 Transfer of the Correction for the Coarse Grid to the Intermediate Grid (Prolongation)

The errors calculated for the coarser grid in the previous phase must now be transferred to the finer grid. For this purpose, any interpolation method can be used. Therefore, using linear interpolation to transfer the errors from the coarser $4h$ grid to the intermediate $2h$ grid, it is

$$\begin{aligned} e_1'^{2h} &= 0.75e_1'^{4h}, \\ e_2'^{2h} &= 0.75e_1'^{4h} + 0.25e_2'^{4h}, \\ e_2'^{2h} &= 0.25e_1'^{4h} + 0.75e_2'^{4h}. \end{aligned}$$

For the first three points of the intermediate grid, the single quote mark has been used to distinguish these errors from those obtained for the same grid during the restriction phase. The error at the edges has also been considered null where the boundary condition is imposed. Therefore, considering the value of the sought quantity at those points as known, it is

$$\mathbf{e}'^{2h} = \begin{bmatrix} e_1'^{2h} \\ e_2'^{2h} \\ e_3'^{2h} \\ \vdots \\ e_9'^{2h} \\ e_{10}'^{2h} \end{bmatrix} = \begin{bmatrix} 17.556 \\ 31.514 \\ 47.726 \\ \vdots \\ 24.062 \\ 12.261 \end{bmatrix}.$$

It is now possible to compute the correct value of the error on the intermediate grid, considering that

$$\mathbf{e}_{corr}^{2h} = \mathbf{e}^{2h} + \mathbf{e}'^{2h}$$

and therefore

$$\begin{bmatrix} e_1^{2h} \\ e_2^{2h} \\ e_3^{2h} \\ \vdots \\ e_9^{2h} \\ e_{10}^{2h} \end{bmatrix} = \begin{bmatrix} 36.713 \\ 89.725 \\ 143.775 \\ \vdots \\ 182.654 \\ 67.612 \end{bmatrix}.$$

4.7.5.4 Smoothing of the Correction for the Intermediate Grid

It is useful to recall here that the Gauss-Seidel algorithm is the chosen method for solving systems of linear equations. Since the intermediate grid is not the initial fine grid, only two iterations of the Gauss-Seidel algorithm are applied to the system formed by the equation in correction form, $\mathbf{A}^{2h}\mathbf{e}^{2h} = \mathbf{r}^{2h}$. This results in an updated value for the correction on the intermediate grid:

$$\begin{bmatrix} e_1^{2h} \\ e_2^{2h} \\ e_3^{2h} \\ \vdots \\ e_9^{2h} \\ e_{10}^{2h} \end{bmatrix} = \begin{bmatrix} 32.639 \\ 95.749 \\ 152.494 \\ \vdots \\ 188.283 \\ 65.248 \end{bmatrix}.$$

4.7.5.5 Transfer of the Correction for the Intermediate Grid to the Fine Grid (Prolongation)

Similarly to the transfer of errors from the coarse grid to the intermediate grid, the errors just calculated for the intermediate grid are transferred to the fine grid, resulting in

$$\begin{bmatrix} e_1^{h} \\ e_2^{h} \\ e_3^{h} \\ \vdots \\ e_1 9^{h} \\ e_{20}^{h} \end{bmatrix} = \begin{bmatrix} 24.479 \\ 48.416 \\ 79.971 \\ \vdots \\ 96.007 \\ 48.936 \end{bmatrix}.$$

4.7.5.6 Updating the Solution on the Fine Grid

In this phase, the initial solution for the fine grid is updated by considering the correction obtained from the coarse grid:

$$\mathbf{y}_{corr} = \mathbf{y} + \mathbf{e}^h$$

which in this case becomes

$$\mathbf{y}^h = \begin{bmatrix} y_1 \\ y_2 \\ y_3 \\ \vdots \\ y_{19} \\ y_{20} \end{bmatrix} = \begin{bmatrix} 116.755 \\ 141.994 \\ 160.427 \\ \vdots \\ 394.392 \\ 468.130 \end{bmatrix} + \begin{bmatrix} 24.479 \\ 48.416 \\ 79.971 \\ \vdots \\ 96.007 \\ 48.936 \end{bmatrix} = \begin{bmatrix} 141.235 \\ 190.411 \\ 240.399 \\ \vdots \\ 490.399 \\ 517.067 \end{bmatrix}.$$

4.7.5.7 Checking the Level of Convergence and Possible Repetition of the Cycle

The phases of residual (restriction) and error (prolongation) transfer involve the application of interpolation at various levels, which can introduce numerical errors. In most practical cases, these errors prevent the solution obtained after a single multigrid cycle from being characterised by a residual value sufficient to consider the solution acceptable. For this reason, the entire multigrid cycle is repeated, using the final solution of the previous cycle as the initial solution, until the required level of convergence is reached.

Chapter 5
Pressure-Velocity Coupling

What has been illustrated so far about the finite volume method (see Chap. 3) assumed the velocity field as known. Some of the techniques for determining this field will be illustrated in this chapter. In this regard, it is useful to highlight that the link—the coupling—existing between pressure and velocity, together with the non-linearities resulting from the presence of advective terms, represents one of the major difficulties in solving the conservation equations of mass and momentum. As seen previously, the total number of equations to be solved depends both on the number of cells in the computational domain and on the conservation equations considered: for example, in the case of two-dimensional laminar compressible flow, it will be necessary to consider the conservation equation of mass, the two components of momentum conservation, energy conservation, and the state equation—five equations for each cell.

A first categorisation of the various approaches with which the finite volume method can be used is based on the way all these equations are grouped into linear systems to be solved. In the case where you want to solve a single system containing all the equations, we speak of the *coupled* or, more accurately, *simultaneous* approach. When solving separately and sequentially the systems resulting from the application of the various conservation equations, the approach is called *segregated*. In the segregated approach, it is said that each quantity "owns its own equation", meaning that each single quantity is associated with the corresponding matrix conservation equation. Considering that, even in the segregated case, the equations are solved depending on each other, the term *simultaneous* may be clearer than *coupled*. Among the main advantages of the segregated approach is the reduced need for computing resources in terms of memory occupied for the solution of linear systems that are characterised by smaller dimensions given the smaller number of unknowns; the greater difficulty associated with the segregated approach is manifested in the need to implement a specific coupling algorithm between pressure and velocity. Correspondingly, the main advantage of the simultaneous approach is that the coupling between pressure and velocity is verified at the very moment in which all the equations are

G. Caramia and E. Distaso, *A Practical Approach to Computational Fluid Dynamics Using OpenFOAM®*, https://doi.org/10.1007/978-3-031-88957-8_5

solved in a single system: it is said in this case that no quantity "owns" its specific equation.

In the case where the segregated approach is chosen to solve a flow in which the variations in density and temperature permit the use of the state equation to calculate pressure, each quantity has its own differential equation: density is obtained from the equation of conservation of mass, velocity from the conservation of momentum, and temperature from the energy equation. This approach is referred to as the *compressible formulation* or, in some cases, the *density-based formulation*. High-speed gas flows necessitate the compressible formulation, where the simultaneous approach is preferred over the segregated one, as the latter is less efficient in cases involving strong density variations, such as those caused by shock waves.

When pressure variations are negligible, density may vary solely due to temperature changes. Although the flow remains strictly compressible, the state equation cannot be used to derive pressure as a function of density and temperature. Small errors in the density calculation through the equation of conservation of mass would result in significant errors in pressure computation via the state equation, potentially leading to unacceptable inaccuracies in the velocity field or, in extreme cases, the failure of the entire iterative process. Consequently, for flows with negligible pressure variations, an equation of state is employed in which density is a function of temperature alone, which is determined using the equation of conservation of energy. In this case, the pressure evolution equation is no longer required, and a pressure field must be computed to determine a corresponding velocity field that satisfies the conservation of both momentum and mass. In other words, the fact that pressure is independent of density introduces significant challenges in handling pressure-velocity coupling. Both for constant-density flows and for flows where density is a function of temperature alone, the *incompressible formulation* is adopted, also referred to in some cases as the *pressure-based formulation*. Among the methods used to compute pressure in the incompressible formulation, the so-called *projection methods* involve solving a Poisson equation for pressure. One of the most well-known and widely used projection methods is the **SIMPLE** (Semi-Implicit Method for Pressure-Linked Equations) algorithm, which, for clarity of exposition, will be presented below in its application on staggered grids, as initially developed by Spalding and Patankar in 1972.

5.1 The Staggered Grid

Although not used in most cases of general interest, the analysis of this type of grid is useful for understanding numerous concepts at the base of modern pressure-velocity coupling algorithms. Considering what has already been seen previously, the application of the finite volume method involves considering all quantities positioned in the cell centre. In Fig. 5.1, a pressure field is shown with a distribution of values. s that gives rise to the so-called "checkerboard problem". Considering the momentum conservation Eq. 5.6 relative to the first component of velocity, it will be necessary to determine the pressure gradient $\partial p/\partial x$, considering the pressure values on the faces w and e. By using linear interpolation to obtain the values of p_w and p_e, we obtain

$$\frac{\partial p}{\partial x} = \frac{p_e - p_w}{\delta x} = \frac{\frac{p_E + p_P}{2} - \frac{p_P + p_W}{2}}{\delta x} = \frac{p_E - p_W}{2\delta x}$$

where the equi-spacing of the computational grid has been considered.

Similarly, considering the momentum conservation Eq. 5.7 for the second component of velocity, we obtain

$$\frac{\partial p}{\partial y} = \frac{p_N - p_S}{2\delta y}.$$

In both cases, in determining the pressure gradient, the pressure value in the cell centre, P, does not appear. Therefore, in the case of a checkerboard pressure distribution, as shown in Fig. 5.1, the algorithm cannot account for the correct pressure distribution: in this specific case, the algorithm would perceive a uniform pressure field instead of a real pressure field with a checkerboard distribution. Obviously, an incorrect pressure field would lead, when used in the solution of the conservation equations of momentum, to incorrect velocity fields. A possible remedy for the checkerboard problem is to consider different control volumes (cells) for the various quantities involved. Specifically, it is possible to consider a control volume for scalar quantities (e.g. pressure) and a different control volume for each of the velocity components considered-two in the two-dimensional case, three in the three-dimensional case.

In order to simplify the understanding of this strategy, a uniform two-dimensional Cartesian grid is considered, where the indices I and J (uppercase indices) are used to indicate the centres of the cells for scalar quantities, and the indices i and j (lowercase indices) are used to indicate the corresponding faces. The control volume used for the scalar physical quantities is the one shown in Fig. 5.2; the control volume used for the x-component of the velocity (u) is the one shown in Fig. 5.3; the control volume used for the y-component of the velocity (v) is the one shown in Fig. 5.4.

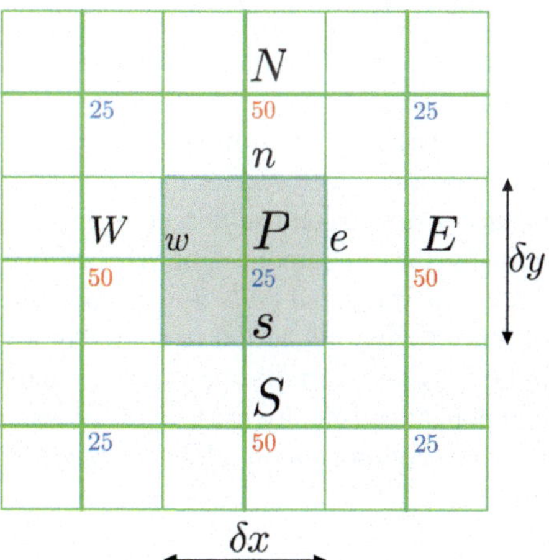

Fig. 5.1 Checkerboard pressure field

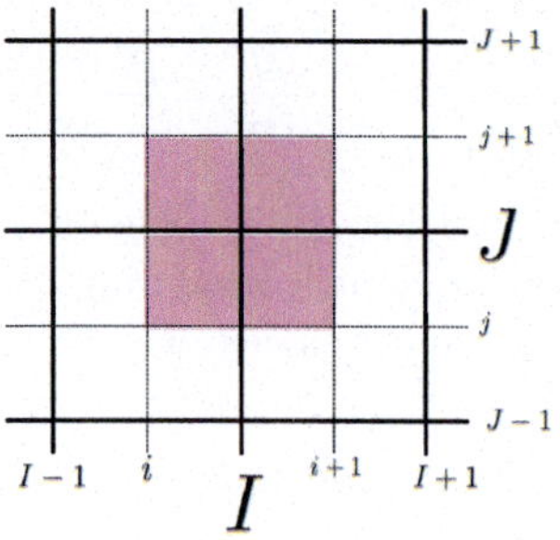

Fig. 5.2 Control volume used for scalars

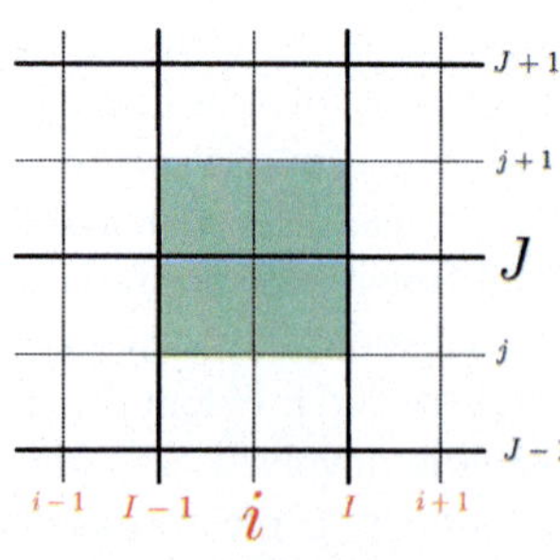

Fig. 5.3 Control volume used for the x-component of the velocity

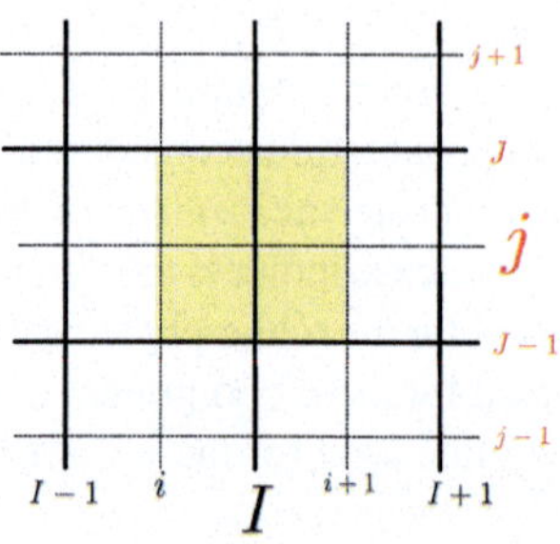

Fig. 5.4 Control volume used for the y-component of the velocity

From Fig. 5.2, and more clearly from the following Figs. 5.3 and 5.4, it is evident how the centres of the faces can be identified using both uppercase and lowercase indices. The centre i, J of face i of the control volume in Fig. 5.2 is also the centre of the control volume used for the u component of velocity in Fig. 5.3. The centre I, j of face j of the control volume in Fig. 5.2 is also the centre of the control volume used for the y-component of velocity in Fig. 5.4. In this case, the grids are called *backward staggered velocity grid* because the grids used for the two components of the velocity are shifted respectively towards the lower value of each of the two indices used to identify the centre of the scalar cell: the centre of the cell for the u component of velocity is placed between indices $I - 1$ and I; the centre of the cell for the y-component of velocity is placed between indices $J - 1$ and J, as shown in Fig. 5.5.

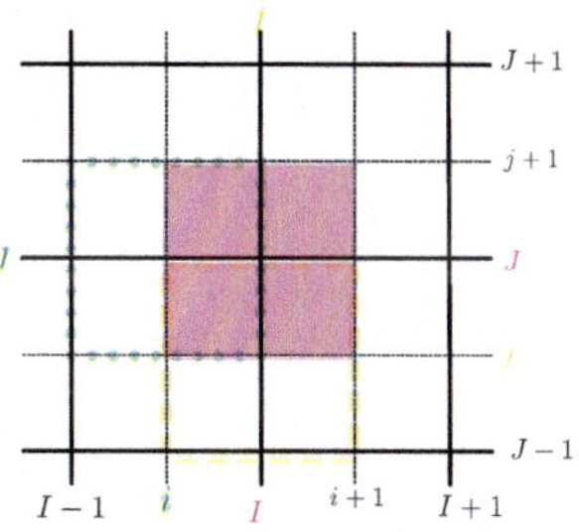

Fig. 5.5 Backward staggered grids for velocity

5.2 Conservation of Momentum

Having defined the new coordinate system, it is possible to write the discretised equation of conservation of momentum for each of the two velocity components with reference to the new staggered grids. Initially, considering the u component of velocity, in Fig. 5.6, the cell used for writing the discretised momentum conservation equation is shown. In detail, Fig. 5.6 shows:

- the cell whose centre is P with indices (i, J);
- the centres e, w, n, s—with indices respectively (I, J), $(I-1, J)$, $(i, j+1)$, (i, j)—of the faces that delimit the cell;
- the centres E, W, N, S—with indices respectively $(i+1, J)$, $(i-1, J)$, $(i, J+1)$, $(i, J-1)$—of the cells adjacent to the considered cell.

In Chap. 3, it was seen that it is possible to write, for each cell of the computational domain, the conservation equation of momentum for the component u of the velocity in the following discretised form

$$a_P u_P = \sum a_{nb} u_{nb} + S_u \tag{5.1}$$

where the subscript P indicates the value at the centre of the considered cell; the symbol $\sum$ indicates the summation extended to the centres of all the cells that share a face with the considered cell (of centre P); the subscript nb indicates all the

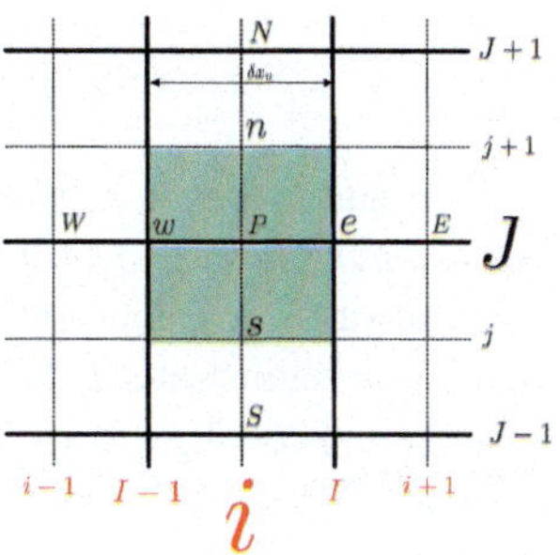

Fig. 5.6 Control volume used for the u component of the velocity on backward staggered grid

neighbouring cells that share a face with the considered cell (of centre P). In the two-dimensional case, it will be

$$\sum a_{nb}u_{nb} = a_E u_E + a_W u_W + a_N u_N + a_S u_S. \tag{5.2}$$

S_u is the constant term of the linearised form $S_u + S_P u_P$ of the component u of the source term $\overline{S}\Delta V_u$ (cf., Eq. 3.4) in which ΔV_u is the volume (the surface in the two-dimensional case) of the cell. By highlighting the contribution of the pressure from the source term and considering the cells shown in Fig. 5.6, Eq. 5.1 can be expressed as

$$a_{i,J}u_{i,J} = \sum a_{nb}u_{nb} - \frac{p_{I,J} - p_{I-1,J}}{\delta x_u}\Delta V_u + \overline{S}\Delta V_u$$

or, equivalently

$$a_{i,J}u_{i,J} = \sum a_{nb}u_{nb} + \left(p_{I-1,J} - p_{I,J}\right)A_{i,J} + b_{i,J}. \tag{5.3}$$

Notice that: (1) the pressure term has been expressed through linear interpolation between the values at the faces of the control volume for the u velocity component, (2) it was set that $b_{i,J} = \overline{S}\Delta V_u$, (3) $A_{i,J}$ is the surface (length in the two-dimensional case) of the faces w and e of the cell. Now, Eq. 5.2 can be written as

$$\sum a_{nb}u_{nb} = a_{i+1,J}u_{i+1,J} + a_{i-1,J}u_{i-1,J} + a_{i,J+1}u_{i,J+1} + a_{i,J-1}u_{i,J-1}. \tag{5.4}$$

The values of the coefficients $a_{i,J}$ and a_{nb} can be calculated using any of the discretisation schemes for convective-diffusive flows presented in Chap. 3. Using the centred scheme, it will be

$$\begin{aligned}
a_E &= D_e - \frac{F_e}{2},\\
a_W &= D_w - \frac{F_w}{2},\\
a_N &= D_n - \frac{F_n}{2},\\
a_S &= D_s - \frac{F_s}{2},\\
a_P &= a_W + a_E + a_N + a_S + (F_e - F_w + F_n - F_s)
\end{aligned}$$

where F and D are the convective and diffusive mass flows respectively that cross the faces e, w, n, and s. Regardless of the interpolation scheme applied, the coefficients a will always be a combination of the fluxes F and D, and for this reason, their calculation is illustrated below for the cell used in the discretisation of the momentum conservation equation in its u component.

The convective fluxes are as follows:

$$F_w = (\rho u)_w = \frac{F_{i,J} + F_{i-1,J}}{2} = \frac{1}{2}\left[\frac{\rho_{I,J} + \rho_{I-1,J}}{2} u_{i,J} + \frac{\rho_{I-1,J} + \rho_{I-2,J}}{2} u_{i-1,J}\right],$$

$$F_e = (\rho u)_e = \frac{F_{i+1,J} + F_{i,J}}{2} = \frac{1}{2}\left[\frac{\rho_{I+1,J} + \rho_{I,J}}{2} u_{i+1,J} + \frac{\rho_{I,J} + \rho_{I-1,J}}{2} u_{i,J}\right],$$

$$F_n = (\rho v)_n = \frac{F_{I,j+1} + F_{I-1,j+1}}{2} = \frac{1}{2}\left[\frac{\rho_{I,J+1} + \rho_{I,J}}{2} v_{I,j+1} + \frac{\rho_{I-1,J+1} + \rho_{I-1,J}}{2} v_{I-1,j+1}\right],$$

$$F_s = (\rho v)_s = \frac{F_{I,j} + F_{I-1,j}}{2} = \frac{1}{2}\left[\frac{\rho_{I,J} + \rho_{I,J-1}}{2} v_{I,j} + \frac{\rho_{I-1,J} + \rho_{I-1,J-1}}{2} v_{I-1,j}\right].$$

The diffusive fluxes are:

$$D_w = \frac{\Gamma_{I-1,J}}{x_i - x_{i-1}},$$

$$D_e = \frac{\Gamma_{I,J}}{x_{i+1} - x_i},$$

$$D_n = \frac{\Gamma_{I-1,J+1} + \Gamma_{I,J+1} + \Gamma_{I-1,J} + \Gamma_{I,J}}{4\,(y_J - y_{J-1})},$$

$$D_s = \frac{\Gamma_{I-1,J} + \Gamma_{I,J} + \Gamma_{I-1,J-1} + \Gamma_{I,J-1}}{4\,(y_{J+1} - y_J)}.$$

It can be noted that, when computing the flows, if a scalar value or a velocity component is not available on the faces of the control volume, the calculation of an appropriate average between the values of the quantity at the closest points for which the quantity itself is known has been used. The values of the two components u and v of the velocity used for the diffusive fluxes are those resulting from the initial conditions, or, if it is not the first iteration, those resulting from the previous iteration of the solution algorithm: these must therefore be distinguished from the unknown values that appear in the discretised Eq. 5.3 and in Eq. 5.4. What has been done so far for the u component of the velocity can be similarly reapplied to the v component with reference to Fig. 5.7.

The discretised form of the conservation equation of momentum for the v component of the velocity is

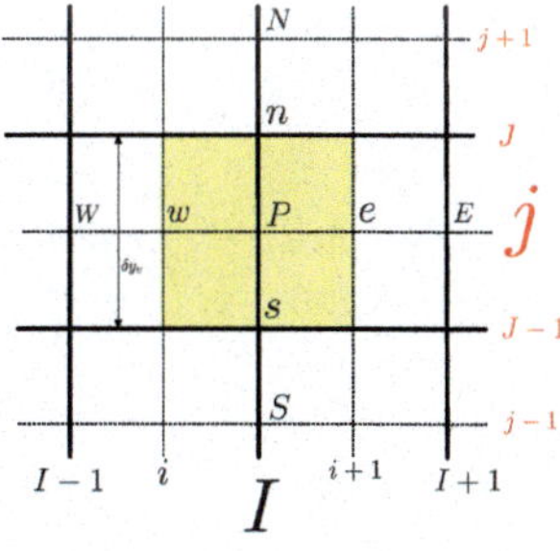

Fig. 5.7 Control volume used for the v component of the velocity on a backward staggered grid

$$a_{I,j}v_{I,j} = \sum a_{nb}v_{nb} + \left(p_{I,J-1} - p_{I,J}\right)A_{I,j} + b_{I,j}. \tag{5.5}$$

The convective fluxes will be:

$$F_w = (\rho u)_w = \frac{F_{i,J} + F_{i,J-1}}{2} = \frac{1}{2}\left[\frac{\rho_{I,J} + \rho_{I-1,J}}{2}u_{i,J} + \frac{\rho_{I-1,J-1} + \rho_{I-1,J-1}}{2}u_{i,J-1}\right],$$

$$F_e = (\rho u)_e = \frac{F_{i+1,J} + F_{i+1,J-1}}{2} = \frac{1}{2}\left[\frac{\rho_{I+1,J} + \rho_{I,J}}{2}u_{i+1,J-1} + \frac{\rho_{I,J-1} + \rho_{I+1,J-1}}{2}u_{i+1,J-1}\right],$$

$$F_n = (\rho v)_n = \frac{F_{I,j} + F_{I,j+1}}{2} = \frac{1}{2}\left[\frac{\rho_{I,J} + \rho_{I,J-1}}{2}v_{I,j} + \frac{\rho_{I,J+1} + \rho_{I,J}}{2}v_{I,j+1}\right],$$

$$F_s = (\rho v)_s = \frac{F_{I,j-1} + F_{I,j}}{2} = \frac{1}{2}\left[\frac{\rho_{I,J-1} + \rho_{I,J-2}}{2}v_{I,j-1} + \frac{\rho_{I,J} + \rho_{I,J-1}}{2}v_{I,j}\right].$$

The diffusive flows will be:

$$D_w = \frac{\Gamma_{I-1,J-1} + \Gamma_{I,J-1} + \Gamma_{I-1,J} + \Gamma_{I,J}}{4\left(x_I - y_{I-1}\right)},$$

$$D_e = \frac{\Gamma_{I,J-1} + \Gamma_{I+1,J-1} + \Gamma_{I,J} + \Gamma_{I+1,J}}{4\left(x_{I+1} - x_I\right)},$$

$$D_n = \frac{\Gamma_{I,J}}{y_{j+1} - y_j},$$

$$D_s = \frac{\Gamma_{I,J-1}}{y_j - y_{j-1}}.$$

Also in this case, the values of the components u and v used to determine the convective flows are those resulting from the initial conditions or, if it is not the first iteration, those resulting from the previous iteration of the solution algorithm. These values must therefore be distinguished from those—unknown—appearing in the discretised Eq. 5.5.

5.3 The SIMPLE Algorithm

From the general transport equation it is possible to derive (see Sect. 2.9) the conservation equation of the momentum in the x direction

$$\frac{\partial}{\partial x}(\rho u u) + \frac{\partial}{\partial y}(\rho u v) = \frac{\partial}{\partial x}\left(\mu\frac{\partial u}{\partial x}\right) + \frac{\partial}{\partial y}\left(\mu\frac{\partial u}{\partial y}\right) - \frac{\partial p}{\partial x} + S_u, \tag{5.6}$$

the conservation equation of momentum in the y direction

$$\frac{\partial}{\partial x}(\rho uv) + \frac{\partial}{\partial y}(\rho vv) = \frac{\partial}{\partial x}\left(\mu\frac{\partial v}{\partial x}\right) + \frac{\partial}{\partial y}\left(\mu\frac{\partial v}{\partial y}\right) - \frac{\partial p}{\partial y} + S_v, \tag{5.7}$$

the continuity equation

$$\frac{\partial}{\partial x}(\rho u) + \frac{\partial}{\partial y}(\rho v) = 0 \tag{5.8}$$

in the case of a two-dimensional laminar stationary flow of a Newtonian fluid. Note that in the two conservation equations of the momentum, the contribution of the pressure has been expressed as a source term to highlight its importance, which will be shown below.

In Chap. 3, the finite volume method is illustrated for the solution of the general transport equation, always considering the velocity field of the fluid transporting a generic quantity ϕ. This velocity field $\mathbf{u}$, with its three components u, v, w, is governed by the three Eqs. 5.6, 5.7, 5.8 whose solution presents three main problems:

- the convective terms contains non-linear quantities (e.g. ρuu in Eq. 5.6);
- each component of the velocity is linked to the others because in each equation all the three components of the velocity appear;
- for the generic control volume, it is not possible to write a transport equation for the pressure. The pressure appears in the two momentum conservation equations but not in the continuity equation.

Regarding the pressure, in the case of compressible flows the continuity equation can be used as a transport equation for density and, in addition to Eqs. 5.6 and 5.7, the conservation equation of energy can be used as a transport equation for temperature. Finally, the pressure can be obtained from the equation of state, given the density and temperature.

In the case of incompressible flows, being constant, the density is not linked to the pressure. The pressure value, if exact, leads to, with the two equations of momentum, a velocity field also satisfying the continuity equation. Specifically, to solve the conservation equation of momentum in the two-dimensional case, there are three unknowns (the two components of the velocity and the pressure) against only two transport equations (one momentum equation for each of the two components of the velocity). The use of the continuity equation is difficult in this case because, as seen, it does not contain the pressure term. The SIMPLE algorithm is one of the techniques used to solve this problem, which is better known as the *pressure-velocity coupling problem*.

Before showing the details of this algorithm, it is worth noting that, given the exact pressure field p, it is possible to solve the conservation equation of momentum for each of the two (in the two-dimensional case) components of the velocity; the velocity field obtained will satisfy both the conservation of momentum equation and the continuity equation.

The SIMPLE algorithm will be illustrated for a two-dimensional laminar stationary flow. The iterative process begins with the imposition of a pressure field p^* as a first guess and the subsequent solution of the discretised momentum conservation equation for each of the two components of velocity:

$$a_{i,J} u^*_{i,J} = \sum a_{nb} u^*_{nb} + \left(p^*_{I-1,J} - p^*_{I,J}\right) A_{i,J} + b_{i,J},$$
$$a_{I,j} v^*_{I,j} = \sum a_{nb} v^*_{nb} + \left(p^*_{I,J-1} - p^*_{I,J}\right) A_{I,j} + b_{I,j}.$$

At this stage, the p' field for pressure correction is defined as the difference between the exact p and the initially imposed p^* pressure field:

$$p' = p - p^*. \tag{5.9}$$

Similarly, a correction field can be defined for both components of the velocity:

$$u = u^* + u', \tag{5.10}$$
$$v = v^* + v'. \tag{5.11}$$

Moreover, the discretised equation of conservation of momentum, when applied to the exact pressure field p, will yield a velocity field that satisfies the continuity equation. By subtracting the discretised equation of momentum for the pressure field p^* from that for the exact pressure field p, the following two equations are obtained, one for each component:

$$a_{i,J}\left(u_{i,J} - u^*_{i,J}\right) = \sum a_{nb}\left(u_{nb} - u^*_{nb}\right) + \left[\left(p_{I-1,J} - p^*_{I-1,J}\right) - \left(p_{I,J} - p^*_{I,J}\right)\right] A_{i,J},$$
$$a_{I,j}\left(v_{I,j} - v^*_{I,j}\right) = \sum a_{nb}\left(v_{nb} - v^*_{nb}\right) + \left[\left(p_{I,J-1} - p^*_{I,J-1}\right) - \left(p_{I,J} - p^*_{I,J}\right)\right] A_{I,j}.$$

Furthermore, using the correction formulas (5.9), (5.10), and (5.11), it becomes

$$a_{i,J} u'_{i,J} = \sum a_{nb} u'_{nb} + \left(p'_{I-1,J} - p'_{I,J}\right) A_{i,J},$$
$$a_{I,j} v'_{I,j} = \sum a_{nb} v'_{nb} + \left(p'_{I,J-1} - p'_{I,J}\right) A_{I,j}.$$

At this stage, the greatest approximation is introduced by the SIMPLE algorithm. The terms $\sum a_{nb} u'_{nb}$ and $\sum a_{nb} v'_{nb}$ are disregarded in order to obtain

$$u'_{i,J} = \left(p'_{I-1,J} - p'_{I,J}\right)\frac{A_{i,J}}{a_{i,J}} = \left(p'_{I-1,J} - p'_{I,J}\right) d_{i,J},$$
$$v'_{I,j} = \left(p'_{I,J-1} - p'_{I,J}\right)\frac{A_{I,j}}{a_{I,j}} = \left(p'_{I,J-1} - p'_{I,J}\right) d_{I,j}.$$

Specifically, the value of the corrections to be inserted in Eqs. 5.10 and 5.11 to obtain velocity values that will no longer be the exact ones, even though these approximated values will be treated as exact during this phase to facilitate comprehension. In summary,

$$u_{i,J} = u^*_{i,J} + \left(p'_{I-1,J} - p'_{I,J}\right) d_{i,J}, \tag{5.12}$$

$$v_{I,j} = v^*_{I,j} + \left(p'_{I,J-1} - p'_{I,J}\right) d_{I,j} \tag{5.13}$$

and, similarly

$$u_{i+1,J} = u^*_{i+1,J} + \left(p'_{I,J} - p'_{I+1,J}\right) d_{i+1,J} \quad with \quad d_{i+1,J} = \frac{A_{i+1,J}}{a_{i+1,J}}, \tag{5.14}$$

$$v_{I,j+1} = v^*_{I,j+1} + \left(p'_{I,J} - p'_{I,J+1}\right) d_{I,j+1} \quad with \quad d_{I,j+1} = \frac{A_{I,j+1}}{a_{I,j+1}}. \tag{5.15}$$

Considering that, in addition to the conservation equations of momentum, the velocity field must also satisfy the continuity equation. Referring to Fig. 5.8, the discretised form of the continuity equation can be expressed as

$$\left[(\rho u A)_{i+1,J} - (\rho u A)_{i,J}\right] + \left[(\rho u A)_{I,j+1} - (\rho u A)_{I,j}\right] = 0.$$

Substituting the values of $u_{i,J}$, $v_{I,j}$, $u_{i+1,J}$, and $v_{I,j+1}$ given respectively by Eqs. 5.12, 5.13, 5.14, and 5.15, the *discretised equation for pressure correction* is obtained:

$$a_{I,J}\, p'_{I,J} = a_{I+1,J}\, p'_{I+1,J} + a_{I-1,J}\, p'_{I-1,J} + a_{I,J+1}\, p'_{I,J+1} + a_{I,J-1}\, p'_{I,J-1} + b'_{I,J} \tag{5.16}$$

in which

$$a_{I+1,J} = (\rho d A)_{i+1,J}\,,$$
$$a_{I-1,J} = (\rho d A)_{i,J}\,,$$
$$a_{I,J+1} = (\rho d A)_{I,j+1}\,,$$

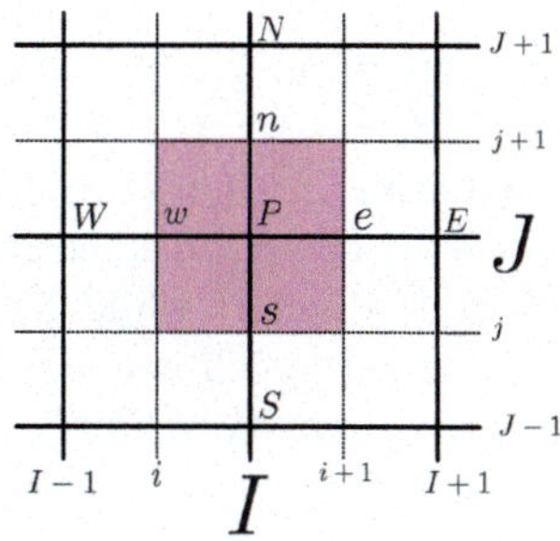

Fig. 5.8 Control volume used for the discretisation of the continuity equation on a staggered backward grid (see also Fig. 5.2)

$$
\begin{aligned}
a_{I,J-1} &= (\rho d A)_{I,j}\,, \\
a_{I,J} &= a_{I+1,J} + a_{I-1,J} + a_{I,J+1} + a_{I,J-1}\,, \\
b'_{I,J} &= \left(\rho u^* A\right)_{i,J} - \left(\rho u^* A\right)_{i+1,J} + \left(\rho u^* A\right)_{I,j} - \left(\rho u^* A\right)_{I,j+1}\,.
\end{aligned}
$$

It can be noted that the source term $b'_{I,J}$ derives from the initial imposition of a velocity field, whose components u^* and v^* are not the exact values, but are either first-guess values or those resulting from the preceding iteration. Solving Eq. 5.16 for each cell in the computational domain results in the p' pressure correction field. Using the correction formula (5.9), the pressure correction field determines the p pressure field, certainly respecting the continuity equation. Similarly, the correction formulas (5.10) and (5.11) are used for the two components of velocity. The velocity field thus obtained certainly respects the conservation of mass equation but does not necessarily satisfy the two conservation equations of momentum (due to the approximation made by neglecting the terms $\sum a_{nb}u'_{nb}$ and $\sum a_{nb}v'_{nb}$).

1. The velocity and pressure fields obtained in this way are referred to as *corrected*;
2. the procedure is iteratively repeated until the conservation equations—mass and momentum—are simultaneously satisfied.

It is important to note that neglecting the terms $\sum a_{nb}u'_{nb}$ and $\sum a_{nb}v'_{nb}$ does not affect the final solution (the one obtained at the end of the iterative procedure) because, once convergence is achieved, the pressure correction term p' and the velocity correction terms become very small—effectively zero. The flowchart of the SIMPLE algorithm is shown in Fig. 5.9. In this figure, the reader can anticipate the implicit solution approach—which enhances numerical stability—as well as the explicit approach. This explains the 'Semi-Implicit' part of the name of this algorithm.

5.3.1 Numerical Example of Application of the Pressure Equation of Correction

The case of a steady unidimensional incompressible flow inside a duct of constant section will be analysed in this case. It is immediately evident that the solution is a constant velocity field throughout the duct. This example demonstrates that by solving the pressure correction equation, an initial velocity field with a non-uniform distribution can still lead to the exact solution that respects the mass conservation equation.

From Fig. 5.10, it is evident that a staggered backward grid is employed. The pressure is computed at the cell centres $I = A, B, C, D$ (as seen in the cell bounded by points 1 and 2), while the velocity is computed at the cell centres $i = 1, 2, 3, 4$ (as seen in the cell bounded by points A and B). The purpose of this example is to demonstrate the validity of the procedure that underlies the SIMPLE algorithm.

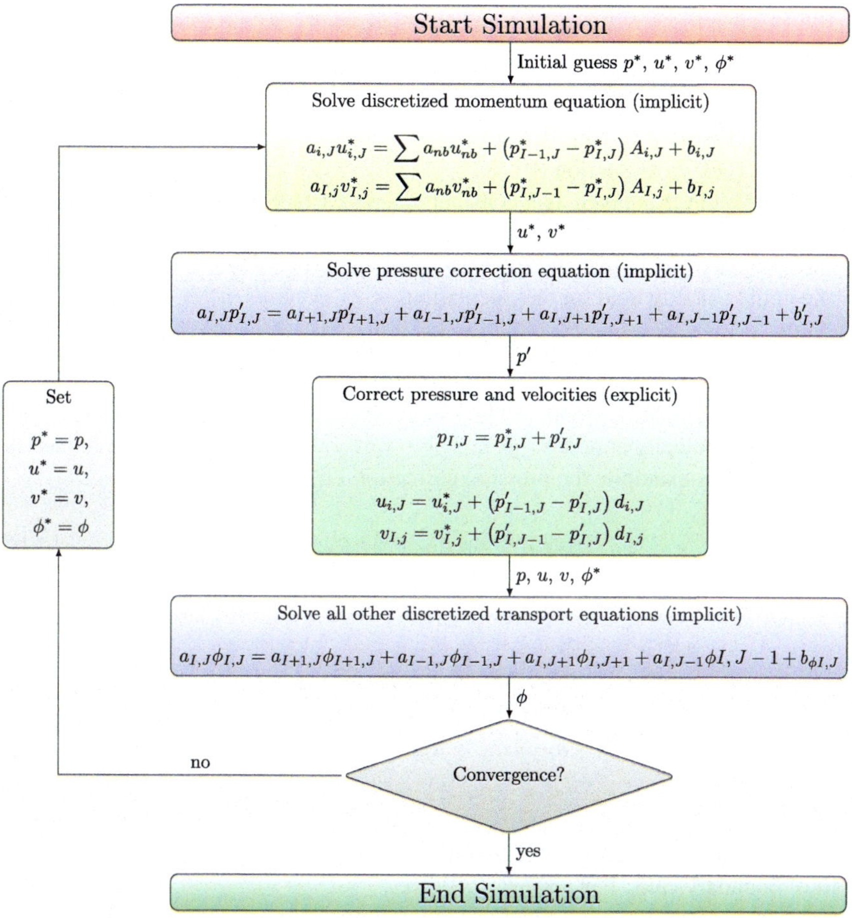

Start Simulation

Initial guess p^*, u^*, v^*, ϕ^*

Solve discretized momentum equation (implicit)

$$a_{i,J} u^*_{i,J} = \sum a_{nb} u^*_{nb} + \left(p^*_{I-1,J} - p^*_{I,J}\right) A_{i,J} + b_{i,J}$$

$$a_{I,j} v^*_{I,j} = \sum a_{nb} v^*_{nb} + \left(p^*_{I,J-1} - p^*_{I,J}\right) A_{I,j} + b_{I,j}$$

u^*, v^*

Solve pressure correction equation (implicit)

$$a_{I,J} p'_{I,J} = a_{I+1,J} p'_{I+1,J} + a_{I-1,J} p'_{I-1,J} + a_{I,J+1} p'_{I,J+1} + a_{I,J-1} p'_{I,J-1} + b'_{I,J}$$

p'

Correct pressure and velocities (explicit)

$$p_{I,J} = p^*_{I,J} + p'_{I,J}$$

$$u_{i,J} = u^*_{i,J} + \left(p'_{I-1,J} - p'_{I,J}\right) d_{i,J}$$

$$v_{I,j} = v^*_{I,j} + \left(p'_{I,J-1} - p'_{I,J}\right) d_{I,j}$$

p, u, v, ϕ^*

Solve all other discretized transport equations (implicit)

$$a_{I,J} \phi_{I,J} = a_{I+1,J} \phi_{I+1,J} + a_{I-1,J} \phi_{I-1,J} + a_{I,J+1} \phi_{I,J+1} + a_{I,J-1} \phi I, J-1 + b_{\phi I,J}$$

ϕ

Convergence?

no → Set

$p^* = p,$
$u^* = u,$
$v^* = v,$
$\phi^* = \phi$

yes

End Simulation

Fig. 5.9 Flowchart of the SIMPLE algorithm

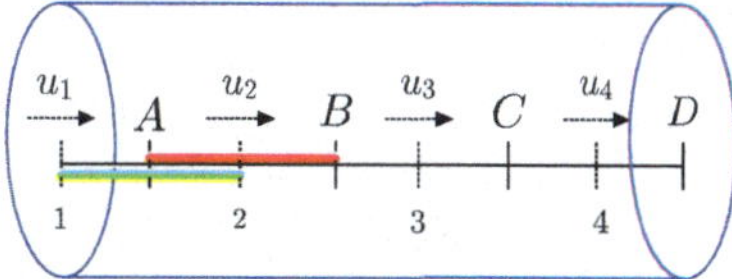

Fig. 5.10 Discretisation diagram of the flow inside a duct with constant section

Specifically, the equation of pressure correction (5.16) is utilised to compute the pressure correction field p', followed by the velocity correction field according to the

$$u' = d\left(p'_I - p'_{I+1}\right) \tag{5.17}$$

from which the corrected velocity field is derived according to the

$$u = u^* + u'.$$

The data for the problem are as follows:

- constant density $\rho = 1\ \text{kg/m}^3$;
- constant duct cross-section A;
- the coefficient d is taken as a constant value of 1 (see Eqs. 5.12, 5.14, 5.13, and 5.15 in Eq. 5.17);
- the boundary conditions are: $u_1 = 10$ m/s and $p_D = 0$ Pa;
- the initial velocity field for the calculations is: $u_2^* = 8$ m/s, $u_3^* = 11$ m/s, and $u_4^* = 7$ m/s.

The solution to this problem is a constant velocity field of 10 m/s. The SIMPLE algorithm is applied to compute the pressure correction at nodes $I = A, B, C, D$ and the corrected velocity at nodes $i = 2, 3, 4$ to verify the correctness of the numerical solution. In this example, the pressure correction Eq. 5.16 is

$$a_P p'_P = a_W p'_W + A_E p'_E + b' \tag{5.18}$$

with

$$\begin{aligned} a_W &= (\rho d A)_w\,, \\ a_E &= (\rho d A)_e\,, \\ a_P &= a_W + a_E, \\ b' &= \left(\rho u^* A\right)_w - \left(\rho u^* A\right)_e. \end{aligned}$$

Equation 5.18 is applied to nodes $I = A, B, C, D$, beginning with nodes that are not part of boundary cells. For node B, this gives:

$$\begin{aligned} a_W &= (\rho d A)_w = (\rho d A)_2 = 1 \times 1 \times A = A, \\ a_E &= (\rho d A)_e = (\rho d A)_3 = 1 \times 1 \times A = A, \\ a_P &= a_W + a_E = A + A = 2A, \\ b' &= \left(\rho u^* A\right)_w - \left(\rho u^* A\right)_e = \left(\rho u^* A\right)_2 - \left(\rho u^* A\right)_3 = (1 \times 8 \times A) - (1 \times 11 \times A) = -3A \end{aligned}$$

thus, the pressure correction equation for node B is expressed as

$$(2A) p'_B = (A) p'_A + A p'_C + (-3A).$$

Considering that section A does not change, it is

$$2p'_B = p'_A + p'_C - 3.$$

For node C we have:

$$
\begin{aligned}
a_W &= (\rho d A)_w = (\rho d A)_3 = 1 \times 1 \times A = A, \\
a_E &= (\rho d A)_e = (\rho d A)_4 = 1 \times 1 \times A = A, \\
a_P &= a_W + a_E = A + A = 2A, \\
b' &= \left(\rho u^* A\right)_w - \left(\rho u^* A\right)_e = \left(\rho u^* A\right)_3 - \left(\rho u^* A\right)_4 = (1 \times 11 \times A) - (1 \times 7 \times A) = 4A
\end{aligned}
$$

thus, the pressure correction equation for node C is expressed as

$$(2A)p'_C = (A)p'_B + Ap'_D + 4A.$$

Given that the cross-section A remains constant, it follows that

$$2p'_C = p'_B + p'_D + 4.$$

For the centre A, it is the centre of a cell that lacks a neighbouring cell to the left. In this case, a_W is set to 0, and the boundary condition is considered by incorporating its contribution as a source term:

$$
\begin{aligned}
a_W &= 0, \\
a_E &= (\rho d A)_e = (\rho d A)_2 = 1 \times 1 \times A = A, \\
a_P &= a_W + a_E = 0 + A = A, \\
b' &= \left(\rho u^* A\right)_w - \left(\rho u^* A\right)_e + (\rho u_1 A) = \\
&\quad \left(\rho u^* A\right)_2 - (\rho u A)_1 = -(1 \times 8 \times A) + (1 \times 10 \times A) = 2A
\end{aligned}
$$

thus, the pressure correction equation for node A is expressed as

$$Ap'_A = 0 + Ap'_B + 2A.$$

Given that the cross-section A remains constant, it follows that

$$p'_A = p'_B + 2.$$

For node D, the pressure correction is not calculated, as $p'_D = 0$ due to the pressure at D being set as a boundary condition.

The system of four equations to determine the pressure corrections p'_A, p'_B, p'_C, and p'_D will be

$$
\begin{aligned}
p'_A &= p'_B + 2 \\
2p'_B &= p'_A + p'_C - 3 \\
2p'_C &= p'_B + p'_D + 4 \\
p'_D &= 0
\end{aligned}
$$

that is, by imposing $p'_D = 0$,

$$\begin{aligned} p'_A &= p'_B + 2 \\ 2p'_B &= p'_A + p'_C - 3 \\ 2p'_C &= p'_B + 4 \end{aligned}$$

which can be written in matrix form as

$$\begin{bmatrix} 1 & -1 & 0 \\ -1 & 2 & -1 \\ 0 & -1 & 2 \end{bmatrix} \begin{bmatrix} p'_A \\ p'_B \\ p'_C \end{bmatrix} = \begin{bmatrix} 2 \\ -3 \\ 4 \end{bmatrix}.$$

The solution to this system of equations gives $p'_A = 4$, $p'_B = 2$, and $p'_C = 3$. Once the pressure corrections are known, the corrected velocities can be obtained using the relation $u = u^* + d(p'_I - p'_{I+1})$.

For node 2:

$$u_2 = u_2^* + d(p'_A - p'_B) = 8 + 1 \times (4 - 2) = 10 \text{ m/s}.$$

For node 3:

$$u_3 = u_3^* + d(p'_B - p'_C) = 11 + 1 \times (2 - 3) = 10 \text{ m/s}.$$

For node 4:

$$u_4 = u_4^* + d(p'_C - p'_D) = 7 + 1 \times (3 - 0) = 10 \text{ m/s}.$$

The solution obtained matches the exact solution. Due to its simplicity, this example has enabled (a) obtaining the exact solution in a single iteration, and (b) not requiring the calculation of the momentum conservation equation to determine the velocity field used in the pressure correction equation. In more general cases, the coefficient d cannot be treated as constant, and the conservation of momentum equation must be solved at each iteration to determine the coefficients of the pressure correction equation in all cells of the computational domain.

5.3.2 *Example of Application of the SIMPLE Algorithm*

In this example, a non-viscous, incompressible and stationary flow within a converging duct is considered. As shown in Fig. 5.11, the computational domain is discretised using the forward staggered grid method, with five equidistant nodes for pressure calculation and four for velocity calculation. The grid is forward staggered because the centre of the cell bounded by points A and B (labelled 1 in Fig. 5.11), initially

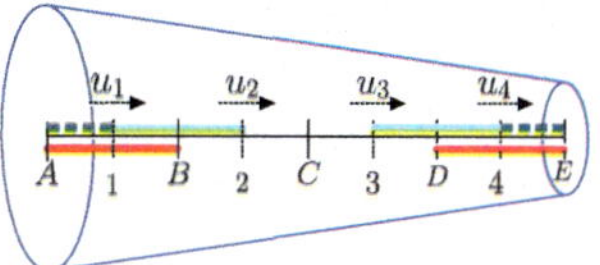

Fig. 5.11 Discretisation diagram of the flow within a variable section duct

Fig. 5.12 Value of the duct area and the exact pressure calculated using Bernoulli's equation at cell centres A, B, C, D and E

Node	$A\ (m^2)$	$p\ (Pa)$
A	0.5	9.60000
B	0.4	9.37500
C	0.3	8.88889
D	0.2	7.50000
E	0.1	0

used for velocity calculation, is shifted in the increasing direction (from left to right) relative to the centre of the cell bounded by points 1 and 2 (labelled A in Fig. 5.11), which is initially used for pressure calculation. The aim of the example is to calculate the pressure in the cell centres A, B, C, D, E and the velocity in the cell centres $1, 2, 3, 4$.

The data given for solving the problem are:

- the fluid density is constant and equal to 1 kg/m^3;
- the length of the duct is 2 m, implying that the extension of a single cell is $\Delta x = 2/4 = 0.5$ m;
- the inlet section at centre A has $A_A = 0.5$ m^2 and the outlet section at centre E has $A_E = 0.1$ m^2. Assuming a linear variation of the duct section between the inlet and outlet, Figs. 5.12 and 5.13 show the areas of the sections at the locations of the centres of the cells used for pressure and velocity calculations;
- the boundary conditions are:
  - total pressure $p_0 = 10$ Pa at the inlet section in correspondence with the centre of cell A;
  - static pressure $p_E = 0$ Pa at the outlet section in correspondence with the centre of cell E.
- For velocity field initialisation, an initial guess of the flow rate $\dot{m} = 1$ kg/s is assumed, and based on this, the velocity is calculated using $u = \dot{m}/(\rho A)$. Thus, the following initial velocity values[1] for nodes $1, 2, 3, 4$ are computed:

[1] Although only 5 decimal places are shown here and in the following, the calculations performed to obtain the results are in double precision.

Fig. 5.13 Value of the duct area and the exact speed calculated using Bernoulli's equation at cell centres 1, 2, 3, 4 and 5

Node	$A\ (m^2)$	$u\ (m/s)$
1	0.45	0.99381
2	0.35	1.27775
3	0.25	1.78885
4	0.15	2.98142

$$u_1 = \dot{m}/(\rho A_1) = 1/(1 \times 0.45) = 2.22222 \text{ m/s},$$
$$u_2 = \dot{m}/(\rho A_2) = 1/(1 \times 0.35) = 2.85714 \text{ m/s},$$
$$u_3 = \dot{m}/(\rho A_3) = 1/(1 \times 0.25) = 4.00000 \text{ m/s},$$
$$u_4 = \dot{m}/(\rho A_4) = 1/(1 \times 0.15) = 6.66666 \text{ m/s};$$

- the initial pressure field is imposed assuming a linear variation of the pressure between nodes A and E: $p_A^* = p_0 = 10$ Pa, $p_B^* = 7.5$ Pa, $p_C^* = 5$ Pa, $p_D^* = 2.5$ Pa, $p_E^* = 0$ Pa (as per boundary condition).

The exact solution to this stationary, incompressible, non-viscous problem can be derived using Bernoulli's equation:

$$p_0 = p_N + \frac{1}{2}\rho\left(\frac{\dot{m}}{\rho A_N}\right)^2$$

where the subscript N denotes the generic pressure node. Considering node E, where $A_E = 0.1$ m^2, and the boundary conditions $p_0 = 10$ Pa and $p_E = 0$ Pa, the flow rate is $\dot{m} = 0.44721$ kg/s. Knowing the flow rate, the exact values of pressure and velocity can be determined, as shown in Figs. 5.12 and 5.13, respectively, for the centres A, B, C, D, E and for nodes 1, 2, 3, 4, and 5.

The governing equations for this stationary, one-dimensional, incompressible, and non-viscous problem are as follows: the conservation of mass equation

$$\frac{d}{dx}(\rho A u) = 0$$

and the conservation of momentum equation

$$\rho A u \frac{du}{dx} = -A\frac{dp}{dx}. \tag{5.19}$$

It should be noted that, although the problem is incompressible, the density term has not been removed to avoid complicating aspects related to dimensional analysis.

In this case, the discretised form of the momentum conservation Eq. 5.19 is

$$(\rho A u)_e \, u_e - (\rho A u)_w \, u_w = \frac{\Delta p}{\Delta x} \Delta V \tag{5.20}$$

where $\Delta p = p_w - p_e$, and ΔV is the volume (area in the two-dimensional case, length in the one-dimensional case) of the considered cell. Using the standard notation introduced earlier, Eq. 5.20, applied to the generic cell of centre P, can be written as

$$a_P u_P^* = a_W u_W^* + a_E u_E^* + S_u. \tag{5.21}$$

In the case where the upwind scheme is applied, the coefficients of Eq. 5.21 can be expressed as

$$\begin{aligned} a_W &= D_w + max(F_w, 0), \\ a_E &= D_e + max(0, -F_e), \\ a_P &= a_W + a_E + (F_e - F_w). \end{aligned} \tag{5.22}$$

Since the flow is inviscid, the terms D_w and D_e will be zero. For the terms F_w and F_e, the area values reported in Fig. 5.13 will be used; in this example, the velocity values for the calculation of the flows F_e and F_w will be obtained by averaging the values at the centres of the cells that share the considered face. At the first iteration, the velocity values are those of the first guess. For the subsequent iterations, the velocity values are those obtained after solving the equation of pressure correction.

The source term S_u will be

$$S_u = \frac{\Delta p}{\Delta x} \times \Delta V = \frac{\Delta p}{\Delta x} \times A_{av} \Delta x = \Delta p \times \frac{1}{2} (A_w + A_e)$$

where the approximation given by setting $A_{av} = 1/2 \, (A_w + A_e)$ has an order of accuracy consistent with the upwind scheme.

Finally, the expression for the coefficients of the discretised momentum conservation equation will be

$$\begin{aligned} F_w &= \rho A_w u_w, \\ F_w &= \rho A_w u_w, \\ a_W &= F_w, \\ a_E &= 0, \\ a_P &= a_W + a_E + (F_e - F_w) \\ S_u &= \Delta p \times \frac{1}{2} (A_w + A_e) = \Delta p \times A_P. \end{aligned}$$

In which, with reference to Eq. 5.22, $a_E = 0$ is set.

The parameter d, required for the calculation of velocity and pressure corrections, can be defined here as

$$d = \frac{A_{av}}{a_P} = \frac{a_w + A_e}{2a_P}. \tag{5.23}$$

Discretisation of the Pressure Correction Equation

The discretised form of the mass conservation equation for the central cell P is

$$(\rho A u)_e - (\rho A u)_w = 0$$

and the corresponding pressure correction equation is

$$a_P p'_P = a_W p'_W + a_E p'_E + b'$$

with

$$\begin{aligned} a_W &= (\rho d A)_w, \\ a_E &= (\rho d A)_e, \\ b' &= (F^*_w - F^*_e) \end{aligned}$$

where Eq. 5.23 provides the expression for the parameter d. In the SIMPLE algorithm, the pressure correction is used to determine the corrected pressure and velocity:

$$\begin{aligned} u' &= d\left(p'_I - p'_{I+1}\right), \\ p &= p^* + p', \\ u &= u^* + u'. \end{aligned}$$

Numerical Values of Momentum Conservation Equation Coefficients

Initially, the internal nodes 2 and 3 are considered.

- Node 2

$$\begin{aligned} F_w &= \rho A_w u_w = 1 \times 0.4 \times \frac{u_1 + u_2}{2} = 1 \times 0.4 \times \frac{2.2222 + 2.8571}{2} = 1.01587, \\ F_w &= \rho A_w u_w = 1 \times 0.3 \times \frac{u_2 + u_3}{2} = 1 \times 0.3 \times \frac{2.8571 + 4}{2} = 1.02857, \\ a_W &= F_w = 1.01587, \\ a_E &= 0, \\ a_P &= a_W + a_E + (F_e - F_w) = 1.01587 + 0 + (1.02857 - 1.01587) = 1.02857, \end{aligned}$$

$$S_u = \Delta p \times \frac{1}{2}(A_w + A_e) = \Delta p \times A_P$$
$$= \Delta p \times A_2 = (p_B - p_C)A_2 = (7.5 - 5)0.35 = 0.875.$$

Therefore, for node 2, the discretised momentum conservation equation is

$$1.02857u_2 = 1.01587u_1 + 0.875.$$

For this node, the value of the parameter d is

$$d_2 = \frac{A_2}{a_P} = \frac{0.35}{1.02857} = 0.34027.$$

- Node 3
  By repeating the same procedure, the result is:

$$1.06666u_3 = 1.02857u_2 + 0.625$$

and

$$d_3 = \frac{A_3}{a_P} = \frac{0.25}{1.06666} = 0.23437.$$

- Node 1
  The cell centred at node 1 must be treated specially because one of its faces is a boundary face. In particular, on the face centred at node A, a fixed total pressure of 10 Pa is imposed. This pressure corresponds to the static pressure of the fluid at rest inside the tank to which the entrance of the duct is connected. At node A, since the velocity is non-zero, the static pressure will be lower than the total pressure. Indicating with u_A the velocity at the centre of section A, it is possible to use Bernoulli's equation to obtain the value of the static pressure at A as a function of the total pressure:

$$p_A = p_0 - \frac{1}{2}\rho u_A^2. \tag{5.24}$$

Considering the continuity equation, the result is

$$u_A = u_1 A_1 \frac{1}{A_A} \tag{5.25}$$

and substituting into Eq. 5.24

$$p_A = p_0 - \frac{1}{2}\rho u_1^2 \left(\frac{A_1}{A_2}\right)^2. \tag{5.26}$$

Considering the expression for p_A, the discretised momentum conservation equation for the cell centred at node 1, using the upwind scheme, is (see Eq. 5.22):

$$F_e u_1 - F_w u_A = (p_A - p_B) A_1 \tag{5.27}$$

where the term $F_w = \rho u_A A_A$ is calculated using the continuity equation (Eq. 5.25):

$$F_w = \rho u_A A_A = \rho u_1 A_1.$$

At this stage, Eq. 5.27 can be written as

$$F_e u_1 - F_w u_1 \frac{A_1}{A_A} = \left[p_0 - \frac{1}{2} \rho u_1^2 \left(\frac{A_1}{A_2} \right)^2 - p_B \right] A_1$$

and rearranging

$$\left[F_e - F_w \frac{A_1}{A_A} + F_w \frac{1}{2} \left(\frac{A_1}{A_2} \right)^2 \right] u_1 = (p_0 - p_B) A_1 \tag{5.28}$$

From this, it is clear that the term in square brackets multiplying u_1 corresponds to the coefficient a_P for this cell. To stabilise the iterative process, the term $F_w \frac{1}{2} \left(\frac{A_1}{A_2} \right) u_1$ is moved to the right-hand side of Eq. 5.28, replacing the current value of u_1 with the value from the previous iteration (see Sect. 3.1.6 in relation to the *deferred correction strategy*):

$$\left[F_e + F_w \frac{1}{2} \left(\frac{A_1}{A_2} \right)^2 \right] u_1 = (p_0 - p_B) A_1 + F_w \frac{A_1}{A_A} u_1^{old}. \tag{5.29}$$

Turning to the numerical values, we have:

$$u_a = u_1 \frac{A_1}{A_2} = 2.2222 \times \frac{0.45}{0.5} = 2,$$
$$F_w = (\rho u A)_w = \rho u_A A_A = 1 \times 2 \times 0.5 = 1.$$

The flow F_e is calculated as for the face of an interior point:

$$F_e = (\rho u A)_e = 1 \times \frac{u_1 + u_2}{2} \times 0.4 = 1 \times \frac{2.2222 + 2.8571}{2} \times 0.4 = 1.01587,$$
$$a_W = 0,$$
$$a_E = 0,$$
$$a_P = F_e + F_w \frac{1}{2} \left(\frac{A_1}{A_2} \right)^2 = 1.01587 + 1 \times 0.5 \times \left(\frac{0.45}{0.5} \right)^2 = 1.42087.$$

By setting $p_0 = 10$ Pa and $u_1^{old} = 2.2222$ m/s, the source term S_u for this cell can be written as

$$S_u = (p_0 - p_B)\, A_1 + F_w \frac{A_1}{A_A} u_1^{old} = (10 - 7.5) \times 0.45 + 1 \times \frac{0.45}{0.5} \times 2.22222 = 3.125.$$

Thus, for node 1, the discretised momentum conservation equation is

$$1.42087 u_1 = 3.125.$$

The value of the parameter d at this node is

$$d_1 = \frac{A_1}{a_P} = \frac{0.45}{1.4209} = 0.31670.$$

- Node 4

$$F_w = (\rho u A)_w = 1 \times \frac{u_3 + u_4}{2} \times 0.2 = 1.06666.$$

As for the east face of the cell with this node at its centre, velocity values are not available at the centre of one of the two cells it belongs to, as the pressure, not the velocity, is imposed as a boundary condition on this face. The mass flux F_e on this face is thus calculated, assuming it coincides with the flow rate passing through the duct. Therefore,

$$F_e = (\rho u A)_4 .$$

Given the initially assumed flow rate of 1 kg/s, it follows that

$$\begin{aligned}
a_W &= F_w = 1.06666, \\
a_E &= 0, \\
a_P &= a_W + a_E + (F_e - F_w) = 1.06666 + 0 + (1 - 1.06666) = 1, \\
S_u &= \Delta p \times A_{av} = (p_D - P_E) \times A_4 = (2.5 - 0) \times 0.15 = 0.375
\end{aligned}$$

where the boundary condition $p_E = 0$ Pa has been applied. Finally, the discretised momentum conservation equation for node 4 becomes

$$1 u_4 = 1.0666 u_3 + 0.375.$$

The value of the parameter d at this node is

$$d_4 = \frac{A_4}{a_P} = \frac{0.15}{1} = 0.15.$$

In conclusion, the discretised momentum conservation equations for the four velocity nodes are

$$
\begin{aligned}
1.42087u_1 &= 3.125,\\
1.02857u_2 &= 1.01587u_1 + 0.875,\\
1.06666u_3 &= 1.02857u_2 + 0.625,\\
1.00000u_4 &= 1.0666u_3 + 0.375.
\end{aligned}
$$

The solution of this system is

$$
\begin{aligned}
u_1 &= 2.19935 \text{ m/s},\\
u_2 &= 3.02289 \text{ m/s},\\
u_3 &= 3.50087 \text{ m/s},\\
u_4 &= 4.10926 \text{ m/s}.
\end{aligned}
$$

These are the first-attempt speeds in the pressure correction calculation procedure, that is, the speeds marked with the superscript * when the SIMPLE algorithm was illustrated.

5.3.2.1 Numerical Values of the Coefficients of the Pressure Correction Equation

The centres of the cells used for pressure calculation are now used. Nodes B, C, and D are internal nodes.

- Node B
  For this node it will be

$$
\begin{aligned}
a_W &= (\rho d A)_1 = 1 \times 0.3167 \times 0.45 = 0.14251,\\
a_E &= (\rho d A)_2 = 1 \times 0.34027 \times 0.35 = 0.11909,\\
F_w^* &= (\rho u^* A)_1 = 1 \times 2.199352 \times 0.45 = 0.98971,\\
F_e^* &= (\rho u^* A)_2 = 1 \times 3.022894 \times 0.35 = 1.05801,\\
a_P &= a_W + a_E = 0.14251 + 0.11909 = 0.26161,\\
b' &= F_w^* - F_e^* = 0.98971 - 1.05801 = -0.06830.
\end{aligned}
$$

  Therefore, the discretised pressure correction equation $a_C p'_C = a_W p'_W + a_E p'_E + b'$, for node B, becomes

$$0.26161 p'_B = 0.14251 p'_A + 0.11909 p'_C - 0.06830.$$

- Node C
  Similarly to what was done for node B, for this node, it is

$$0.17769 p'_C = 0.11909 p'_B + 0.058593 p'_D + 0.18279.$$

- Node D
  Similarly to what was done for node B, for this node it is

$$0.081093 p'_D = 0.058593 p'_C + 0.25882.$$

- Node E
  On this node the boundary condition that sets the pressure to zero is imposed. Since the pressure is known, the value of the pressure correction will be zero: $p'_E = 0$.
- Node A
  On this node the boundary condition that sets the total pressure is imposed. For simplicity, Eq. 5.26 is reported here.

$$p_A = p_0 - \frac{1}{2}\rho u_1^2 \left(\frac{A_1}{A_2}\right)^2 \tag{5.30}$$

  Knowing the pressure and speed values, the total pressure value is obtained. In the SIMPLE algorithm, the available velocity value before the solution of the pressure correction equation is the one marked with the superscript *, resulting from the solution of the momentum conservation equation at the previous iteration. Nevertheless, the value of static pressure, if calculated with Eq. 5.30, is consistent with the current velocity value. For this reason, at node A, the value of the pressure correction is set to zero: $p'_A = 0$.

In conclusion, the discretised pressure correction equations for the three pressure nodes are:

$$\begin{aligned}
0.26161 p'_B &= 0.11909 p'_C - 0.06830,\\
0.17769 p'_C &= 0.11909 p'_B + 0.058593 p'_D + 0.18279,\\
0.081093 p'_D &= 0.058593 p'_C + 0.25882.
\end{aligned}$$

The solution of this system is

$$\begin{aligned}
p'_A &= 0,\\
p'_B &= 1.63935,\\
p'_C &= 4.17461,\\
p'_D &= 6.20805,\\
p'_E &= 0.
\end{aligned}$$

Knowing the value of the pressure correction, it is possible to obtain the value of the corrected pressure

$$p_B = p_B^* + p_B' = 7.5 + 1.63935 = 9.13935,$$
$$p_C = p_C^* + p_C' = 5 + 4.17461 = 9.17461,$$
$$p_D = p_D^* + p_D' = 2.5 + 6.20805 = 8.70805$$

and, therefore, the corrected speeds after the first iteration.

$$u_1 = u_1^* + d_1(p_A' - p_B') = 2.19935 + 0.31670 \times (0 - 1.63935) = 1.68015 \text{ m/s},$$
$$u_2 = u_2^* + d_2(p_B' - p_C') = 3.02289 + 0.34027 \times (1.63935 - 4.17461) = 2.16020 \text{ m/s},$$
$$u_3 = u_3^* + d_3(p_C' - p_D') = 3.50087 + 0.23437 \times (4.17461 - 6.20805) = 3.02428 \text{ m/s},$$
$$u_4 = u_4^* + d_4(p_D' - p_E') = 4.10926 + 0.15 \times (6.20805 - 0) = 5.04047 \text{ m/s}.$$

The value of the static pressure at node A is also known after the first iteration:

$$p_A = p_0 - \frac{1}{2}\rho u_1^2\left(\frac{A_1}{A_2}\right)^2 = 10 - \frac{1}{2} \times 1 \times (1.68015 \times \frac{0.45}{0.5})^2 = 8.85671.$$

With the known velocity values, it is possible to calculate the value of the flow rate at each of the four velocity nodes:

$$\rho u_1 A_1 = 0.75607,$$
$$\rho u_2 A_2 = 0.75607,$$
$$\rho u_3 A_3 = 0.75607,$$
$$\rho u_4 A_4 = 0.75607.$$

Observing the values obtained for the flow rate after the first iteration, it can be deduced that:

- since the values are the same for each of the nodes in this numerical example, one of the aspects that makes the SIMPLE algorithm so widely used and popular is highlighted: at each iteration, the continuity equation is always respected, even if the velocity field does not satisfy the momentum conservation equation.
- the flow rate value obtained at the first iteration differs by 69% from that determined with the Bernoulli equation, equal to 0.44721 kg/s: this is mainly due to the fact that the coefficients of the discretised momentum conservation equation were obtained based on the first guess values for the velocity. To obtain a velocity and pressure field that simultaneously satisfy both the mass and momentum conservation equations, further iterations will need to be performed, ideally until the perfect satisfaction of the two equations is achieved. In reality, given the finite number

of digits with which computers can store numbers and the consequent limited precision, the perfect balance between the mass conservation equation (i.e., the pressure correction equation) and the momentum conservation equation will be impossible to achieve. As a consequence, the iterative procedure will be stopped when further iterations produce negligible variations in pressure and velocity.

Chapter 6
OpenFOAM®

In general, a solver based on the finite volume method applies the discretised transport equations to all the cells in the computational domain. This results in a number of algebraic equations whose system, once solved, provides the solution for all transported quantities. For this to be possible, it is necessary to provide very precise information:

- the computational grid (mesh);
- the boundary and initial conditions;
- physical properties such as density, viscosity, diffusion coefficient, etc.;
- the spatial discretisation scheme for each term (convective, diffusive, source) of the conservation equations;
- the time discretisation scheme;
- the strategy for solving the linear system constituted by the discretised conservation equations applied to each element of the grid;
- the value of the parameters (under-relaxation factors, stop condition, etc.) that regulate the execution of the resolution process.

In the case of OpenFOAM®, this information is provided through text files contained in the folder of the case to be executed or in sub-folders:

- the computational grid is provided through the files contained in the directory `constant/polymesh`;
- the initial and boundary conditions are contained in the files located in the folder `0`;
- the discretisation schemes definition is contained in the file `fvSchemes` inside the folder `system`;
- the information related to the solution of the system of differential algebraic equations and the under-relaxation factors are contained in the file `fvSolution` inside the folder `system`;

G. Caramia and E. Distaso, *A Practical Approach to Computational Fluid Dynamics Using OpenFOAM®*, https://doi.org/10.1007/978-3-031-88957-8_6

- the value of the parameters that regulate the execution of the process resolution is contained in the file `controlDict` inside the folder `system`.

6.1 Discretisation Schemes

In the file `fvSchemes`, the selected discretisation schemes are contained, one for each term of the general transport equation. Figure 6.1 shows an extract of this file.

- the area identified by the word **ddtSchemes** is the one in which the time discretisation is defined;
- the area identified by the word **gradSchemes** is the one in which it is defined how to calculate, in the cell centre, the gradient of the transported quantity;
- the area identified by the word **divSchemes** is the one in which the discretisation of the convective terms is defined;
- the area identified by the word **laplacianSchemes** is the one where the discretisation of the diffusive terms is defined;
- the area identified by the word **interpolationSchemes** is the one where the type of interpolation chosen to obtain the value of the transported quantity on the centroid of the face from that in the centroid of the adjacent cells is defined;
- the area identified by the word **snGradSchemes** is the one where the scheme for computing the normal to the face component of the gradient at the centroid of the face is defined. The calculation of this quantity, limited to the case of diffusive terms, is specified in the area **laplacianSchemes**.

To know, for each item, what the possible values are, simply modify the word originally present in the file so that it is incorrect (for example, backward -> backward) and launch the solver. The program will stop, providing the list of acceptable values.

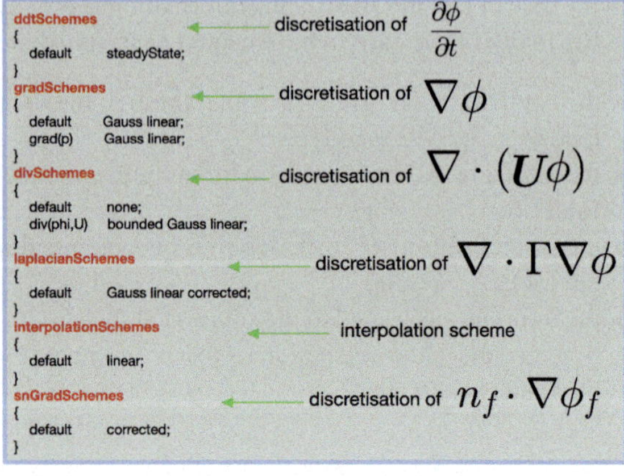

Fig. 6.1 Extract from the file `fvSchemes`

6.1.1 *Temporal Discretisation Schemes*

The source code of the available time discretisation schemes is contained in the folder `WM_PROJECT_DIR/src/finiteVolume/finiteVolume/ddtSchemes`. The most frequently used schemes are:

- **steadyState**, used for stationary calculations;
- **Euler**, first-order accurate and used to perform non-stationary calculations;
- **backward**, second-order accurate and used to perform non-stationary calculations;
- **CrankNicolson**, second-order accurate and used to perform non-stationary calculations.

First-order accurate schemes are stable and do not cause non-physical oscillations, but the low level of accuracy may be inadequate. Second-order accurate schemes, while more accurate, can give rise to non-physical oscillations. In the case of the Euler scheme, the time integration takes place according to the following formula:

$$\int_V \frac{\partial \phi}{\partial t} dV \approx \left(\phi - \phi^0\right) \frac{V}{\Delta t} \tag{6.1}$$

where ϕ^0 is the column vector whose terms represent the value of the considered quantity in all the cells of the computational domain at the previous time. Equation 6.1 modifies the system of equations resulting from the application of the conservation equations, represented in matrix form, as follows:

$$\begin{bmatrix} \square & * & & * \\ * & \square & * & \\ & * & \square & * \\ * & & * & \square \end{bmatrix} [\phi] = \begin{bmatrix} * \\ * \\ * \\ * \end{bmatrix}. \tag{6.2}$$

Specifically, the main diagonal of the coefficients matrix is modified by the presence of the term $\frac{V}{\Delta t}$: as the time integration interval decreases, there will be an increase in the value of this term, which will contribute to improve the diagonal dominance of the matrix. The term $\phi^0 \frac{V}{\Delta t}$ instead modifies the vector of known terms on the right hand side of Eq. 6.2.

In the case of the backward scheme, the time integration takes place according to the following formula

$$\int_V \frac{\partial \phi}{\partial t} dV \approx \left(3\phi - 4\phi^0 + \phi^{00}\right) \frac{V}{2\Delta t} \tag{6.3}$$

where ϕ^{00} is the column vector whose terms represent the value of the considered quantity in all the cells of the computational domain two time steps before the current step. The implementation of the Crank-Nicolson scheme (see also Sect. 3.5.2) involves the use of a blending factor ψ that allows a gradual transition from a first-order accurate scheme, coinciding with the Euler scheme, to a second-order accurate

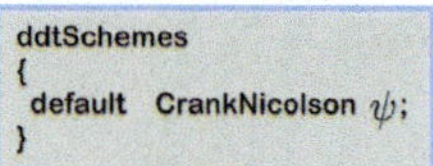
```
ddtSchemes
{
 default  CrankNicolson ψ;
}
```

Fig. 6.2 Syntax of the terms to be inserted in the file `fvSchemes` to set the Crank-Nicolson scheme

scheme, coinciding with the Crank-Nicolson scheme. The syntax of the terms to be inserted in the `fvSchemes` file is shown in Fig. 6.2. By setting the value of ψ to zero, the Euler scheme (stable but not very accurate) is obtained. By setting ψ equal to 1, the pure Crank-Nicolson scheme (very accurate but less stable and with possible non-physical oscillations in the results) is obtained. In most cases, $\psi = 0.7 \div 0.9$ is set to achieve a good compromise between stability and accuracy. The word "default" indicates that all the time derivative terms present in the treated equations will be discretised using the scheme specified on the same line after this word. In the case of non-stationary simulations, the word `bounded` must not appear in any of the schemes for the discretisation of the convective terms (`divSchemes`).

6.1.2 Discretisation Schemes of the Convective Terms

The source code of the discretisation schemes for the convective terms is contained in the folder `WM_PROJECT_DIR/src/finiteVolume/interpolation/surfaceInterpolation`. The most frequently used schemes are:

- **upwind**, first-order accurate;
- **linear**, second-order accurate but with the possibility of producing non-physical over- or under-estimates;
- **limitedLinear**, this scheme requires specifying a number between 0 and 1 next to the word `limitedLinear`:
  - if the number is equal to 0, the interface value is calculated using the linear scheme for all cells in the domain;
  - if the number is equal to 1, the scheme with a flow limiter is used (see Sect. 3.1.7) with $\psi(r) = \min(2r, 1)$;
  - if the number is between 0 and 1, each cell will have an intermediate value between those obtained for the two extreme values;
- **limitedLinearV**, in the case of vector fields, instead of calculating a limiter value for each component of the vector field, a single limiter value is applied to all components, choosing the one related to the component exhibiting the highest gradient value. This increases stability (by eliminating possible wiggles) at the expense of accuracy;
- **linearUpwind**, second-order accurate and without non-physical over- or under-estimates;

- **linearUpwindV**, in the case of vector fields, the greatest of the gradients of the vector components is used as the gradient for the linear corrective term;
- **Minmod**, second-order accurate TVD scheme without non-physical over- or under-estimates;
- **vanLeer**, second-order accurate TVD scheme without non-physical over- or under-estimates;
- **LUST (Linear Upwind STabilised)**, blended scheme: 75% linear and 25% linearUpwind.

Recalling what was already seen in Sect. 3.1.7, schemes with flux limiters aim to optimise the blending factor (the *flux limiter*) ψ between the value of the quantity ϕ on the face obtained with the simple upwind scheme and the same value obtained by applying the centred scheme:

$$\phi_f = [1 - \psi(r)]\,\phi_{UW} + \psi(r)\phi_{LI}$$

in which it can be noted that (i) for $\psi = 0$ the upwind scheme is obtained, (ii) for $\psi = 1$ the centred scheme is obtained, (iii) the limiter ψ is a function of the ratio r between the value of the gradient $\nabla\phi$ of the quantity ϕ at the upwind cell centre and the value of the component $\nabla_n \phi_f$ normal to the face of the gradient of the same quantity at the centroid of the considered face. In the case where ϕ is a scalar quantity, the ratio r is calculated in OpenFOAM® as

$$r = \max\left[2\frac{\Delta\mathbf{d}\cdot\nabla\phi}{|\Delta\mathbf{d}|\nabla_n\phi_f} - 1, 0\right]$$

where $\Delta\mathbf{d}$ is the vector connecting the centres of the two cells that share the considered face.

In the case where ϕ is a vector quantity, it is possible to use the schemes indicated as `limitedLinearV`, `linearUpwindV`, etc. In these cases, the ratio r is calculated in OpenFOAM® as

$$r = 2\frac{\nabla_n\phi_f\cdot\Delta\mathbf{d}\cdot\nabla\phi}{|\Delta\mathbf{d}|\nabla_n\phi_f\cdot\nabla_n\phi_f} - 1.$$

Various expressions exist for the flux limiter function. Here is the implementation in OpenFOAM® for some of them:

- $\psi(r) = \min(2r, 1)$ for the limited linear scheme;
- $\psi(r) = \min(r, 1)$ for the minmod scheme;
- $\psi(r) = \dfrac{r + |r|}{1 + |r|}$ for the van Leer scheme;
- $\psi(r) = \dfrac{r^2 + r}{1 + r^2}$ for the van Albada scheme.

In the case where LUST or linearUpwind schemes (see Sects. 3.1.4 and 3.1.5) are used, it is necessary to know the value of the gradient of ϕ at the cell centre because, in the case of linearUpwind, the value of the quantity considered at the face is calculated as (Figs. 6.3 and 6.4)

$$\begin{cases} \phi_f = \phi_P + \nabla\phi_P \cdot \mathbf{r} & for \quad (\mathbf{u} \cdot \mathbf{n})_f \geq 0, \\ \phi_f = \phi_N + \nabla\phi_N \cdot \mathbf{r} & for \quad (\mathbf{u} \cdot \mathbf{n})_f < 0 \end{cases}$$

where $\mathbf{r}$ is the cell centre-to-face centre distance vector. It is clear the necessity to specify the way in which this gradient (for example, the velocity gradient **grad(U)**) is calculated. As indicated in Fig. 6.5, the gradient calculation setting can be done explicitly or implicitly by inserting the word **default** instead of **grad(U)** in the Section **gradSchemes** and removing **grad(U)** in the Section **divSchemes**. It should be noted that in this second case, the gradient of all quantities will be calculated in the same way. To differentiate the way in which the gradient is calculated based on the quantity considered, it is possible to proceed as done in Fig. 6.1, in which the method to calculate the gradient of the pressure is specified separately from that used for all other quantities.

Still, in the case where LUST or linearUpwind schemes are used, the value of the gradient, calculated as specified in Section **gradSchemes**, can lead to unacceptable

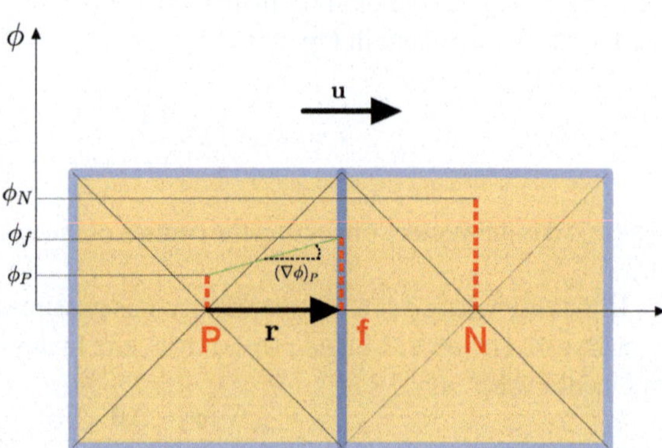

Fig. 6.3 Value of ϕ_f for $(\mathbf{u} \cdot \mathbf{n})_f \geq 0$

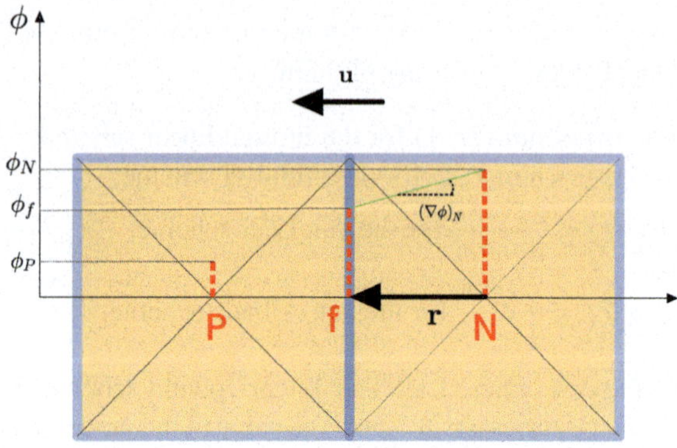

Fig. 6.4 Value of ϕ_f for $(\mathbf{u} \cdot \mathbf{n})_f < 0$

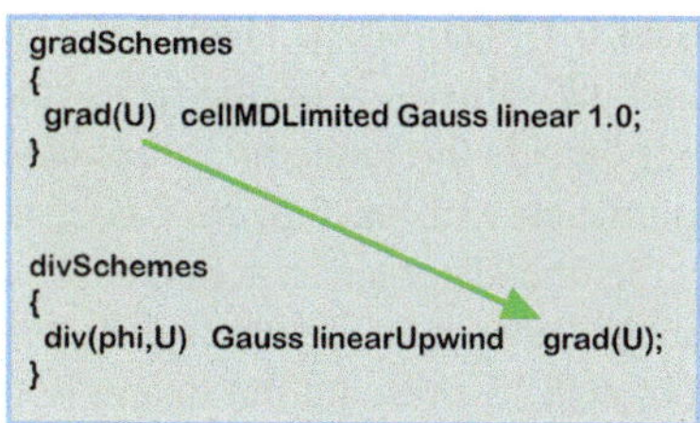

Fig. 6.5 Explicit setting of the gradient calculation method of velocity

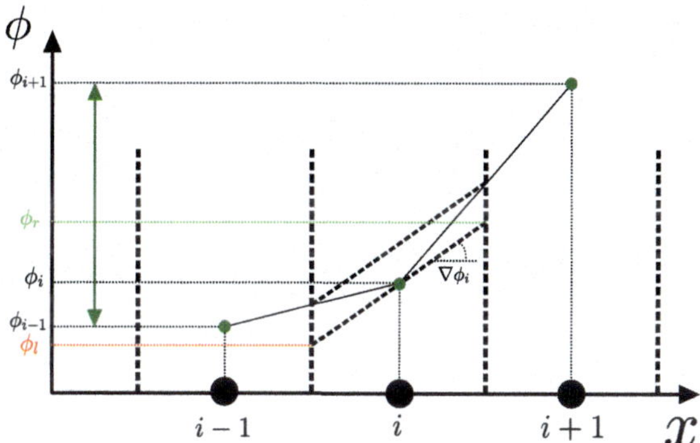

Fig. 6.6 Calculation of ϕ on the face

values of the considered quantity at the faces of the generic cell. For simplicity, and with reference to Fig. 6.6, a one-dimensional grid is considered. In this case, the gradient is indicated by the slope of the line representative of the variation of ϕ.

If the cell-based strategy (i.e., linear interpolation, see Sect. 3.4) is adopted to calculate the value of the gradient in cell i, one should consider the value of ϕ at the centres of the two cells $i-1$ and $i+1$ to calculate the values ϕ_l and ϕ_r respectively. ϕ_l and ϕ_r are the values of the quantity at the two faces that delimit the cell. The two values thus obtained determine the value of the gradient (i.e., the inclination of the two dashed and parallel segments in Fig. 6.6) at the centre of the cell. This gradient value is used for the subsequent calculation of the face values of the quantity ϕ when discretising the convective fluxes (Section **divSchemes**).

It is evident from Fig. 6.6 that ϕ_{i-1} and ϕ_{i+1} are the values of ϕ at the centres of the two cells adjacent to the one considered. Considering the interval on the ordinate axis between ϕ_{i-1} and ϕ_{i+1}, the application of this gradient calculation procedure can lead to values of ϕ on the faces outside this interval. In Fig. 6.6, the value ϕ_l is lower than ϕ_{i-1}.

This phenomenon is known as the "unboundedness" of the scheme. Such a phenomenon can make the entire simulation process unstable. If, for example, one considers the turbulent viscosity calculation, the correctness of the velocity gradient is fundamental: excessive values would lead to an incorrect increase in viscosity, which in turn would artificially raise the value of the velocity gradient, triggering a self-feeding process that could lead to the destabilisation of the entire simulation. To avoid the occurrence of such situations, the so-called "*gradient limiters*" (see

Sect. 6.1.3) are used. Thanks to gradient limiters, the gradient is modified to ensure that the value of ϕ on the faces (the left face of the cell i in the case of Fig. 6.6) is not outside the range of values determined by the cells close to the one considered (in the case of Fig. 6.6, $\phi_l < \phi_{i-1}$ is avoided).

In the divergence schemes area of the file `fvSchemes` (for historical reasons), advective or transport terms of quantities other than velocity appear. Figure 6.15 presents the example of the advection scheme setting for turbulence-related quantities such as k and ω. In this area, some very specific terms of a non-precisely advective type also appear. For example, one can observe the diffusive term in the last line of the divSchemes area of Fig. 6.15:
`div((nuEff*dev2(T(grad(U))))) Gauss Linear;`
in which explicit reference is made to turbulent kinematic viscosity with the term `nuEff`. This diffusive term requires a type of discretisation different from the one used for advective terms. The heterogeneity of the nature of the terms present in this area requires that, unlike other areas, the word "none" must be used to explicitly state the chosen scheme for each term.

It is finally noted that, in the case of non-stationary simulations, none of the chosen schemes should include the word `bounded`. Normally, convective terms are identified by terms that begin with `div(phi,...)`, in which `phi` generally indicates

- the volumetric flux through the faces of cells for incompressible cases;
- the mass flux for compressible calculations.

As an example, `div(phi,U)` indicates the advection term of velocity, `div(phi,e)` the advection term of internal energy, and so on.

Schemes such as `linearUpwindV` or `limitedLinearV` differ from the corresponding versions for scalar fields in that the only limiting factor value for all components is calculated based on the component that presents the highest gradients. Consequently, this scheme is more stable, although less accurate.

The `bounded` version of the schemes for the discretisation of convective terms refers to the treatment of the material derivative, which can be expressed in terms of the time derivative and advection. In the case of internal energy e, it is:

$$\frac{De}{Dt} = \frac{\partial e}{\partial t} + \mathbf{U} \cdot \nabla e = \frac{\partial e}{\partial t} + \nabla \cdot (\mathbf{U}e) - (\nabla \cdot \mathbf{U})\, e.$$

For incompressible cases, the `bounded` version is used to improve convergence and maintain the *boundedness* of the scheme, including the term $(\nabla \cdot \mathbf{U})\, e$, which becomes null once convergence is reached.

6.1.3 Gradient Discretisation Schemes

The available cell centre gradient calculation methods are:

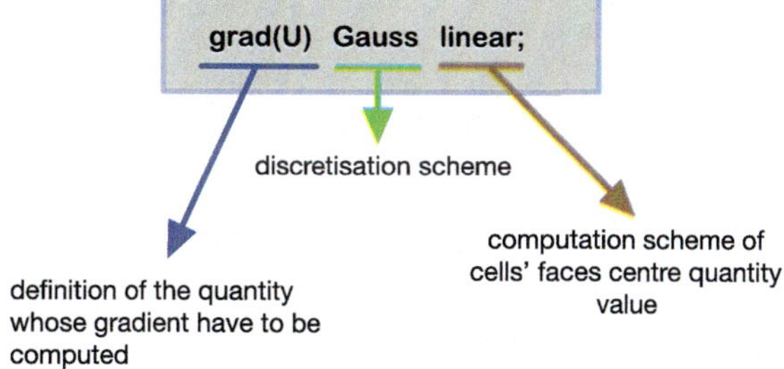

Fig. 6.7 Setting for the calculation of the velocity gradient at the centroid of the cell using the Gauss method and the cell based approach to calculate the velocity at the centroid of the faces

- **edgeCellsLeastSquares**;
- **fourth**;
- **Gauss**;
- **leastSquares**;
- **pointCellsLeastSquares**.

The source code for these schemes—all accurate to the second order—is contained in the folder `WM_PROJECT_DIR/src/finiteVolume/finiteVolume/gradSchemes`. For some gradient calculation methods, it is necessary to specify the strategy used to obtain the value of the quantity at the centroid of the generic face. Considering, for example, the Gauss method for calculating the velocity gradient, it is necessary to specify whether to use the cell- or node-based approach (see Sect. 3.4) to calculate the velocity at the centroid of the faces of the cell. If the cell-based strategy is chosen, the word `linear` must be specified (see Fig. 6.7); if the node-based method is chosen, the word `pointLinear` should be specified instead of `linear`. In Fig. 6.7, the gradient of the referred quantity is specified first, followed by the method for calculating the gradient at the centroid of the cell. The final word specifies the method for calculating the quantity at the centroid of the face. There are gradient calculation schemes for which it is possible to use a limiter. In ascending order of numerical diffusivity, they are:

- **cellMDLimited**;
- **cellLimited**;
- **faceMDLimited**;
- **faceLimited**.

The **cell*** type limiters restrict the value of the gradient in its component along the direction identified by two adjacent cell centres. The limiters of the **face*** type restrict the value of the gradient in its component along the direction identified by the face and the cell centroid. The multidimensional limiters (**cellMDLimited** and **faceMDLimited**) restrict the value of the gradient only along the direction normal to the face considered. The other limiters restrict all components of the gradient, not just the one along the normal to the face direction. The limiter used by default is **minmod**, but if the **cellLimited** option is used, it is possible to use two other limiters: **cubic** and **Venkatakrishnan**. For these two, the syntax to use is **cellLimited<cubic>**

Fig. 6.8 Syntax for the definition of gradient calculation

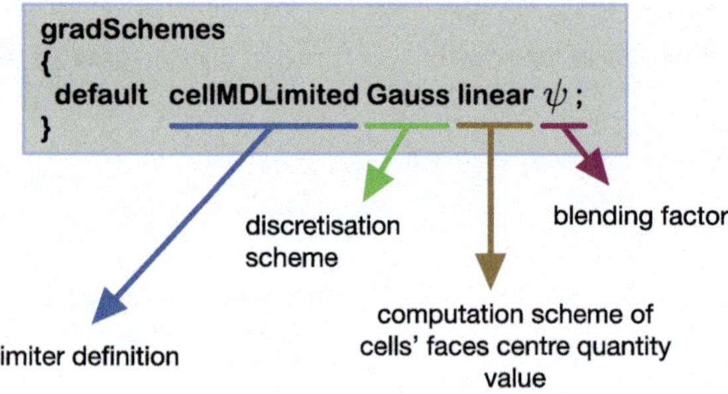

and **cellLimited<Venkatakrishnan>**, respectively. In OpenFOAM®, the implementation of limiters involves the use of a blending factor ψ, as shown in Fig. 6.8. This coefficient can vary between 0 and 1. By setting this coefficient to 0, the limiter is disabled; by setting it equal to 1, it is enabled. Intermediate values will result in a weighted average of the two extreme cases. The most commonly used gradient discretisation schemes are **Gauss**, which involves the use of interpolation, and **leastSquares**, which does not involve interpolation. The most commonly used limiter schemes are **cellLimited** and **cellMDLimited**. Except in special cases, these limiters are applied only to certain quantities, including velocity and quantities related to the modelling of turbulence (k, ϵ, ω), in order to avoid results affected by excessive numerical dissipation and excessively high residual values.

6.1.4 Discretisation Schemes of Laplacian or Diffusive Terms

The available schemes for calculating the component of the gradient normal to the face, in the context of the discretisation of Laplacian terms (see Sect. 3.3), are:

- **corrected**;
- **faceCorrected**;
- **limited**;
- **linearFit**;
- **orthogonal**;
- **quadraticFit**;
- **uncorrected**.

The source code for these schemes is contained in the folder `WM_PROJECT_DIR/src/finiteVolume/finiteVolume/snGradSchemes`. The choice of scheme depends on the geometric characteristics of the computational grid. Figures 6.9, 6.10, 6.11, and 6.12 show some possible types of grid. The most frequently used schemes are:

- **orthogonal**: used in the case of perfectly orthogonal, non-deformed hexagonal grids (see Fig. 6.9). This scheme is second-order accurate. It does not include

Fig. 6.9 Orthogonal grid: **orthogonal** schemes should be used

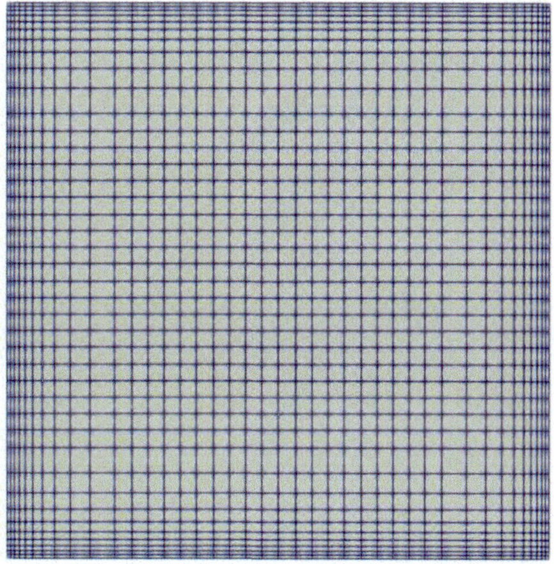

Fig. 6.10 Stretched orthogonal grid: **corrected** or **limited** with $\psi = 1$ schemes should be used

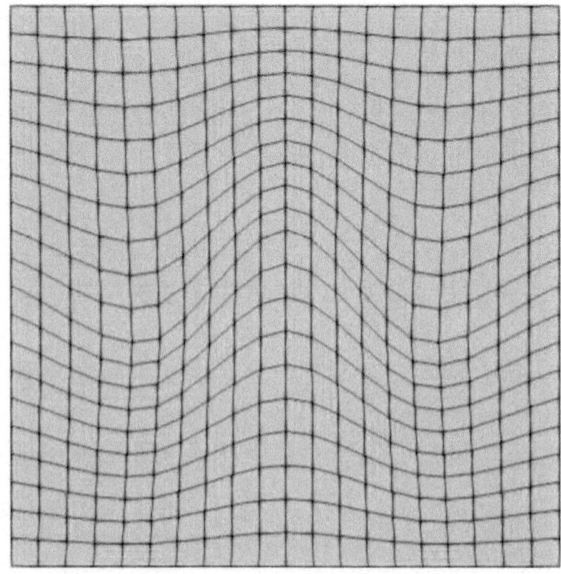

Fig. 6.11 Deformed grid with medium/low degree of non-orthogonality: **limited** with $\psi = 1$ or $\psi = 0.5$ scheme should be used

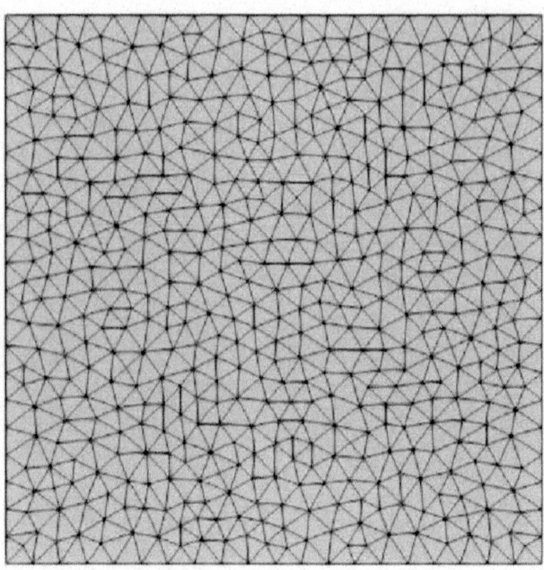

Fig. 6.12 Unstructured grid: **limited** with $\psi = 0.5$ scheme should be used

corrections for non-orthogonality. Definition (3.37) is rewritten here for the sake of convenience.

$$\mathbf{S} \cdot (\nabla \phi)_f = |\mathbf{S}| \frac{\phi_N - \phi_P}{|\mathbf{d}|};$$

- **uncorrected**: used in the case of non-deformed hexagonal grids with low non-orthogonality (see Fig. 6.10), this scheme is second-order accurate. It does not include corrections for non-orthogonality. Definition (3.39) is used, truncated of the corrective term for non-orthogonality:

$$\mathbf{S} \cdot (\nabla \phi)_f = \underbrace{|\Delta| \frac{\phi_N - \phi_P}{|\mathbf{d}|}}_{orthogonal\ contribution}$$

  where the term Δ is calculated according to the over-relaxed method (see Sect. 3.3);
- **corrected**: used in the case of grids (see Fig. 6.11) with high non-orthogonality, this scheme is second-order accurate and applies corrections for non-orthogonality (see Sect. 3.3). Definition (3.39) is rewritten here for the sake of simplicity.

$$\mathbf{S} \cdot (\nabla \phi)_f = \underbrace{|\Delta| \frac{\phi_N - \phi_P}{|\mathbf{d}|}}_{orthogonal\ contribution} + \overbrace{\mathbf{k} \cdot (\nabla \phi)_f}^{non\ orthogonal\ contribution} \quad ;$$

- **limited**: derived from the **corrected** scheme, this scheme provides the blending factor ψ, which limits the weight of the non-orthogonal contribution to a fraction of the orthogonal one. Highly non-orthogonal grids produce coefficient matrices with scarce diagonal dominance. When dealing with grids with high non-orthogonality (see Fig. 6.12), this approach prevents divergence of the solution process due to excessively high values of the non-orthogonal contribution. Some considerations regarding the blending factor ψ follow.
  - By setting $\psi = 0$ (no limitation), the **uncorrected** scheme is obtained. This option ensures greater stability but lower accuracy;
  - By setting $\psi = \frac{1}{3} = 0.333$, the contribution of the orthogonal part will be $1 - \frac{1}{3} = \frac{2}{3} = 0.666$, and the ratio (blending factor)/(orthogonal part) will be equal to $\frac{1}{2} = 0.5$. Therefore, the correction cannot exceed 50% of the value of the orthogonal part;
  - By setting $\psi = \frac{1}{2} = 0.5$, the contribution of the orthogonal part will be $1 - \frac{1}{2} = \frac{1}{2} = 0.5$, and the ratio (blending factor)/(orthogonal part) will be equal to $\frac{1}{1} = 1$. Therefore, the correction cannot exceed 100% of the value of the orthogonal part;
  - By setting $\psi = 1$, the **corrected** scheme is obtained, which ensures greater accuracy but lower stability.

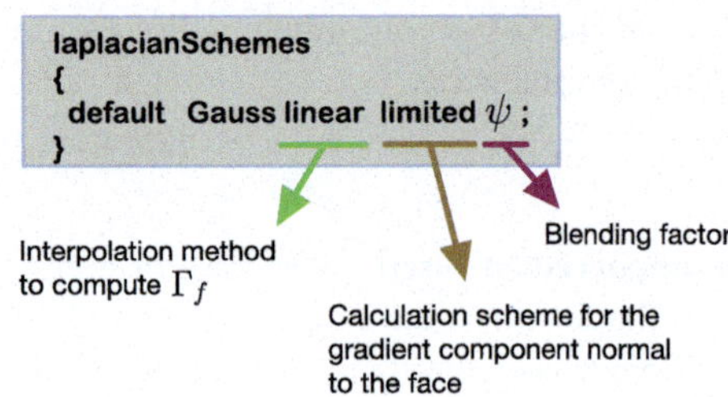

Fig. 6.13 Syntax for defining of the scheme **limited**

In general, for grids with non-orthogonality less than 70°, $\psi = 1$ can be set (no limitation). For a non-orthogonality factor between 70° and 80°, $\psi = 0.5$ can be set, simultaneously increasing the number of non-orthogonal corrections (see further on). For a non-orthogonality factor greater than 80°, $\psi = 0.33$ can be set, increasing the number of non-orthogonal corrections (see further on). Grids with a non-orthogonality factor greater than 85° should be discarded due to insufficient quality. The syntax to use in the case of the **limited** scheme is indicated in Fig. 6.13, where the word **Gauss** indicates that the Gauss theorem is being used to switch from volume to surface integrals. The word **Gauss** is the only option in the case of discretisation of Laplacian terms. The presence of this word responds to code implementation needs. The word **linear** indicates the interpolation method used to calculate the value Γ_f of the diffusion coefficient on the face based on that at the cell centre.

The available interpolation methods for the diffusion coefficient are:

- **cubic**;
- **harmonic**;
- **linear**;
- **midPoint**;
- **pointLinear**;
- **reverseLinear**.

The one used in most cases is **linear**. As for the terms related to the **snGradSchemes** entry, the same value inserted for the **laplacianSchemes** entry is normally used, as shown in Fig. 6.14.

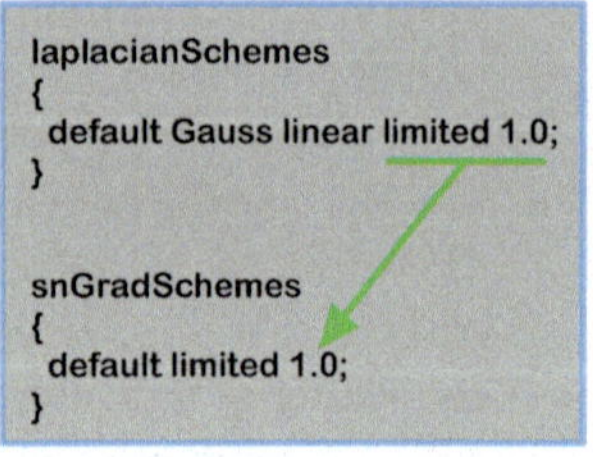

Fig. 6.14 Syntax for defining the scheme for the normal-to-the-face component of gradient

6.2 Examples of Discretisation Scheme Settings

Below are some examples of settings for the file `fvSchemes`.

6.2.1 Generic Setting

These settings (see Fig. 6.15) are valid in most cases and are very similar to those preset in commercial solvers. They are second order accurate, and depending on the quality of the grid, it may be necessary to reduce the value of the blending factor related to the **laplacianSchemes** and **snGradSchemes** entries. If turbulent quantities (i.e. k and ω) assume unacceptable values, it is advisable to change the corresponding entry in `divSchemes` from linearUpwind to upwind, thus lowering the order of accuracy from second to first for these quantities. To keep the simulation stable and accurate over time, the value of the CFL (in the file `controlDict`) should be less than 2 for implicit solvers. As for the calculation of the gradient, although the leastSquares method is generally more accurate, non-physical oscillations may occur in the presence of tetrahedral grids because the leastSquares method is not conservative.

Fig. 6.15 General settings

```
ddtSchemes
{
 default   CrankNicolson  0;
}
gradSchemes
{
 default   cellLimited Gauss linear  0.5;
 grad(U)   cellLimited Gauss linear  1.0;
}
divSchemes
{
 default   none;
 div(phi,U)  Gauss linearUpwindV grad(U);
 div(phi,omega)  Gauss linearUpwind default;
 div(phi,k)  Gauss linearUpwind default;
 div(nuEff*dev(T(grad(U))) Gauss linear;
}
laplacianSchemes
{
 default   Gauss linear limited 1.0;
}
interpolationSchemes
{
 default   linear;
}
snGradSchemes
{
 default   limited 1;
}
```

6.2.2 Accurate Setting

Figure 6.16 shows settings that make the calculation very accurate, and therefore more exposed to the risk of non-physical oscillations. These settings can be used for high-quality grids (low skewness and non-orthogonality) and for LES (Large Eddy Simulation) or RANS (Reynolds Averaged Navier–Stokes) simulations involving not too complex physical phenomena. To mitigate the effects of low-quality grids, lower values of the blending factor for **laplacianSchemes** and **snGradSchemes** can be used.

6.2.3 Stable Setting

Figure 6.17 shows settings that make the calculation very stable, more diffusive, and therefore less accurate (see also Fig. 3.17). These settings can be used for grids with poor quality (high skewness and non-orthogonality) and for simulations involving strong discontinuities or in cases of divergence in the calculations, leading to an unexpected interruption of the solution process. To mitigate the effects of low-quality grids, one can lower the values of the blending factor for **laplacianSchemes** and **snGradSchemes**. This type of setting is also used to perform a limited number of initial iterations, followed by the necessary number of iterations to reach convergence with settings that make the calculation more accurate.

6.3 Linear Solvers

The solvers for linear systems available in OpenFOAM® are:

- **PCG (Preconditioned Conjugate Gradient)**: a solver based on the preconditioned gradient method, suitable for symmetric coefficient matrices;
- **PBiCGStab (Preconditioned Bi-conjugate Gradient Stabilised)**: a solver based on the preconditioned bi-conjugate gradient method, stabilised for both symmetric and asymmetric coefficient matrices;
- **PBiCG (Preconditioned Bi-conjugate Gradient)**: a solver based on the preconditioned bi-conjugate gradient method, valid for asymmetric coefficient matrices;
- **smoothSolver**: a solver that requires the specification of a smoother;
- **GAMG (Geometric-Algebraic Multi-Grid)**: a multi-grid solver;
- **diagonal**: a diagonal solver valid for both symmetric and asymmetric coefficient matrices.

The symmetry of the coefficient matrix depends on the terms of the equation being considered: time discretisation and Laplacian/diffusive terms introduce symmetric elements, whereas the discretisation of advective terms introduces asymmetric elements. The preconditioners (see also Sect. 4.6.1) available in OpenFOAM® are:

- diagonal: preconditioner valid for symmetric and asymmetric coefficient matrices;
- DIC (Diagonal Incomplete Cholesky preconditioner): preconditioner valid for symmetric coefficient matrices;
- DILU (Diagonal Incomplete LU preconditioner): preconditioner valid for asymmetric coefficient matrices;
- FDIC (Fast Diagonal Incomplete Cholesky preconditioner): preconditioner valid for symmetric coefficient matrices;
- GAMG (Geometric Algebraic MultiGrid preconditioner): preconditioner valid for symmetric and asymmetric coefficient matrices. In this case, the GAMG method is used as a preconditioner.

Finally, in the case where a smoother needs to be specified for the chosen solver, the options available in OpenFOAM® are:

- GaussSeidel: the Gauss-Seidel method is applied to symmetric and asymmetric coefficients matrices;

Fig. 6.16 Settings for accurate but unstable calculation

```
ddtSchemes
{
 default   CrankNicolson 0.7;
}
gradSchemes
{
 default   leastSquares;
}
divSchemes
{
 default   none;
 div(phi,U) Gauss linear;
 div(phi,omega) Gauss linear;
 div(phi,k) Gauss linear;
 div(nuEff*dev(T(grad(U))) Gauss linear;
}

laplacianSchemes
{
 default   Gauss linear limited 1.0;
}

interpolationSchemes
{
 default   linear;
}
snGradSchemes
{
 default   limited 1;
}
```

Fig. 6.17 Settings for less accurate (diffusive) but stable calculation

```
ddtSchemes
{
 default   Euler;
}
gradSchemes
{
 default   cellLimited Gauss linear 1.0;
}

divSchemes
{
 default   none;
 div(phi,U)  Gauss upwind;
 div(phi,omega)  Gauss upwind;
 div(phi,k)  Gauss upwind;
 div(nuEff*dev(T(grad(U)))  Gauss linear;
}

laplacianSchemes
{
 default   Gauss linear limited 0.5;
}

interpolationSchemes
{
 default   linear;
}
snGradSchemes
{
 default   limited 0.5;
}
```

- symGaussSeidel: the Gauss-Seidel method is applied to symmetric coefficients matrices;
- DIC: in the case of symmetric matrices, the Diagonal Incomplete-Cholesky method is applied;
- DILU: in the case of non-symmetric matrices, the Diagonal Incomplete-LU method is applied;
- DICGaussSeidel: in this case, if the matrices are symmetric, an iteration performed with the DIC method is followed by an iteration with the Gauss-Seidel method that eliminates any high-frequency errors resulting from the iteration performed with the DIC method; if the matrices are non-symmetric, an iteration performed with the DILU method is followed by an iteration with the Gauss-Seidel method, which eliminates any high-frequency errors resulting from the iteration performed with the DILU method.

A smoother is a solver for systems of equations whose application eliminates high-frequency errors. It is applied to the system of equations before the actual solver, and the number of times it is applied is equal to the value set for the optional parameter `nSweeps`, with a default value of 1.

6.3.1 Geometric-Algebraic Multi-grid (GAMG)

As mentioned in Sect. 4.7.4, this algorithm begins with the assembly phase of the grids that will be used in the subsequent phases. In OpenFOAM®, the strategy for generating the coarser grids is specified by the parameter `agglomerator`. If the values of the elements of the coefficients matrix are used to perform the agglomeration phase, the value to set for `agglomerator` is `algebraicPair`. If the chosen strategy for merging the cells of the grid is based on geometric parameters, the value to set for `agglomerator` is `faceAreaPair`. This latter setting is considered more efficient than the former. The final option is `MGridGen`, which allows the use of the MGridGen library to perform the agglomeration process based on geometric considerations. In this case, it will be necessary to add the following line to the `controlDict` file: `geometricGamgAgglomerationLibs ("libMGridGenGamgAgglomeration.so")`. Other settings related to the agglomeration process are as follows.

- `nCellsInCoarsestLevel`: This parameter sets the total number of cells for the coarsest grid used in the multigrid process. The default value is 10.
- `directSolveCoarsest`: This parameter specifies whether to use a direct method to solve the linear system associated with the coarsest grid in the multigrid process. The default value is `false`.
- `mergeLevels`: This parameter controls the number of grids used in the multigrid process by selectively excluding some of the grids generated during the agglomeration phase. By setting this parameter to 2, only half of the generated grids will be used in the process. In most cases, this parameter is set to 1, and the value 2 is used only in the case of very simple starting grids.

It is possible to set the solver (`smoother`) used to solve the linear systems associated with the various grid levels through the parameter `smoother`. The possible values for this parameter are:

- `GaussSeidel`: this method applies the Gauss-Seidel approach to matrices that do not have zero values on the main diagonal. It guarantees convergence only for diagonal-dominant matrices that are symmetric and positive definite;
- `symGaussSeidel`: this method applies the Gauss-Seidel approach specifically to symmetric coefficient matrices;
- `DIC/DILU`: the method of incomplete diagonal decomposition (DIC) is applied to symmetric matrices, while the method of incomplete LU decomposition (DILU) is used for non-symmetric matrices;
- `DICGaussSeidel`: for symmetric matrices, an iteration performed with the DIC method is followed by an iteration with the Gauss-Seidel method to eliminate high-frequency errors caused by the DIC iteration. For non-symmetric matrices, an iteration with the DILU method is followed by an iteration with the Gauss-Seidel method to eliminate high-frequency errors resulting from the DILU iteration.

The number of iterations to be performed using the chosen smoother is determined by the value of the following optional parameters:

- `nPreSweeps`: number of iterations to be performed on the grids during the coarsening phase (default 0);
- `nPostSweeps`: number of iterations to be performed on the grids during the refining phase (default 2);
- `nFinestSweeps`: number of iterations to be performed on the finest grid.

6.4 Pressure-Velocity Coupling

The projection methods implemented in OpenFOAM® are:

- **SIMPLE** (Semi-Implicit Method for Pressure-Linked Equations);
- **SIMPLEC** (SIMPLE Corrected/Consistent);
- **PISO** (Pressure Implicit with Splitting Operators);
- **PIMPLE** (hybrid between SIMPLE and PISO).

The **SIMPLE** and **SIMPLEC** methods are implemented for steady-state simulations, while **PISO** and **PIMPLE** are implemented for transient simulations.

6.4.1 Implementation of SIMPLE and PISO in OpenFOAM®

As previously mentioned, in the continuity equation for incompressible flows,

$$\nabla \cdot \mathbf{U} = 0 \tag{6.4}$$

the pressure term is not present. Therefore, in this case, it is impossible to link the continuity equation with the conservation of momentum Eq. 2.43 here reported in vector form:

$$\frac{\partial \mathbf{U}}{\partial t} + \nabla \cdot (\mathbf{U}\mathbf{U}) - \nu \nabla^2 \mathbf{U} = -\nabla p \tag{6.5}$$

whose three components are

$$\frac{\partial U_x}{\partial t} + \nabla \cdot (\mathbf{U} U_x) - \nu \nabla^2 U_x = -\frac{\partial p}{\partial x}, \tag{6.6}$$

$$\frac{\partial U_y}{\partial t} + \nabla \cdot (\mathbf{U} U_y) - \nu \nabla^2 U_y = -\frac{\partial p}{\partial y}, \tag{6.7}$$

$$\frac{\partial U_z}{\partial t} + \nabla \cdot (\mathbf{U}U_z) - \nu \nabla^2 U_z = -\frac{\partial p}{\partial z}. \tag{6.8}$$

U is the fluid velocity, while U_x (or U_y or U_z) is the transported quantity (the ϕ of Eq. 2.38). It should be noted that here the pressure refers to the kinematic pressure, p/ρ. ν is the kinematic viscosity, which is related to dynamic viscosity by the relation $\nu = \mu/\rho$. Equations 6.6, 6.7, and 6.8 can each be applied to every cell of the computational domain, resulting in the following matrix equations.

$$\mathbf{M}_x\mathbf{U}_x = \mathbf{b}_x, \tag{6.9}$$

$$\mathbf{M}_y\mathbf{U}_y = \mathbf{b}_y,$$

$$\mathbf{M}_z\mathbf{U}_z = \mathbf{b}_z$$

where $\mathbf{M}_x$, $\mathbf{M}_y$, and $\mathbf{M}_z$ are the coefficient matrices; $\mathbf{U}_x$, $\mathbf{U}_y$, and $\mathbf{U}_z$ are the column matrices containing the considered component of the velocity corresponding to each of the N cells of the computational domain; $\mathbf{b}_x$, $\mathbf{b}_y$, and $\mathbf{b}_z$ are the column matrices containing the considered component of the pressure gradient corresponding to each of the N cells of the computational domain. Considering only the x component of the velocity, Eq. 6.9 is

$$\begin{bmatrix} M_{11} & M_{12} & \dots & M_{1N} \\ M_{21} & M_{22} & \dots & M_{2N} \\ \vdots & \vdots & \dots & \vdots \\ M_{N1} & M_{N2} & \dots & M_{NN} \end{bmatrix} \begin{bmatrix} U_{x1} \\ U_{x2} \\ \vdots \\ U_{xN} \end{bmatrix} = \begin{bmatrix} (\partial p/\partial x)_1 \\ (\partial p/\partial x)_2 \\ \vdots \\ (\partial p/\partial x)_N \end{bmatrix}$$

where the terms M_{ij} of the coefficient matrix are known.

It is important to note that the discretisation of the various terms of Eq. 6.5 contributes to determining the value of the elements M_{ij} of the coefficient matrix. For simplicity, we will refer only to the component U_x of the velocity from now on. As seen in Sect. 3, for the discretisation of the convective term, we can refer to Eq. 3.2. If the transported quantity is U_x, Eq. 3.2 becomes

$$\oint_{\partial V_P} d\mathbf{S} \cdot (\rho \mathbf{U} U_x) \approx \sum_f \mathbf{S}_f \cdot (\rho \mathbf{U} U_x)_f = \sum_f \mathbf{S}_f \cdot \rho_f \mathbf{U}_f (U_x)_f = \sum_f \phi_f (U_x)_f \tag{6.10}$$

where the mass flow rate $\phi_f = \mathbf{S}_f \cdot \rho_f \mathbf{U}_f$ becomes a volumetric flow rate, $\phi_f = \mathbf{S}_f \cdot \mathbf{U}_f$, in the case of incompressible flow, where the density ρ is absent.

For the discretisation of the diffusive term, we refer to Eq. 3.3, which, when the diffused quantity is U_x, becomes:

$$\oint_{\partial V_P} d\mathbf{S} \cdot (\rho \nu \nabla U_x) \approx \sum_f \mathbf{S}_f \cdot (\rho \nu \nabla U_x)_f = \sum_f S_f \rho_f \nu_f \nabla_n (U_x)_f$$

S_f denotes the magnitude of the vector $\mathbf{S}_f$, and the term $\nabla_n(U_x)_f$ is the normal component of the gradient of U_x at the centre of the face f. Considering that

- $\nabla \cdot \nabla^2 \mathbf{U} \equiv \nabla \cdot (\nabla \cdot \nabla \mathbf{U}) \equiv \nabla^2 (\nabla \cdot \mathbf{U})$;
- the kinematic viscosity is considered constant;
- the semi-discretised form (i.e., discretised only with respect to time) of the momentum Eq. 6.5 is used;
- the continuity Eq. 6.4 must be satisfied.

A new equation can be written when calculating the divergence of the momentum Eq. 6.5. In this new equation, the pressure term appears:

$$\cancel{\nabla \cdot \frac{\partial \mathbf{U}}{\partial t}} + \nabla \cdot (\nabla \cdot (\mathbf{UU})) - \cancel{\nabla \cdot (\nu \nabla^2 \mathbf{U})} = -\nabla^2 p$$

that is

$$\nabla^2 p + \nabla \cdot [\nabla \cdot (\mathbf{UU})] = 0 \tag{6.11}$$

known as the *Poisson pressure equation* (see also Sect. 1.3.5).
The Poisson equation can be better understood by noting that the symbol **UU** represents a second-order tensor:

$$\mathbf{UU} = \begin{bmatrix} U_x U_x & U_x U_y & U_x U_z \\ U_y U_x & U_y U_y & U_y U_z \\ U_z U_x & U_z U_y & U_z U_z \end{bmatrix}.$$

Note that, since the transport term $\nabla \cdot (\mathbf{UU})$ is present in the Poisson pressure equation, its discretisation necessarily leads to the calculation of the flux ϕ_f, as seen for the same term in the discretised form of the momentum equation. The system composed of Eqs. 6.5 and 6.11 is equivalent to the system of the Navier-Stokes equations in their incompressible formulation and can be solved by imposing appropriate boundary and initial conditions. Specifically, by setting an initial pressure field, a velocity field can be calculated through Eq. 6.5 (*momentum predictor step*), which has a non-zero divergence. Through Eq. 6.11, this velocity field can be used to calculate a new pressure field (*pressure corrector step*). The pressure field thus obtained is used to "correct" the velocity field (*momentum corrector step*) and update the values of the fluxes ϕ_f and the coefficients of matrices $\mathbf{M}_x$, $\mathbf{M}_y$, and $\mathbf{M}_z$. This sequence is executed iteratively until a velocity field with zero divergence is reached, that is, until the continuity equation is satisfied with an acceptable error. This procedure is also known as *pressure-velocity coupling* (see also Chap. 5).

It is now necessary to describe the notation underlying the implementation of the SIMPLE algorithm. The terms on the left-hand side of Eq. 6.5 can be represented as follows:

$$\frac{\partial \mathbf{U}}{\partial t} + \nabla \cdot (\mathbf{UU}) - \nu \nabla^2 \mathbf{U} \equiv A\mathbf{U} - \mathbf{H}(\mathbf{U})$$

in which A is a constant and $\mathbf{H}(\mathbf{U})$ is a vector which depends on $\mathbf{U}$ and any source term not considered here for simplicity. Considering the three components of the velocity, it is

$$AU_x - H_x(\mathbf{U}) \equiv \frac{\partial U_x}{\partial t} + \nabla \cdot (\mathbf{U}U_x) - \nu \nabla^2 U_x,$$

$$AU_y - H_y(\mathbf{U}) \equiv \frac{\partial U_y}{\partial t} + \nabla \cdot (\mathbf{U}U_y) - \nu \nabla^2 U_y,$$

$$AU_z - H_z(\mathbf{U}) \equiv \frac{\partial U_z}{\partial t} + \nabla \cdot (\mathbf{U}U_z) - \nu \nabla^2 U_z$$

in which $H_x(\mathbf{U})$, $H_y(\mathbf{U})$, and $H_z(\mathbf{U})$ are the three components of the vector $\mathbf{H}(\mathbf{U})$.

The momentum conservation Eq. 6.5 can now be written as

$$A\mathbf{U} - \mathbf{H}(\mathbf{U}) = -\nabla p \tag{6.12}$$

and, considering the three components of the velocity,

$$\begin{aligned} AU_x - H_x(\mathbf{U}) &= -\frac{\partial p}{\partial x}, \\ AU_y - H_y(\mathbf{U}) &= -\frac{\partial p}{\partial y}, \\ AU_z - H_z(\mathbf{U}) &= -\frac{\partial p}{\partial z}. \end{aligned} \tag{6.13}$$

Applying each of Eq. 6.13 to all cells of the computational domain, three matrix equations are obtained:

$$\begin{aligned} \mathbf{A}_x\mathbf{U}_x - \mathbf{H}_x(\mathbf{U}) &= \mathbf{b}_x, \\ \mathbf{A}_y\mathbf{U}_y - \mathbf{H}_y(\mathbf{U}) &= \mathbf{b}_y, \\ \mathbf{A}_z\mathbf{U}_z - \mathbf{H}_z(\mathbf{U}) &= \mathbf{b}_z \end{aligned}$$

in which $\mathbf{A}_x$, $\mathbf{A}_y$, and $\mathbf{A}_z$ are diagonal matrices whose terms are the diagonal elements (constants) of the coefficient matrices $\mathbf{M}_x$, $\mathbf{M}_y$, and $\mathbf{M}_z$, respectively. $\mathbf{U}_x$, $\mathbf{U}_y$, and $\mathbf{U}_z$ are column matrices containing the corresponding velocity components for each cell of the computational domain. $\mathbf{H}_x(\mathbf{U})$, $\mathbf{H}_y(\mathbf{U})$, and $\mathbf{H}_z(\mathbf{U})$ are matrices containing the corresponding components of the vector $\mathbf{H}(\mathbf{U})$ for each cell of the computational domain. $\mathbf{b}_x$, $\mathbf{b}_y$, and $\mathbf{b}_z$ are column matrices containing the corresponding components of the pressure gradient for each of the N cells of the computational domain.

Equation 6.12 can now be rewritten as

$$\mathbf{U} = \frac{\mathbf{H}(\mathbf{U})}{A} - \frac{1}{A}\nabla p \tag{6.14}$$

and, applying the three component equations of this vector equation to all the cells of the computational domain, three matrix equations are obtained:

$$\begin{aligned} \mathbf{U}_x &= \frac{\mathbf{H}_x(\mathbf{U})}{\mathbf{A}_x} + \frac{1}{\mathbf{A}_x}\mathbf{b}_x, \\ \mathbf{U}_y &= \frac{\mathbf{H}_y(\mathbf{U})}{\mathbf{A}_y} + \frac{1}{\mathbf{A}_y}\mathbf{b}_y, \\ \mathbf{U}_z &= \frac{\mathbf{H}_z(\mathbf{U})}{\mathbf{A}_z} + \frac{1}{\mathbf{A}_z}\mathbf{b}_z \end{aligned} \tag{6.15}$$

where $\mathbf{b}_x$, $\mathbf{b}_y$, and $\mathbf{b}_z$ are still the column matrices containing the considered components of the pressure gradient corresponding to each of the N cells of the computational domain. Equation 6.15 has been written in this way to reflect the implementation in OpenFOAM® programming language. Nevertheless, Eq. 6.15 has been written in this way to highlight the fact that the calculation of the terms $\frac{1}{\mathbf{A}_x}$, $\frac{1}{\mathbf{A}_y}$, and $\frac{1}{\mathbf{A}_z}$ is very simple, since the inverse of a diagonal matrix is a diagonal matrix whose elements are each the inverse of the corresponding elements of the initial matrix. Referring only to the component in the x direction, it is

$$\frac{1}{\mathbf{A}_x} = \mathbf{A}_x^{-1} = \begin{bmatrix} \frac{1}{M_{11}} & 0 & \dots 0 & 0 \\ 0 & \frac{1}{M_{22}} & \dots 0 & 0 \\ \vdots & \vdots & \dots \vdots & \vdots \\ 0 & 0 & \dots 0 & \frac{1}{M_{NN}} \end{bmatrix}_x .$$

Referring to (6.14) and noting the definition $\phi_f = \mathbf{S}_f \cdot \mathbf{U}_f$ of volumetric flow rate at the interface, we can write

$$\phi_f = \mathbf{S}_f \cdot \left(\frac{\mathbf{H}(\mathbf{U})}{A}\right)_f - \left(\frac{S_f}{A}\right)_f \nabla_n p_f . \tag{6.16}$$

Referring to Eq. 6.14 and noting the definition $\nabla \cdot \mathbf{U} = 0$ of the mass conservation equation, the Poisson pressure equation can be written as

$$\nabla \cdot \frac{1}{A} \nabla p = \nabla \cdot \left[\frac{\mathbf{H}(\mathbf{U})}{A}\right] .$$

Given that a Laplacian is present in the pressure equation, as occurs in the discretisation of diffusive terms, the presence of non-orthogonal grids results in the additional term $f(\nabla p)$:

$$\nabla \cdot \frac{1}{A} \nabla p = \nabla \cdot \left[\frac{\mathbf{H}(\mathbf{U})}{A} \right] + f\left(\nabla p\right). \tag{6.17}$$

Skewness and non-orthogonality of cells imply *secondary gradients* in every equation where diffusive phenomena are considered. Noting that the non-orthogonal correction term (see Sect. 3.3) is solved as a source term and therefore explicitly, the values of pressure used to calculate the term $f\left(\nabla p\right)$ are always those obtained at the actual iteration. To reduce the errors due to secondary gradients, the pressure value just obtained from the pressure correction equation is substituted back into the same equation, as shown in Figs. 6.21, 6.22, and 6.24, where the *non-orthogonal correctors* decision block is used to obtain a new and more accurate value of the pressure field. By doing so, accuracy and stability are improved at the expense of a significant increase in computational cost. In OpenFOAM®, the SIMPLE algorithm is implemented for steady-state calculations only. For non-steady-state calculations, the PISO or PIMPLE algorithms are used. The control parameters of the SIMPLE cycle are contained in the `fvOptions` file. The number of executions of the non-orthogonal correction cycle is set through the word **nonOrthogonalCorrectors** present in the **SIMPLE** section of the `fvOptions` file. When dealing with orthogonal grids, this value can be set to zero; otherwise, it cannot be less than 1 and may increase depending on the quality of the grid. When there is at least one non-orthogonal correction, the cycle is called SIMPLEC.

6.4.1.1 Field and Equation Under-Relaxation

The use of under-relaxation factors is typical of SIMPLE-type steady-state solvers. As shown in Fig. 6.19, it is possible to distinguish the application of such factors to a specific quantity ("`fields`") from their application to a matrix equation ("`equations`"). In the case of applying them to a generic quantity (for example, pressure), these factors are used to control the rate of change of the quantity between one iteration and the next (Fig. 6.18):

$$\phi_P^n = \phi_P^{n-1} + \alpha \left(\phi_P^{n^*} - \phi_P^{n-1} \right)$$

where ϕ_P^n is the final value of the quantity at iteration n in the cell centred at P, ϕ_P^{n-1} is the value of the quantity at iteration $n-1$, and $\phi_P^{n^*}$ is the value of the quantity at iteration n before the application of the under-relaxation factor α. The values that this factor can assume range between 0 and 1. High values imply greater rapidity but less stability in the calculations, and vice versa. This approach to the application of the under-relaxation factor is somewhat inefficient from a memory usage perspective because it requires storing the value $\phi_P^{n^*}$ in order to calculate ϕ_P^n. This method is known as *field under-relaxation* or *explicit under-relaxation*. It is applied by specifying the quantity for which it is to be used in the "`fields`" section of the `relaxationFactors` zone in the `fvSolution` file (see Fig. 6.19). For numerical reasons, this approach is required for pressure when using the SIMPLE algorithm.

When under-relaxation factors are applied to a matrix equation, it is referred to as *equation under-relaxation* or *implicit under-relaxation*. In this approach, the coefficients matrix resulting from the discretisation of the equation (for example, the conservation equation of the quantity of motion) for the considered quantity, applied to all the cells of the computational domain, is modified. Indicating the elements of the main diagonal of the coefficients matrix with the symbol □ and the remaining elements with the symbol $*$, it can be written:

$$\begin{bmatrix} \square & * & & * \\ * & \square & * & \\ & * & \square & * \\ * & & * & \square \end{bmatrix} [\phi] = \begin{bmatrix} * \\ * \\ * \\ * \end{bmatrix}$$

in which $[\phi]$ is the column vector whose elements are the values of the quantity ϕ corresponding to all the cells of the computational domain. The under-relaxation process involves modifying the elements of the main diagonal as follows:

$$\blacksquare = \max\left(\square, sum\,|*|\right)/\alpha \tag{6.18}$$

in which the symbol $\sum |*|$ represents the sum of the absolute values of all the off-diagonal elements in the row to which the element of the main diagonal belongs. It can be noted that, as the value of α decreases, the values of the elements of the main diagonal increase, along with the diagonal dominance of the coefficients matrix, thus making the inversion process more stable. Subsequently, an additional term is added to the column vector of known terms, obtaining:

$$\begin{bmatrix} \blacksquare & * & & * \\ * & \blacksquare & * & \\ & * & \blacksquare & * \\ * & & * & \blacksquare \end{bmatrix} [\phi] = \begin{bmatrix} * \\ * \\ * \\ * \end{bmatrix} + [\blacksquare - \square]\left[\phi^{n-1}\right] \tag{6.19}$$

in which $\left[\phi^{n-1}\right]$ is the column vector whose elements are the values of the quantity ϕ at the previous iteration. Due to the OpenFOAM® implementation of this approach, when $\alpha = 1$, the elements of the main diagonal are still modified according to Eqs. 6.18 and 6.19 to ensure the diagonal dominance of the coefficients matrix.

For non-stationary cases (PISO or PIMPLE algorithms), it is necessary:

- to delete all the rows in the `fields` section of the `fvSolution` file to avoid incorrect results due to the variation of the elements of the main diagonal, which have already been modified by the presence of elements related to the time derivative (see Sect. 6.1.1);

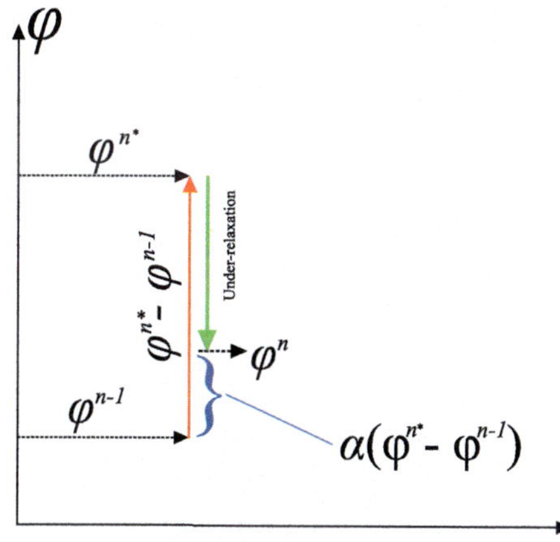

Fig. 6.18 Application of the under-relaxation factor α

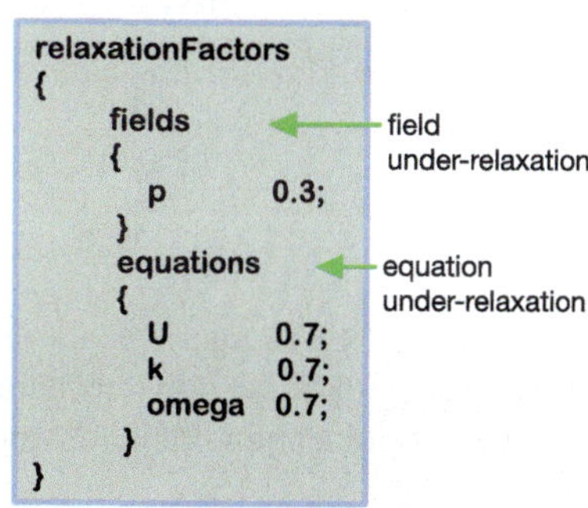

Fig. 6.19 Setting of the under-relaxation factors in the case of a solver based on the SIMPLE algorithm

- to set the value of the under-relaxation factors for the quantities listed in the `equations` section to 1, in order to ensure diagonal dominance. In the `relaxationFactors` section of the `fvOptions` file, it is recommended to have only a single row containing the text: `".*" 1;` which sets the under-relaxation value of all the quantities to 1, thus ensuring diagonal dominance without under-relaxing (see Fig. 6.26).

An example of setting such factors in the `fvSolution` file for stationary calculations with a solver based on the SIMPLE algorithm is shown in Fig. 6.19. Figure 6.20 shows the setting of such factors in the case of a solver based on the SIMPLEC algorithm. In this case, both field under-relaxation and equation under-relaxation approaches have been activated. An example of setting such factors for non-stationary calculations in the `fvSolution` file is shown in Fig. 6.26.

Figure 6.21 shows the SIMPLE algorithm in the form of a flowchart as implemented in OpenFOAM®. The corresponding source code lines are also shown for each block of the flowchart. Given an initial velocity field or one resulting from the previous iteration, a "momentum matrix" is defined for each of the three components

Fig. 6.20 Setting of the under-relaxation factors in the case of a solver based on the SIMPLEC algorithm

```
relaxationFactors
{
    fields
    {
        p           0.9;
    }
    equations
    {
        p           1.0;
        U           0.7;
        k           0.7;
        omega       0.7;
    }
}
```

of the velocity, based on the discretised form of the momentum conservation Eq. 6.5, excluding the pressure term.

$$\begin{aligned}\mathbf{M}_x\mathbf{U}_x &\equiv \nabla\cdot(\mathbf{U}U_x) - \nu\nabla^2 U_x,\\ \mathbf{M}_y\mathbf{U}_y &\equiv \nabla\cdot(\mathbf{U}U_y) - \nu\nabla^2 U_y,\\ \mathbf{M}_z\mathbf{U}_z &\equiv \nabla\cdot(\mathbf{U}U_z) - \nu\nabla^2 U_z.\end{aligned}$$

Subsequently, these matrices are under-relaxed. Given an initial pressure field or one resulting from the previous iteration, the following matrix equations are solved implicitly (momentum predictor step):

$$\begin{aligned}\mathbf{M}_x\mathbf{U}_x &= \mathbf{b}_x,\\ \mathbf{M}_y\mathbf{U}_y &= \mathbf{b}_y,\\ \mathbf{M}_z\mathbf{U}_z &= \mathbf{b}_z\end{aligned} \tag{6.20}$$

in which $\mathbf{b}_x$, $\mathbf{b}_y$ and $\mathbf{b}_z$ are the column matrices containing the considered component of the pressure gradient in each of the N cells of the computational domain. The result is a velocity field $\mathbf{U}$ with non-zero divergence. This velocity field allows defining the term $\mathbf{H}(\mathbf{U})$ present in the pressure Eq. 6.17, here reported for simplicity.

$$\nabla\cdot\frac{1}{A}\nabla p = \nabla\cdot\left[\frac{\mathbf{H}(\mathbf{U})}{A}\right] + f(\nabla p)$$

which, solved implicitly, provides the new pressure field. The new pressure field can be used to update the flow rates on the faces of the cells (flux corrector step) using Eq. 6.16, here reported for simplicity.

$$\phi_f = \mathbf{S}_f\cdot\left(\frac{\mathbf{H}(\mathbf{U})}{A}\right)_f - \left(\frac{S_f}{A}\right)_f \nabla_n p_f.$$

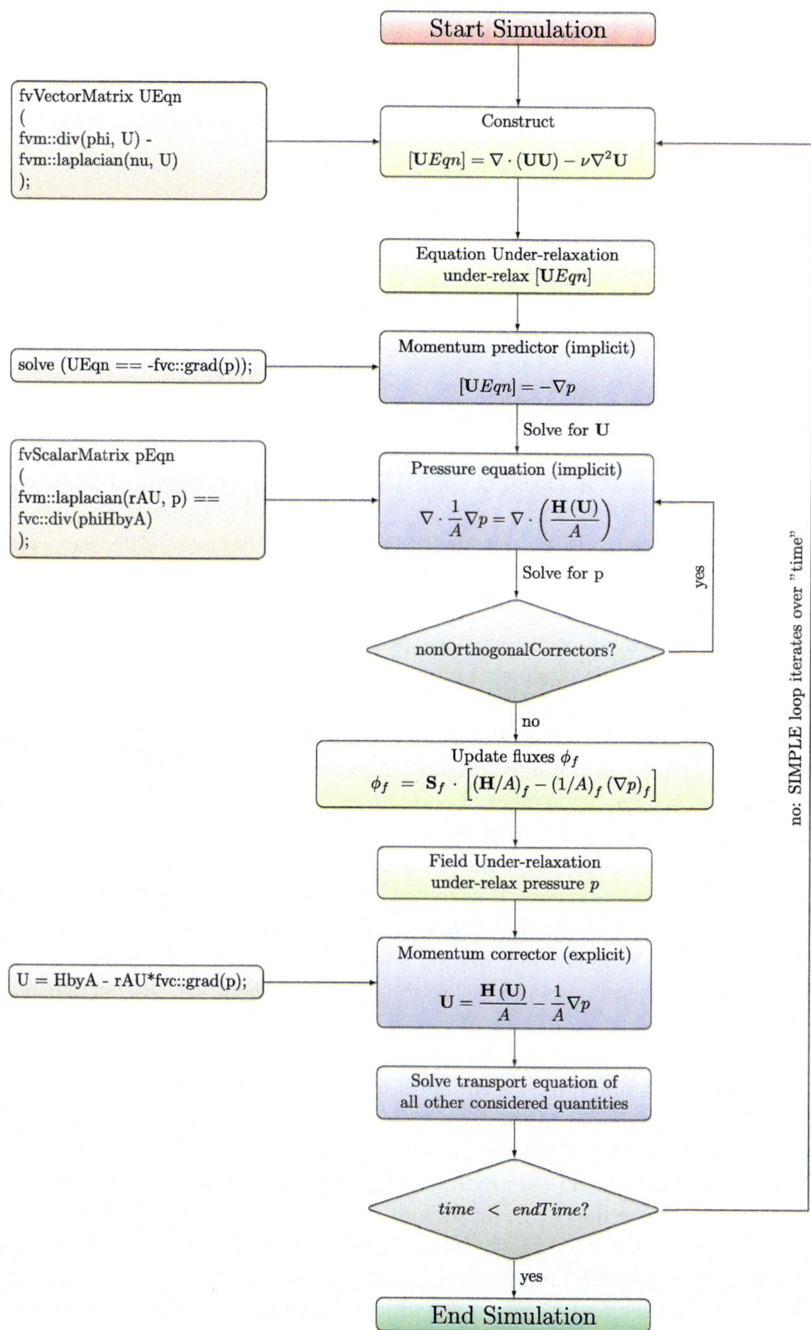

Fig. 6.21 SIMPLE predictor-corrector process with non-orthogonality correction cycle orthogonality

To avoid flow rates ϕ_f not respecting the equation of continuity (represented by the pressure equation), the pressure field under-relaxation is performed after the flow rates update. With the new pressure field and the updated value of $\mathbf{H}(\mathbf{U})$, the new velocity field can be obtained by explicitly solving the three matrix Eq. 6.15.

$$\mathbf{U}_x = \frac{\mathbf{H}_x(\mathbf{U})}{\mathbf{A}_x} + \frac{1}{\mathbf{A}_x}\mathbf{b}_x,$$
$$\mathbf{U}_y = \frac{\mathbf{H}_y(\mathbf{U})}{\mathbf{A}_y} + \frac{1}{\mathbf{A}_y}\mathbf{b}_y,$$
$$\mathbf{U}_z = \frac{\mathbf{H}_z(\mathbf{U})}{\mathbf{A}_z} + \frac{1}{\mathbf{A}_z}\mathbf{b}_z.$$

The entire cycle is repeated until the convergence criteria are met or the maximum number of iterations is reached. For implementation reasons, the iteration number is referred to as "time" in OpenFOAM®. This cycle is often identified as the *outer corrector loop*.

When comparing the PISO and SIMPLE algorithms, the main difference is that, once the divergence-free velocity field is obtained, it is used to update the term $\mathbf{H}(\mathbf{U})$ and solve the pressure equation again. This cycle, also called the *inner corrector loop*, is repeated a finite number of times, as specified by the keyword `nCorrectors` (see Fig. 6.22). Once the number of repetitions is completed, the cycle proceeds as in the SIMPLE case.

The control parameters of the PISO cycle in OpenFOAM® are contained in the file `fvOptions`. This algorithm requires at least one correction; however, to improve stability and accuracy, the number of corrections can be increased by modifying the value associated with the keyword **nCorrectors**, as shown in Fig. 6.23. In the case of a non-orthogonal grid, and to further improve stability, the number of times the non-orthogonal correction cycle is executed can be increased by modifying the value associated with the keyword **nonOrthogonalCorrectors**. For orthogonal grids, this value can be set to zero, although setting it to 1 can help improve stability. For non-orthogonal grids, the value cannot be less than 1 and may need to be higher depending on the quality of the grid.

Through the value associated with the keyword **momentumPredictor**, it is possible to activate or inhibit the execution of the *momentum predictor* step necessary to calculate the first value of the term $\mathbf{H}(\mathbf{U})$ in the pressure equation. This step is activated in the case of flows with a high Reynolds number. Although it helps to stabilise the calculation, it is advisable to inhibit the execution of this step in the case of weakly convective flows (i.e. those with a low Reynolds number). If this step is executed, it is necessary to specify, in the `fvOptions` file, the linear solver to use for all quantities ***.Final**. For the same quantities, the value of the under-relaxation factor should be specified if under-relaxation is necessary.

In the case of non-stationary calculations, it is useful to stabilise the solution process by reducing the time integration step size, possibly also acting on the constraint

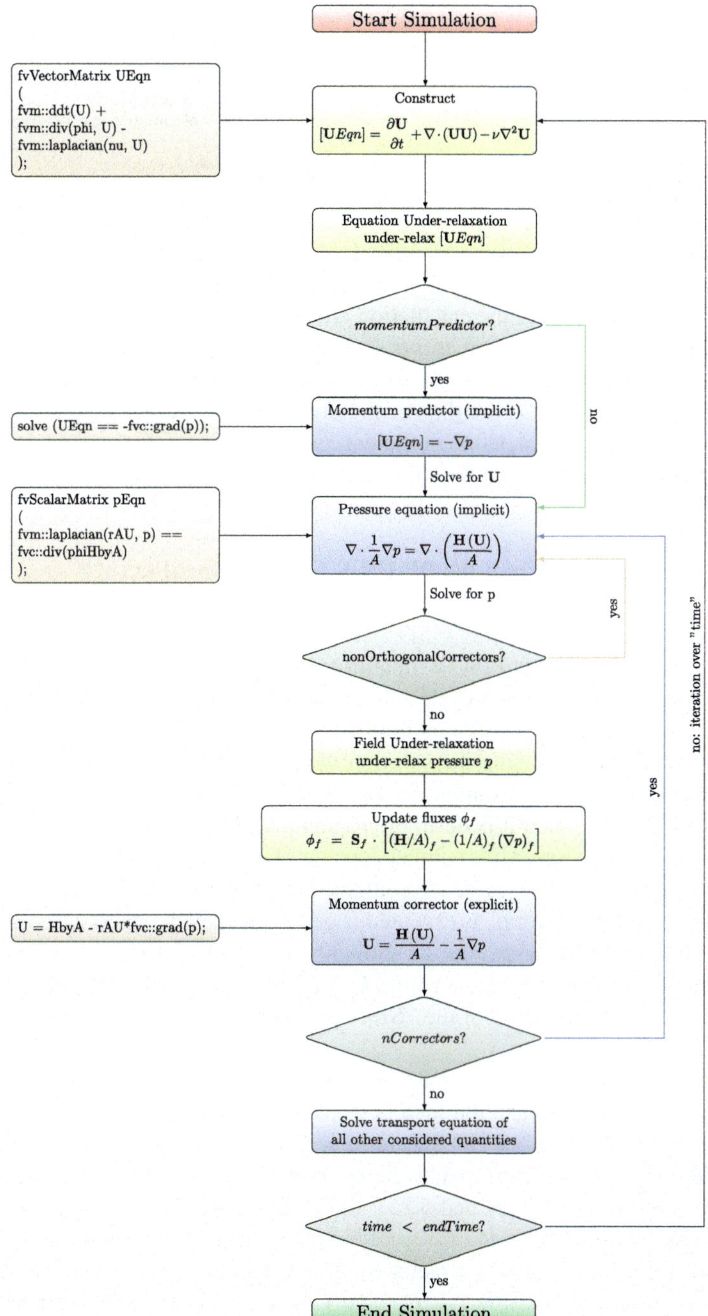

Fig. 6.22 Flowchart of the PISO cycle

Fig. 6.23 Control parameters of the PISO cycle

```
PISO
{
 momentumPredictor yes;
 nCorrectors 2;
 nNonOthogonalCorrectors 1;
}
```

related to the maximum Courant number (see Sect. 6.4.2) achievable during the calculation. Both the value of the time integration step size and the maximum Courant number achievable during the calculation are specified in the `controlDict` file. Reducing these two values increases the value of the time derivative term ($\Delta\phi/\Delta t$), which, in the discretisation process, always appears within the elements of the main diagonal of the coefficients matrix. As already seen, increasing the value of the elements of the main diagonal of the coefficients matrix increases its diagonal dominance and, therefore, the stability of the solution process.

6.4.1.2 Implementation of the PIMPLE Cycle in OpenFOAM®

This algorithm (also known as PISO with iterative marching—PISO-ITA) differs from PISO (also known as PISO with non-iterative marching—PISO-NITA) due to the presence of an additional cycle indicated as the "SIMPLE Loop". Figure 6.24 presents the PIMPLE algorithm as implemented in OpenFOAM®. The additional cycle is useful for stabilising the solution of cases dealing with the simulation of complex phenomena (for example, combustion or discontinuities). The additional cycle is also useful when it is desirable to maintain large time steps, although it is recommended not to have CFL numbers greater than 2. In OpenFOAM®, the control parameters of the PIMPLE cycle are contained in the file `fvOptions`. The number of times the "SIMPLE Loop" is executed is determined by the value associated with the keyword **nOuterCorrectors**, as shown in Fig. 6.25. Setting this value to 1 is equivalent to using the PISO algorithm. Given the considerable computational load resulting from the execution of the "SIMPLE Loop", usually, the value of 3 is not exceeded.

Also for this algorithm, as in the previous two SIMPLE and PISO, it is possible to make use of under-relaxation factors to increase the diagonal dominance of the coefficients matrix. Like in PISO, by the value associated with the keyword **momentumPredictor**, it is possible to activate or inhibit the execution of the *momentum predictor* step (see Figs. 6.24 and 6.25). In the case where such a step is executed, it is necessary to specify, in the `fvOptions` file, the linear solver to use as well as the value of the under-relaxation factor for all ***.Final** quantities. If under-relaxation is not to be used, it is possible to leave empty the area related to the keyword **relaxationFactors**. Otherwise, the syntax to set the same under-relaxation factor for all quantities is shown in Fig. 6.26. Figures 6.27 and 6.28 show acceptable values of under-relaxation factors if SIMPLE or SIMPLEC formulation is used, respectively.

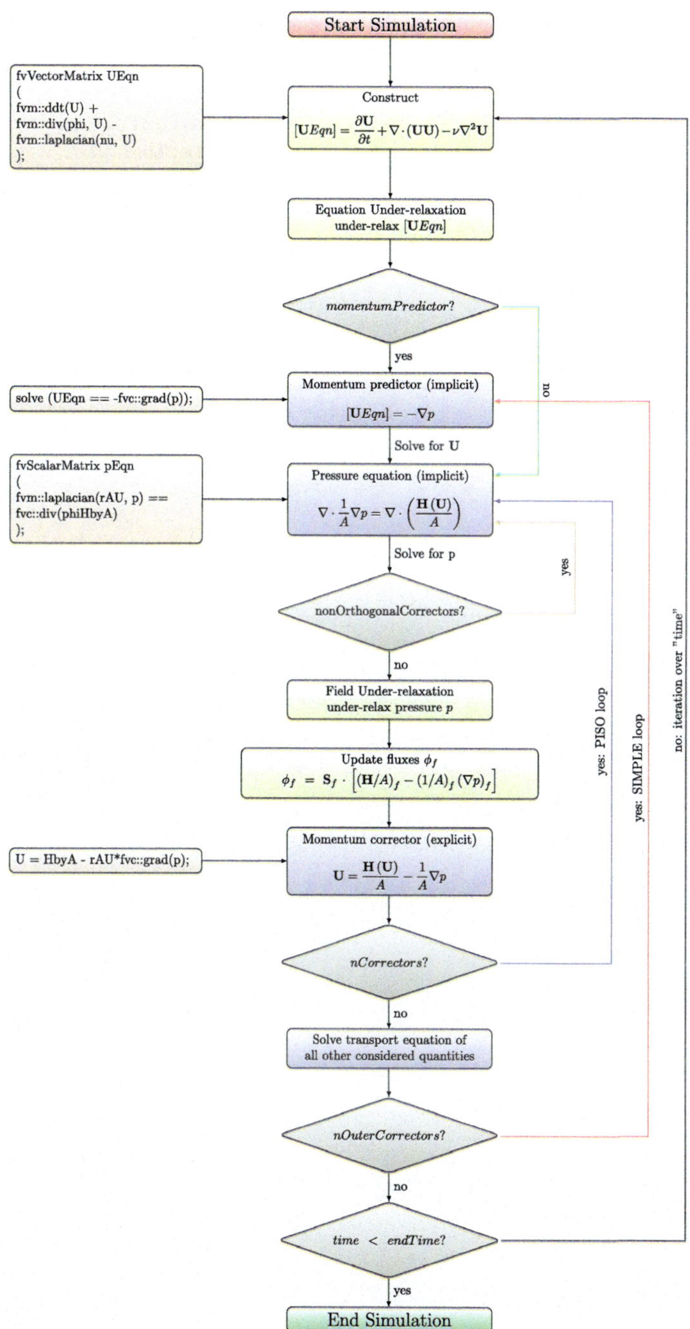

Fig. 6.24 Flow diagram of the PIMPLE cycle

Fig. 6.25 Control parameters of the PIMPLE cycle

```
PIMPLE
{
 momentumPredictor yes;
 nOuterCorrectors 1;
 nCorrectors 2;
 nNonOthogonalCorrectors 1;
}
```

Fig. 6.26 Setting of the under-relaxation factors

```
relaxationFactors
{
        fields
        {
           ".*"          1.0;
        }
        equations
        {
           ".*"          1.0;
        }
}
```

Fig. 6.27 Possible values of the under-relaxation factors for SIMPLE formulation

```
relaxationFactors
{
        fields
        {
           "p.*"      0.3;
        }
        equations
        {
           "U.*"          0.7;
           "k.*"          0.7;
           "omega.*"      0.7;
        }
}
```

Fig. 6.28 Possible values of the under-relaxation factors for SIMPLEC formulation

```
relaxationFactors
{
        fields
        {
           "p.*"      0.3;
        }
        equations
        {
           "p.*"          0.7;
           "U.*"          0.7;
           "k.*"          0.7;
           "omega.*"      0.7;
        }
}
```

6.4.2 *The Courant Number*

In transient simulations, the calculation is conducted at successive time values. It has been seen that the interval defined by two successive times cannot assume any values but must respect the *stability criterion of Courant-Friedrichs-Lewy*, which is a necessary, but not sufficient, condition for the numerical convergence of the solution of some partial differential equations (usually, hyperbolic equations). According to this criterion, to determine the amplitude of a wave that crosses the computational domain by calculating its value at successive times, the interval determined by such two times should not be greater than the time the wave takes to cross the single cell. In this regard, considering for simplicity a one-dimensional computational domain discretised with cells of the same size, it is possible to define the *Courant number* Co as

$$Co = \frac{u\Delta t}{\Delta x}$$

where u is the flow velocity (the wave propagation velocity), Δt is the time interval, and Δx is the extension of the cell. This number expresses the ratio between the space travelled by the flow with velocity u in time Δt and the extension Δx of the cell. The stability criterion of Courant-Friedrichs-Lewy—*CFL condition*—requires that $Co \leq 1$ in the case of explicit schemes, while for implicit schemes, this value can be greater when an accurate description of the transient is not required. Therefore, given a generic computational grid, the CFL condit

6.4.2.1 Expression of the Courant Number

In two- or three-dimensional computational domains, the calculation of the Courant number is defined as follows. The length Δx is calculated as the ratio between the volume V_P—area in the two-dimensional case—of the cell and the sum of the areas A_f—lengths in the two-dimensional case—of the faces that define the cell.

$$\Delta x = \frac{V_P}{\sum_f A_f}.$$

As for the velocity, the only component to consider is the one normal to the face. Adding the velocity contribution of all the faces bounding the cell would result in a zero value due to the validity of the mass conservation equation. For this reason, the absolute value of the contribution of each face is summed, and the resulting value is halved. By doing so, the following formula is obtained for the calculation of the Courant number:

$$C_o = \frac{1}{2}\Delta t \frac{\sum_f |\mathbf{U}_f \cdot \mathbf{n}_f| A_f}{V_P}$$

where $\mathbf{U}_f$ is the velocity vector at the centroid of the generic face, and $\mathbf{n}_f$ is the unit vector normal to the corresponding face.

6.5 Residual and Tolerances

As already seen in Sect. 4.5, residuals are a measure of the error made when iteratively solving a system of equations. Before solving the matrix equation resulting from the application of the discretised form of the considered conservation equation, in OpenFOAM®, the residual is calculated based on the known values of the quantities involved. At the end of each iteration for the solution of the system, the calculation of the residual is performed. Considering the system resulting from the application of the momentum conservation equation for the only component along x to all the cells of the computational domain, the residual is calculated as

$$r_x = \frac{\|\mathbf{M}_x \mathbf{U}_x - \mathbf{b}_x\|}{\|\mathbf{M}_x \mathbf{U}_x - \mathbf{M}_x \overline{\mathbf{U}}_x\| + \|\mathbf{M}_x \overline{\mathbf{U}}_x - \mathbf{b}_x\|}. \tag{6.21}$$

The numerator of Eq. 6.21 includes the norm of the difference between the left and right-hand sides of Eq. 6.20. If the found solution were exact, this value would be zero. At the denominator of Eq. 6.21, there is a dimensionless factor ensuring that the calculated value for the residuals is not dependent on the scale (i.e., geometric dimensions, module of characteristic velocities, etc.) of the generic problem addressed. Indicating with the symbol $\overline{\mathbf{U}}_x$ the average of $\mathbf{U}_x$ calculated considering all the cells of the computational domain, and remembering that the norm $\|\mathbf{b}_x\|$ of the column matrix $\mathbf{b}_x$ is equal to the sum of the absolute values of each of its elements, the use of the dimensionless factor also ensures that the residual is a single number rather than a vector of size equal to the total number of cells with which the computational domain has been discretised. The iterative process of solving the considered system stops in OpenFOAM® if one of the following conditions is verified:

- the value of the residual is less than that set for the parameter `tolerance`;
- the value of the ratio between the current residual and that at the previous iteration is less than that set for the parameter `relTol`;
- the number of iterations performed is greater than that set for the parameter `maxIter`.

As seen before, when using algorithms such as PISO or PIMPLE, the matrix equation associated with a generic quantity can be solved multiple times within the same iteration of the algorithm (PISO/PIMPLE), according to the value assigned to the parameter `nCorrectors`. In these cases, it is necessary to set specific values for the stop parameters for the last step of the solution. An example of such a setting, relating to the pressure solution equation—similarly it proceeds for velocity—is shown in Fig. 6.29, in which the tolerance parameters for the last execution of the iterative process are specified in There is a specific section in the `fvSolution` file

Fig. 6.29 Tolerance parameters for the last execution of the iterative process

```
p
{
    solver PCG;
    preconditioner DIC;
    tolerance  1e-6;
    relTol       0.05;
}
p.Final
{
    $p;
    relTol    0;
}
```

to specify the tolerance values of the residuals for the final iteration solution. The name of this section is the same as the section relating to the intermediate iterations, with the addition of the suffix `Final`. If the value associated with the parameter `nCorrectors` is 4, the instructions in Fig. 6.29 impose that:

- the first three solutions of the matrix equation of the pressure can be stopped at a reduced computational cost for a value of `relTol` equal to 0.05;
- for the last solution, greater accuracy is required and, with higher computational costs, it will necessarily have to reach the value of the residual equal to $1e - 06$ since the value 0 deactivates the parameter `relTol`.

Chapter 7
Boundary Conditions

Incorrect imposition of boundary conditions can lead to computational instability, lack of convergence, and incorrect or inaccurate results. Boundary conditions must be specified consistently with various characteristics of the flow, such as the velocity regime at input and output, interactions with viscous walls, etc. A fundamental aspect is understanding the mathematical characteristics of the equations that describe the flow under consideration, as described in Sect. 1.3. These mathematical characteristics define how flow properties and disturbances are transported both within the computational domain and across its boundaries. Therefore, understanding how information propagates in the flow, which aspects enter and which exit through the boundaries, is fundamental in selecting the most appropriate boundary conditions consistent with the physical characteristics of the flow and the mathematical equations that represent it. In the case of incompressible flow, it has been observed that information propagates equally in all directions. Conversely, to understand the fundamental concepts underlying the correct imposition of boundary conditions in compressible flow, it is necessary to briefly describe the *Riemann problem of gas dynamics*. The Riemann problem is an initial value problem for the Euler equations; in the one-dimensional case, the initial condition consists of a jump in the variables between two states, with a uniform distribution on the left of the discontinuity and another equally uniform distribution on the right (see Fig. 7.1). The solution of the Riemann problem depends on the values of the variables in the left and right states. The solution generally consists of three distinct waves that propagate at specific speeds. In the shock tube problem, the initial discontinuity evolves into three waves. The intermediate wave is a contact discontinuity that propagates at the local fluid velocity u. The other two waves propagate at velocities $u - c$ and $u + c$, where c is the speed of sound in the fluid under consideration. Referring to Fig. 7.2, in the simplest case where the local fluid velocity u is zero, there will be a rarefaction wave propagating in the gas towards the left and a shock wave propagating to the right.

G. Caramia and E. Distaso, *A Practical Approach to Computational Fluid Dynamics Using OpenFOAM®*, https://doi.org/10.1007/978-3-031-88957-8_7

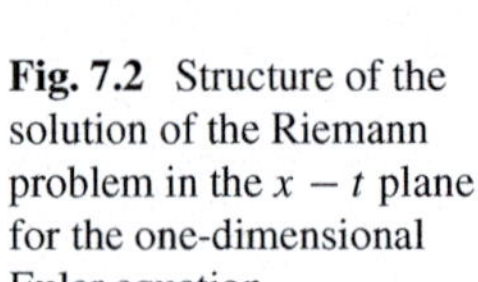

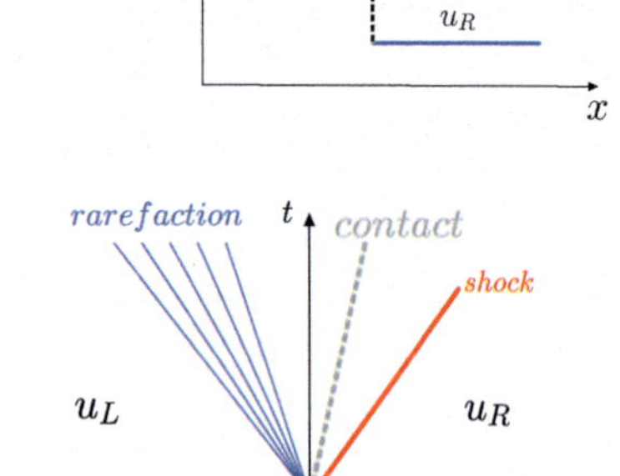

Fig. 7.1 Initial condition of the Riemann problem

Fig. 7.2 Structure of the solution of the Riemann problem in the $x-t$ plane for the one-dimensional Euler equation

Each of these waves follows a characteristic curve (see Sect. 1.3.3): C_0 for the contact discontinuity, C_- for the wave propagating at velocity $u-c$, and C_+ for the wave propagating at velocity $u+c$.

To apply these concepts to boundary condition imposition, a one-dimensional compressible flow is considered. Referring to Figs. 7.3, 7.4, and 7.5, the characteristic C_- is *negative* (residing in the negative semi-plane of the abscissas) if the flow is subsonic and *positive* (residing in the positive semi-plane of the abscissas) if the flow is supersonic. At the inlet, the characteristics C_0 and C_+ have slopes u and $u+c$, respectively, and are always positive for a flow directed in the positive abscissa direction. Therefore, these two characteristics convey information from the exterior to the interior of the computational domain through the boundary. The third characteristic, C_-, has a sign that depends on the Mach number: it is positive for a supersonic inlet flow and negative for a subsonic inlet flow. It follows that, for subsonic inlet flows, information related to the characteristic C_- cannot be specified *a priori* in the boundary condition at the inlet. Similar considerations apply to the outlet boundary, where no conditions need to be imposed for the characteristics C_+ and C_0. For C_-, a condition must be imposed if the exit flow is subsonic, whereas no condition

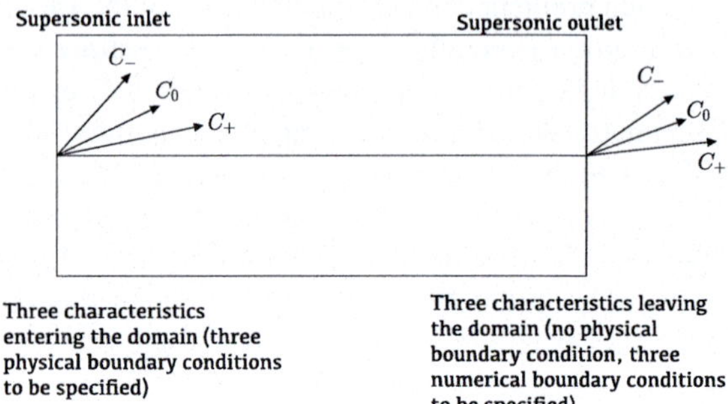

Fig. 7.3 Boundary conditions for one-dimensional supersonic flow

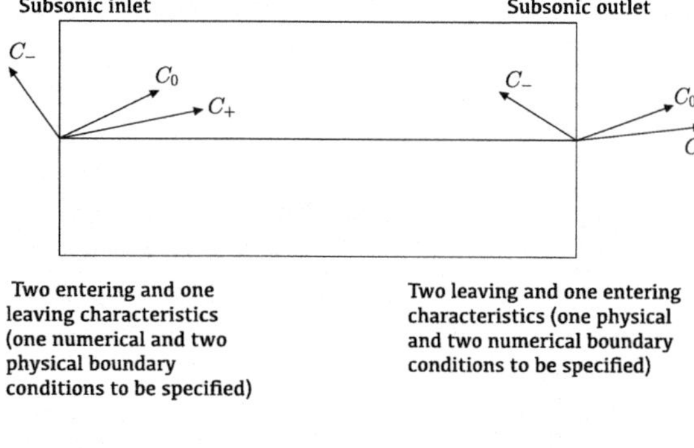

Fig. 7.4 Boundary conditions for one-dimensional subsonic flow

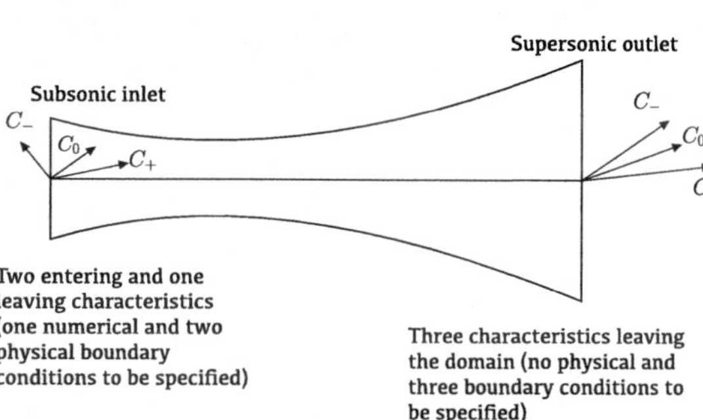

Fig. 7.5 Boundary conditions for variable section unidimensional flow

is required if the flow is supersonic. Each characteristic conveys information about a single variable, and only the variables transported through the boundaries into the domain define a physical boundary condition. The remaining variables, transported out of the computational domain, are determined by the computed flow itself. In this case, numerical boundary conditions are considered, where the necessary information is obtained by extrapolation from the downstream flow (for an inlet boundary) or from the upstream flow (for an outlet boundary). A *boundary cell* is a cell with at least one face positioned on the boundary of the computational domain; such a face is referred to as a *boundary face*.

7.1 Boundary Conditions for Incompressible Flow

The conservation equations of momentum (Eq. 6.5)

$$\frac{\partial \mathbf{U}}{\partial t} + \nabla \cdot (\mathbf{U}\mathbf{U}) - \nu \nabla^2 \mathbf{U} = -\nabla p$$

and pressure (Eq. 6.11)

$$\nabla^2 p + \nabla \cdot [\nabla \cdot (\mathbf{U}\mathbf{U})] = 0$$

as reported for simplicity, describe the motion of an incompressible flow. The pressure equation combines the conservation equations of momentum and mass and does not contain any time derivative term, thus highlighting that, for incompressible flows, the speed of sound is infinite ($c = \infty$). This means that quantities throughout the computational domain are instantaneously influenced by any disturbance at any point in the domain. Specifically, the term $\nabla^2 p$ describes the instantaneous pressure change across the computational domain in response to a change in velocity at any single point. In turn, the pressure change causes an instantaneous change in the velocity field across the domain, which also varies (on finite time scales) due to advection and diffusion.

In transport phenomena, modelling advection terms requires the calculation of the face value (ϕ_f) by interpolating between the values (ϕ) of the same quantity at the centres of the two cells sharing the face. For boundary cells at the inlet, there is no second cell centre through which to determine the value of the quantity on the boundary face. Thus, from a numerical perspective, it is straightforward to set a constant value for the considered quantity at these faces. In this context, for incompressible flow, the behaviour of the pressure is particular because the instantaneous propagation of pressure disturbances results in the absence of advection. As a consequence, it is not necessary to discretise the advective term for pressure (as it is absent) and therefore calculate the pressure at the boundary face of the inlet cell by interpolation. For these reasons, it is preferred not to fix the value of pressure at the inlet of the computational domain. Instead, the pressure gradient is set at the inlet. For the faces of the outlet boundary, the opposite is true: the pressure value is set, while the gradient value is set for all other quantities.

- velocity, and possibly temperature, are imposed at the inlet;
- for boundaries with a symmetry condition, the gradient in the direction normal to the boundary is set to zero for all scalar quantities. Additionally, the component of velocity in the direction parallel to the boundary is set to zero, as is the component of velocity normal to the boundary;
- for boundaries with a wall condition, velocity is set to zero, and the normal stress at the wall, temperature, and heat flow are prescribed as constant;
- pressure is imposed on the exit boundary;
- zero gradient is imposed for the remaining quantities at the outlet.

The imposition of a velocity value at both the inlet and the outlet results in a numerically unstable system.

7.1.1 *The Relative Nature of Pressure*

In incompressible cases where none of the boundaries have a specified pressure value, a difficulty arises due to the relative nature of the pressure. This is because, in the momentum conservation equation, only pressure gradients appear. In this case,

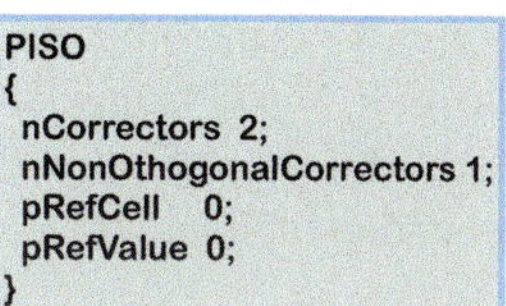

```
PISO
{
 nCorrectors 2;
 nNonOthogonalCorrectors 1;
 pRefCell  0;
 pRefValue 0;
}
```

Fig. 7.6 Example of setting parameters related to the reference pressure

pressure differences are meaningful, whereas absolute pressure values are irrelevant. This indeterminacy results in a singular coefficient matrix. Because the coefficient matrix is singular, it is not invertible, leading to the failure of the simulation process. This problem is resolved by setting the pressure value at a point within the domain. This ensures that the pressure values at all other points in the computational domain are relative to the set value.

In OpenFOAM®, this value is set by inserting the entry `pRefValue` in the `fvSolution` file, within the zone where the parameters of the chosen pressure-velocity coupling strategy are specified. Figure 7.6 shows an example of how this parameter is set.

The parameter `pRefPoint/pRefCell` defines the position inside the computational domain at which the reference pressure is assumed to be equal to the value specified by `pRefValue`.

7.1.2 Inlet

At the inlet of the computational domain, three types of boundary conditions are typically used:

1. imposed velocity (magnitude and direction);
2. static pressure and direction of velocity imposed;
3. total pressure and direction of velocity imposed.

7.1.2.1 Imposed Velocity

The subscript b indicates the quantities referred to the boundary. In this case, it is:

- p_b, the pressure to be determined during the simulation execution based on the values inside the computational domain;
- the flow rate $\dot{m}_b$, uniquely determined;
- the velocity $\mathbf{v}_b$, imposed.

The pressure at the boundary is extrapolated from the centroid C of the considered boundary cell:

$$p_b = p_C + \nabla p_C^{(n)} \cdot \mathbf{d}_{Cb}.$$

Being ∇p_C the pressure gradient calculated at the centre C of the considered boundary cell, $\nabla p_C^{(n)}$ is the component in the direction normal to the considered boundary face. $\mathbf{d}_{Cb}$ is the vector indicating the distance between the centre of the considered boundary cell and the centre of the corresponding boundary face.

7.1.2.2 Static Pressure and Velocity Direction Imposed

In this case, the pressure p_b and the versor $\mathbf{e}_v$ of the velocity vector are known because they are imposed. The magnitude of the velocity can be calculated using the continuity equation since the boundary pressure gradient is known, as the pressure on the boundary face and in the centroid of the actual cell are known. The velocity at the boundary is recalculated at each iteration, consistently modifying the coefficients of the momentum equation as in the previous case.

7.1.2.3 Total Pressure and Velocity Direction Imposed

In this case, the velocity and the pressure at the boundary are not known, although they are linked by the total pressure expression:

$$p_0 = \underbrace{p}_{static\ pressure} + \underbrace{\frac{1}{2}\rho \mathbf{v} \cdot \mathbf{v}}_{dynamic\ pressure} \quad . \tag{7.1}$$

The mass flow is calculated using the continuity equation. Knowing the pressure from the initial condition or from the value obtained at the previous iteration, the velocity is obtained from Eq. 7.1. The velocity is then treated as a condition of fixed velocity (that is, Dirichlet boundary condition) by consistently modifying the coefficients in the momentum equation.

7.1.3 Outlet

Three types of outlet boundary conditions will be considered:

1. imposed static pressure;
2. imposed flow rate;
3. fully developed flow.

7.1.3.1 Imposed Static Pressure

In this case, the pressure at the boundary p_b is known because it is imposed while velocity and flow rate are not known. It is assumed that conditions of fully developed flow and therefore zero velocity gradient in the direction normal to the considered boundary are verified. Therefore, the velocity at the boundary is assumed to be equal to that at the centroid of the considered boundary element.

7.1.3.2 Imposed Flow Rate

In this case, the flow rate is known because it is imposed, while the velocity and the pressure at the boundary p_b are not known. Given the incompressibility of the flow, imposing the flow rate is equivalent to imposing the component of the velocity normal to the boundary. It is assumed that the direction of the velocity is the same as that at the centroid of the actual boundary cell.

7.1.3.3 Fully Developed Flow (Outflow)

In this case, the velocity gradient in the normal direction is assumed to be zero at the outlet, so the velocity at the outlet is assumed to be equal to that at the centroid of the considered boundary element. The value of the pressure at the boundary is extrapolated from the pressure values and pressure gradient in the centroid of the considered boundary element. Particular attention must be paid to the use of this boundary condition in relation to the positioning of the boundary itself with respect to the gradients of the velocity. As shown in Fig. 7.7, the assumption of zero velocity gradient in the direction normal to the outlet would be incorrect in the case where the exit boundary is positioned in correspondence with Sects. 1 or 2.

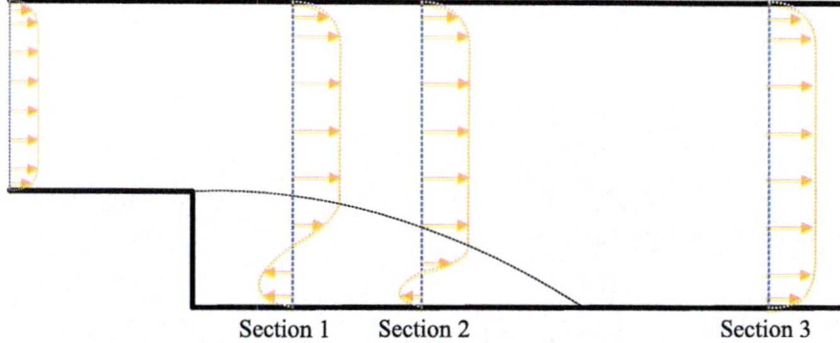

Fig. 7.7 Positioning of the boundary with respect to the gradients of the quantities representative of the considered flow

7.2 Boundary Conditions for Compressible Flow

The flow at the inlet can be subsonic or supersonic; the nature of equations changes depending on the considered case, going from elliptical to hyperbolic respectively. The approach used to solve these equations will be consistently different.

7.2.1 *Subsonic Inlet*

In the case of a subsonic inlet, three types of boundary conditions will be considered:

1. fixed velocity;
2. static pressure and fixed velocity direction;
3. fixed velocity direction and total pressure.

The last type of boundary condition should be used only when the flow within the domain becomes supersonic.

7.2.1.1 Fixed Velocity

In this case it will be: p_b to be calculated, $\dot{m}_b$ imposed, $\mathbf{v}_b$ imposed. Differently from the incompressible case, the density depends on the pressure so the flow rate remains unknown. The pressure at the boundary is therefore firstly calculated by extrapolation as done for the incompressible case and then, from this, the density is deduced.

7.2.1.2 Static Pressure and Imposed Velocity Direction

The implementation is similar to that of the incompressible case.

7.2.1.3 Total Pressure and Imposed Velocity Direction

In this case, the pressure and velocity are unknown although they result linked by the definition of total pressure:

$$p_{0,b} = p_b \left(1 + \frac{\gamma - 1}{2} M_b^2\right)^{\frac{\gamma}{\gamma-1}} \tag{7.2}$$

where the subscript b is used to refer to the conditions on the boundary, while γ is the ratio of specific heats. Finally, M_b is the Mach number at the boundary:

$$M_b = \sqrt{\frac{\mathbf{v}_b \cdot \mathbf{v}_b}{\gamma R T_b}}. \tag{7.3}$$

Similarly to what has already been done in the incompressible case, it is possible to obtain the value of M_b by setting the value of $p_{0,b}$, knowing the value of p_b from the initial conditions or from the previous iteration, and using Eq. 7.2. Equation 7.3 gives the value of $\mathbf{v}_b$ since the value of T_b is obtained from the initial conditions or from the previous iteration. It should be mentioned that the boundary condition for the energy equation at the inlet consists of specifying the static temperature T_b or the total temperature $T_{0,b}$. When the total temperature is specified, at each iteration the value of the static temperature is calculated using the formula that defines the total temperature:

$$T_{0,b} = T_b + \frac{\mathbf{v}_b \cdot \mathbf{v}_b}{2c_p}.$$

7.2.2 *Supersonic Inlet*

In this case, it is necessary to specify the value of all variables: pressure, velocity, and temperature.

7.2.3 *Subsonic Outlet*

In the case of a subsonic exit, two types of boundary conditions are considered:

1. imposed static pressure;
2. imposed flow rate.

7.2.3.1 Fixed Static Pressure

In this case, flow rate and velocity must be determined. Assuming to be zero the value of the velocity gradient, it is possible to extrapolate its value at the boundary from the centroid of the cells inside the domain. Even for the calculation of the density (and therefore the flow rate) a constant gradient (that is, Neumann) condition is applied to the energy equation.

7.2.3.2 Fixed Flow Rate

Given the flow rate, it is possible to obtain pressure, velocity and temperature by applying the Neumann boundary condition of constant gradient to the energy equation.

7.2.4 *Supersonic Outlet*

In this case, the value of none of the variables pressure, density, temperature, velocity has to be specified; such values are extrapolated from the values in the centroids of the boundary cells of the computational domain.

7.3 Boundary Conditions Available in OpenFOAM®

Some of the boundary conditions available in OpenFOAM® will be briefly described.

7.3.1 *Imposition of the Value and Gradient of a Quantity at the Boundary*

As seen in Chap. 4, the finite volume method leads to writing a discretised conservation equation for each grid cell. Grouping the equations of all the cells results in a system that, in its matrix form can be written as $\mathbf{A}\phi = \mathbf{b}$. In Chap. 3, it was seen that the discretised general transport equation contains two elements involving the computation of the value of the quantity considered at the centre of the face shared by two cells:

- the advection term.[1] In the case of constant density, the advection term is $\nabla \cdot (\mathbf{u}\phi)$ which, discretised becomes $\sum_f \mathbf{S}_f \cdot \mathbf{u}_f \phi_f$. When the density is not constant, the advection term is $\nabla \cdot (\rho\mathbf{u}\phi)$ which, discretised becomes $\sum_f \mathbf{S}_f \cdot (\rho\mathbf{u})_f \phi_f$;
- the Laplacian term. In the case of constant density, the Laplacian term is $\nabla \cdot (\Gamma\nabla\phi)$ which, discretised becomes $\sum_f \left|\mathbf{S}_f\right| \Gamma_f \nabla_n \phi_f$. When the density is not constant, the Laplacian term is $\nabla \cdot (\rho\Gamma\nabla\phi)$ and discretised becomes $\sum_f \left|\mathbf{S}_f\right| (\rho\Gamma)_f \nabla_n \phi_f$ in which the symbol $\nabla_n \phi_f$ represents the component normal to the face of the gradient of the quantity ϕ_f.

From what has just been observed, it is clear that, for boundary cells, it is necessary to impose, at the corresponding face, both the value of ϕ_f (to allow the calculation of the advective term) and $\nabla_n \phi_f$ (to allow the calculation of the Laplacian or diffusive term). This process is called *boundary conditions imposition*. The boundary condition that specifies the imposition of a specific value ϕ_b for the considered quantity is called the *Dirichlet condition*. In OpenFOAM®, the Dirichlet boundary condition is indicated by the term *fixed value*. The boundary condition that specifies the imposition of a particular value for $\nabla_n \phi_b$ (the component normal to the face of the gradient of

[1] Volumetric $\mathbf{S}_f \cdot \mathbf{u}_f$ and mass $\mathbf{S}_f \cdot (\rho\mathbf{u})_f$ flows are indicated in OpenFOAM® with the symbol ϕ_f while here the symbol ϕ_f represents the value of the quantity considered at the centre of the face (see also Sect. 6.1.2).

the considered quantity) is called the *Neumann condition*. In OpenFOAM®, it is indicated by the term *fixed gradient*.

For the calculation of the advective term in the case of the *fixed value* boundary condition, $\phi_f = \phi_b$ is set. When the *fixed gradient* boundary condition is imposed, the value of ϕ_f on the boundary face is set by extrapolating the value of the quantity from the cell centre using the gradient $\nabla_n \phi_b$ set in the boundary condition. For the calculation of the Laplacian term, in the case of the *fixed value* boundary condition, the value of $\nabla_n \phi_b$ is based on the value assumed by the quantity in the cell centre and the value imposed for the quantity on the face by the boundary condition. In the case of the *fixed gradient* boundary condition, $\nabla_n \phi_f = \nabla_n \phi_b$ is set. An example of the use of these two types of boundary conditions is the case of incompressible flow within a duct. In this case, the imposed boundary conditions are:

- at the inlet: zero value for the gradient (*zeroGradient*) of the pressure and fixed value for all other quantities;
- at the outlet: fixed value for the pressure and zeroGradient for all other quantities.

7.3.2 Inlet-Outlet

There may be cases where it is not possible to uniquely define the flow as outgoing or incoming at the boundary, as shown in Fig. 7.8, where:

- at the upper boundary, the flow is partly outgoing and partly incoming in the computational domain;
- at the thin boundary on the left, the flow is incoming in the computational domain;
- at the thin boundary on the right, the flow is outgoing from the computational domain;
- the remaining boundaries are walls and are not crossed by flow.

Naming `atmosphere` as the upper boundary in Fig. 7.8, in the `p` configuration file for the boundary and initial conditions of pressure, the lines shown in Fig. 7.9 will appear. This boundary condition will ensure that, for those faces where the flow is outgoing, the pressure value will be fixed and equal to that specified by the word `value`. For those faces where the flow is incoming with velocity **U**, the pressure value is calculated according to the following formula:

$$p = p_0 - \frac{|\mathbf{U}|^2}{2}.$$

Since the flow is incompressible, the software identifies p as the kinematic pressure, which is the ratio between pressure and density, explaining the absence of the density term on the right-hand side of the above formula.

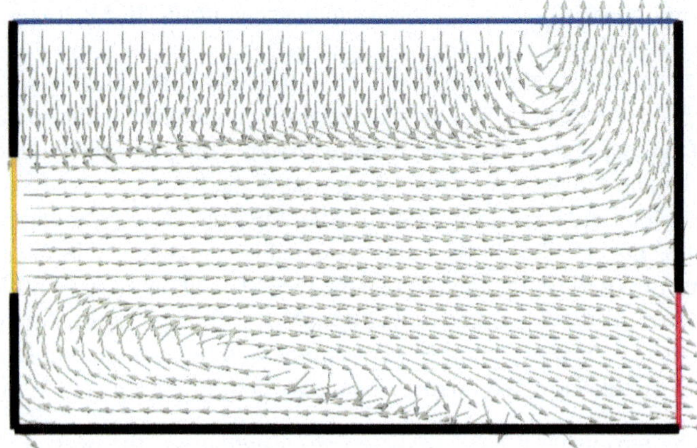

Fig. 7.8 Computational domain with a boundary characterised by partly incoming and partly outgoing flow

Fig. 7.9 Example of total pressure boundary condition configuration

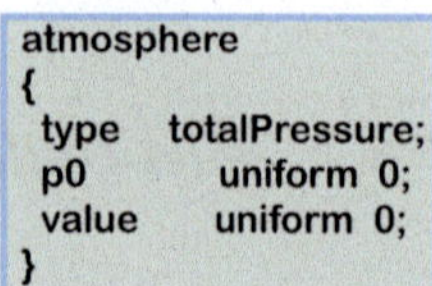

```
atmosphere
{
 type    totalPressure;
 p0         uniform  0;
 value      uniform  0;
}
```

Fig. 7.10 Example of boundary condition configuration for velocity

```
atmosphere
{
 type    pressureInletOutletVelocity;
 value       uniform  (0 0 0);
}
```

In the U configuration file for the boundary and initial conditions of velocity, the lines shown in Fig. 7.10 will appear. This boundary condition will ensure that, for those faces where the flow is outgoing, a zero gradient boundary condition is set for the velocity; for those faces where the flow is incoming, a zero value will be set for the velocity component parallel to the boundary and a zero gradient for the velocity component orthogonal to the boundary.

The tutorial `incompressible/pimpleFoam/RAS/flowWithOpenBoundary`, available in the version downloaded from the website www.openfoam.org, demonstrates how to handle a boundary where the flow is

Fig. 7.11 Example of boundary condition configuration for the temperature

```
atmosphere
{
 type          inletOutlet;
 inletValue  uniform 0;
 value          uniform  0;
}
```

partly outgoing and partly entering the computational domain. In this tutorial, the transport of temperature is also considered. In the `T` configuration file for the boundary and initial conditions of temperature, the lines shown in Fig. 7.11 appear. This boundary condition ensures that, for those faces where the flow is outgoing, a boundary condition with zero gradient (`zeroGradient`) is set for the temperature. For those faces where the flow is incoming, a zero value will be set for the temperature. For any other scalar quantity involved in the simulation, this type of boundary condition must be used.

Chapter 8
Turbulence

Turbulence is a phenomenon that arises from the instability of laminar flow caused by the amplification of disturbances due to strongly non-linear inertial effects.

The universally accepted theory is that developed by Kolmogorov, known as the *energy cascade*. According to this theory, a turbulent flow consists of vortices of various sizes, each associated with a different energy level. Figure 8.1 presents a schematic, referred to as the *energy density spectrum*, which illustrates the energy levels of turbulent vortices as a function of the inverse of their sizes.

The energy level of turbulent vortices is represented by the turbulent kinetic energy density, denoted by the symbol E. The inverse of the size of turbulent vortices is called the *wave number* and is typically denoted by the symbol κ. Although very similar, this symbol differs from k, which represents turbulent kinetic energy. It may be useful here to distinguish between the concept expressed by the word *vortex* and that of the word "*eddy*" (*turbulent vortex*). In the context of turbulence, we generally refer to eddies where turbulent vortices break down into smaller turbulent vortices, giving rise to the phenomenon known as the *energy cascade*. The term vortex is used when describing more stable structures whose physics does not necessarily involve turbulence, meaning they are not expected to decay into smaller vortices.

The size of larger turbulent vortices is typically comparable to the dimensions of the objects that confine the flow. These larger eddies progressively break down into smaller eddies with lower energy levels, continuing this process until they reach scales small enough that molecular viscosity can dissipate kinetic energy into thermal energy. At the smallest scales, local velocity gradients become sufficiently high to generate significant viscous stress, even when fluid viscosity is low (see the concept of turbulent kinetic viscosity discussed below).

The Navier–Stokes equations can describe the energy cascade phenomenon, provided that temporal and spatial integration scales are chosen to resolve phenomena at the smallest relevant scales. This approach is known as *Direct Numerical Simulation* (DNS).

G. Caramia and E. Distaso, *A Practical Approach to Computational Fluid Dynamics Using OpenFOAM®*, https://doi.org/10.1007/978-3-031-88957-8_8

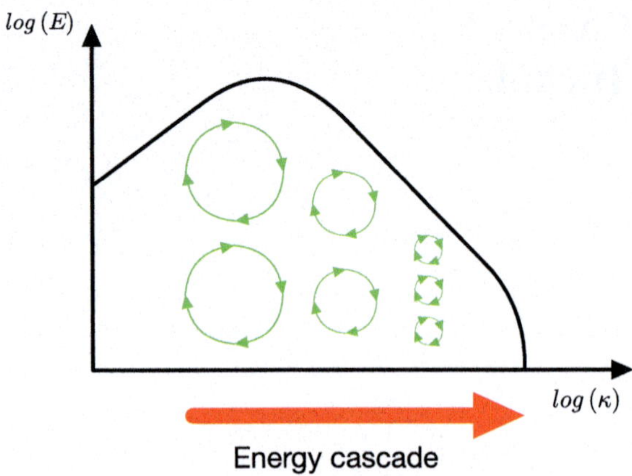

Fig. 8.1 Kolmogorov's energy density spectrum

The DNS approach entails computational costs that are prohibitive in most practical cases due to the need for extremely small temporal and spatial integration intervals.

The random fluctuations inherent in turbulence, combined with the wide range of temporal and spatial scales involved, have necessitated the use of statistical analysis techniques to mitigate the challenges of DNS. One such technique is known as *Large Eddy Simulation* (LES), which identifies larger vortices through spatial statistical analysis and simulates them, while modelling the smaller vortices.

Computational cost can be further reduced using the *Reynolds Averaged Navier–Stokes* (RANS) technique, which applies temporal rather than spatial statistical analysis. This remains the most widely used approach, as it provides acceptable results in most cases while requiring significantly less demanding computational grids and temporal integration intervals compared to DNS or LES.

8.1 Reynolds Averaged Navier–Stokes Approach

This technique is based on using the time-averaged values of velocity, pressure, and temperature in the Navier–Stokes equations. Referring to Fig. 8.2, these quantities are decomposed into their mean ($\overline{\mathbf{v}}$, $\overline{p}$, $\overline{T}$) and fluctuating ($\mathbf{v}'$, p', T') components:

$$
\begin{aligned}
\mathbf{v} &= \overline{\mathbf{v}} + \mathbf{v}', \\
p &= \overline{p} + p', \\
T &= \overline{T} + T', \\
\overline{\mathbf{v}} &= \overline{u}\mathbf{i} + \overline{v}\mathbf{j} + \overline{w}\mathbf{k}, \\
\mathbf{v}' &= u'\mathbf{i} + v'\mathbf{j} + w'\mathbf{k}.
\end{aligned}
$$

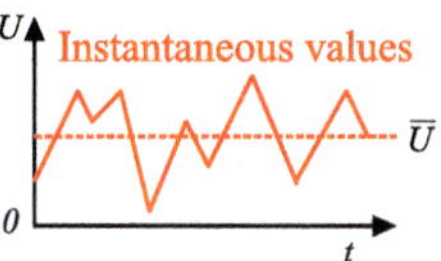

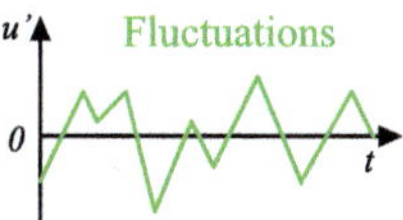

Fig. 8.2 Example of instantaneous velocity values, average velocity and fluctuations

We will now consider the Navier–Stokes equations written for the case of an incompressible flow with constant viscosity, and in the absence of body forces and energy sinks or sources.

$$\begin{aligned} \nabla \cdot (\rho \mathbf{v}) &= 0, \\ \frac{\partial}{\partial t}(\rho \mathbf{v}) + \nabla \cdot (\rho \mathbf{v}\mathbf{v}) &= -\nabla p + \nabla \cdot \boldsymbol{\tau}, \\ \nabla \cdot \left(\rho c_p T\right) + \nabla \cdot \left(\rho c_p \mathbf{v} T\right) &= \nabla \cdot (k \nabla T) + \rho T \frac{D c_p}{D t} \end{aligned}$$

The substitution of time-averaged values of velocity, pressure, and temperature leads to the *Reynolds-averaged Navier–Stokes (RANS) equations*:

$$\begin{aligned} \nabla \cdot (\rho \overline{\mathbf{v}}) &= 0, \\ \frac{\partial}{\partial t}(\rho \overline{\mathbf{v}}) + \nabla \cdot (\rho \overline{\mathbf{v}}\overline{\mathbf{v}}) &= -\nabla \overline{p} + \nabla \cdot \left(\overline{\boldsymbol{\tau}} - \rho \overline{\mathbf{v}'\mathbf{v}'}\right), \\ \nabla \cdot \left(\rho c_p \overline{T}\right) + \nabla \cdot \left(\rho c_p \overline{\mathbf{v}}\overline{T}\right) &= \nabla \cdot \left(k \nabla \overline{T} - \rho c_p \overline{\mathbf{v}'T'}\right) + \rho \overline{T} \frac{D c_p}{D t} \end{aligned} \tag{8.1}$$

These equations are similar to the original Navier–Stokes equations but differ in key aspects. In the case of the momentum conservation equation, an additional term, $\rho \overline{\mathbf{v}'\mathbf{v}'}$, appears alongside the stress tensor $\boldsymbol{\tau}$. This additional term is known as the *Reynolds stress tensor*.

$$\boldsymbol{\tau}^R = -\rho \overline{\mathbf{v}'\mathbf{v}'} = -\rho \begin{pmatrix} \overline{u'u'} & \overline{u'v'} & \overline{u'w'} \\ \overline{u'v'} & \overline{v'v'} & \overline{v'w'} \\ \overline{u'w'} & \overline{v'w'} & \overline{w'w'} \end{pmatrix}$$

which, in the three-dimensional case, introduces six new unknowns. Indeed, considering that

$$\overline{u'v'} = \overline{v'u'} \qquad \overline{u'w'} = \overline{w'u'} \qquad \overline{v'w'} = \overline{w'v'}$$

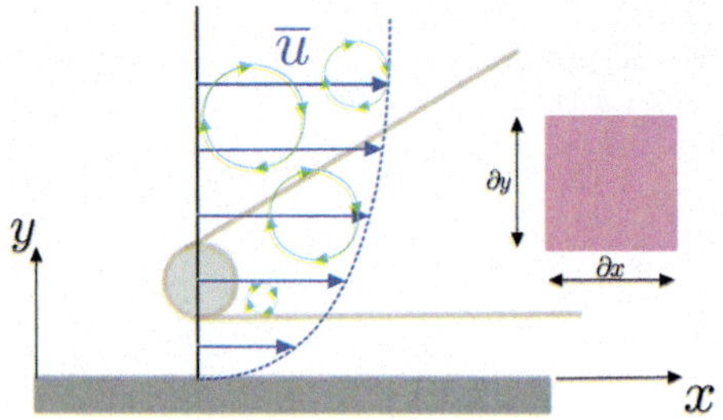

Fig. 8.3 A fluid element within the turbulent boundary layer of momentum

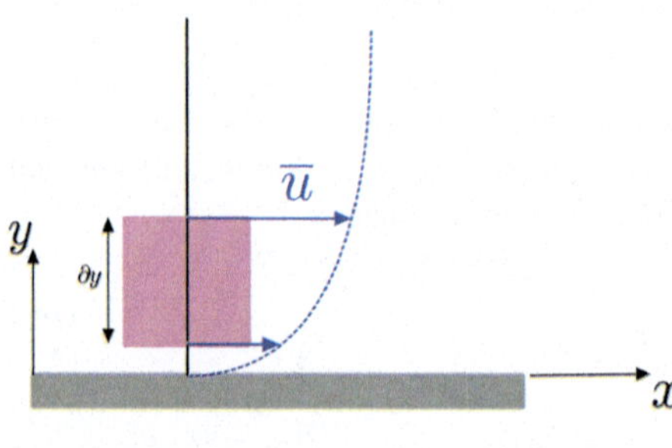

Fig. 8.4 Gradient of mean velocity

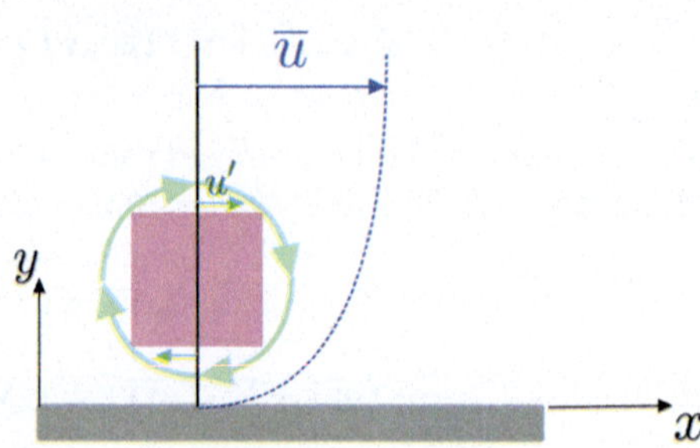

Fig. 8.5 Gradient of the fluctuating component of velocity

we obtain the symmetric tensor

$$\boldsymbol{\tau}^R = -\rho \begin{pmatrix} \overline{u'u'} & \overline{u'v'} & \overline{u'w'} \\ & \overline{v'v'} & \overline{v'w'} \\ & & \overline{w'w'} \end{pmatrix}.$$

To better understand the physical meaning of the Reynolds stress tensor, consider the fluid element within a two-dimensional turbulent flow within the boundary layer, as shown in Fig. 8.3. The shear stress acting on this generic fluid element can be divided into two components: one due to the gradient of the mean velocity and the other due to the gradient of the fluctuating velocity component.

The component of the shear stress due to the gradient of the mean velocity (see Fig. 8.4) is given by the term $\overline{\boldsymbol{\tau}}$ in Eq. 8.1; the component of the shear stress due to the gradient of the fluctuating velocity component (see Fig. 8.5) is given by the Reynolds stress tensor in Eq. 8.1, $\boldsymbol{\tau}^R = -\rho\overline{\mathbf{v}'\mathbf{v}'}$.

In the case of the energy conservation equation, an additional term $\rho c_p \overline{\mathbf{v}'T'}$ appears. This term is known as the *turbulent heat flux vector*

$$\dot{\mathbf{q}}^R = -\rho c_p \begin{pmatrix} \overline{u'T'} \\ \overline{v'T'} \\ \overline{w'T'} \end{pmatrix}$$

which, in the three-dimensional case, introduces three new unknowns.

The calculation techniques used to determine the values of these new unknowns are referred to in the literature as *turbulence modelling*.

Attempting to use the Navier-Stokes conservation equations directly to determine these unknowns would introduce even more unknowns, resulting in a not-closed system where the number of equations is lower than the number of unknowns. For this reason, any turbulence model must be capable of representing the non-linear fluctuation components described by the Reynolds stress tensor, as well as the three components of the turbulent heat flux, in terms of the mean components.

One possibility for obtaining information about the fluctuation components from the mean ones is provided by the Boussinesq hypothesis.[1] In this context, it is important to recall Newton's law of viscosity, which states that the shear stress is proportional to the velocity gradient, with the proportionality constant being μ, known as *dynamic viscosity* (see Sect. 2.5.1). The underlying observation of the Boussinesq hypothesis is that, even in turbulent flows, there is momentum transport between layers of fluid characterised by different mean velocities (i.e., particles in layers with higher mean velocities "drag" or accelerate particles in layers with lower mean velocities, and vice versa). The Boussinesq hypothesis states that the elements of the Reynolds stress tensor are a linear function of the gradient of the mean velocity proportional to the constant μ_t. This relationship is expressed in vector notation as

$$\tau^R = -\rho\overline{\mathbf{v}'\mathbf{v}'} = \mu_t \left[\nabla\overline{\mathbf{v}} + (\nabla\overline{\mathbf{v}})^T - \frac{1}{3}(\nabla \cdot \overline{\mathbf{v}}) \right] - \frac{2}{3}\rho k \mathbf{I} \tag{8.2}$$

or, using tensor notation (see Eq. 2.23), as

$$\tau_{ij}^R = -\rho\overline{v_i' v_j'} = \mu_t \left[\frac{\partial v_i}{\partial x_j} + \frac{\partial v_j}{\partial x_i} - \frac{1}{3}\frac{\partial v_k}{\partial x_k}\delta_{ij} \right] - \frac{2}{3}\rho k \delta_{ij} \tag{8.3}$$

where $i, j = 1, 2, 3$, δ_{ij} is the Kronecker delta ($\delta_{ij} = 1$ when $i = j$, $\delta_{ij} = 0$ when $i \neq j$). In this context we employ the convention of repeated indices (see Sect. 1.1.5) where, when $i = j$,

$$\frac{\partial v_k}{\partial x_k} = \frac{\partial v_1}{\partial x_1} + \frac{\partial v_2}{\partial x_2} + \frac{\partial v_3}{\partial x_3} = \frac{\partial u}{\partial x} + \frac{\partial v}{\partial y} + \frac{\partial w}{\partial z}.$$

[1] Joseph Boussinesq, *Essay on the theory of running waters*, Paris, National Printing Office, 1877.

Referring to the strain rate tensor (see Sect. 1.2.1), defined in tensor notation as

$$S_{ij} = \frac{1}{2}\left(\frac{\partial v_i}{\partial x_j} + \frac{\partial v_j}{\partial x_i}\right)$$

and in particular to its deviatoric part

$$S_{ij}^* = \frac{1}{2}\left(\frac{\partial v_i}{\partial x_j} + \frac{\partial v_j}{\partial x_i} - \frac{1}{3}\frac{\partial v_k}{\partial x_k}\delta_{ij}\right)$$

it is possible to write Eq. 8.3 as

$$\tau_{ij}^R = -\rho\overline{v_i' v_j'} = 2\mu_t S_{ij}^* - \frac{2}{3}\rho k \delta_{ij}.$$

In the incompressible case, Eq. 8.2 becomes

$$\tau^R = -\rho\overline{\mathbf{v}'\mathbf{v}'} = \mu_t\left[\nabla\overline{\mathbf{v}} + (\nabla\overline{\mathbf{v}})^T\right] - \frac{2}{3}\rho k\mathbf{I}$$

or, using tensor notation, as

$$\tau_{ij}^R = -\rho\overline{v_i' v_j'} = \mu_t\left[\frac{\partial v_i}{\partial x_j} + \frac{\partial v_j}{\partial x_i}\right] - \frac{2}{3}\rho k \delta_{ij}$$

where k is the *turbulent kinetic energy* per unit mass defined as

$$k = \frac{1}{2}\overline{\mathbf{v}' \cdot \mathbf{v}'} = \frac{1}{2}\left(\overline{u'u'} + \overline{v'v'} + \overline{w'w'}\right). \tag{8.4}$$

The presence of the term $-\frac{2}{3}\rho k\mathbf{I}$ or equivalently, the term $-\frac{2}{3}\rho k\delta_{ij}$ is necessary to ensure that the expression of the Reynolds stress tensor is consistent with the definition of turbulent kinetic energy. Specifically, this term allows the sum of the normal stresses, or the sum of the elements of the main diagonal of the Reynolds stress tensor, to equal the turbulent kinetic energy as defined in Eq. 8.4.

μ_t is the proportionality constant between Reynolds stresses and the gradients of the mean velocity component. μ_t is called *turbulent dynamic viscosity* or *eddy viscosity*. As μ_t increases, for a given mean velocity gradient, the transport of momentum between fluid particles with different mean velocity values also increases. It is also noted that:

- μ_t is not a characteristic of the fluid itself, but rather of the specific flow being considered;
- μ_t is a mathematical abstraction whose value must nonetheless be determined.

I is the identity matrix of order three. Under the Boussinesq hypothesis, the problem of calculating the Reynolds stress tensor is therefore reduced to determining the kinetic energy and turbulent viscosity.

Similarly, turbulent thermal fluxes are calculated in analogy with Fourier's law:

$$\dot{\mathbf{q}}^R = -\rho c_p \overline{\mathbf{v}' T'} = k_t \nabla \overline{T}$$

where k_t denotes the *turbulent thermal diffusivity*.

It is noted that:

1. there are turbulence models that do not rely on the Boussinesq hypothesis, among which are the Reynolds Stress models;
2. models based on the Boussinesq hypothesis are used in the LES approach, including the Smagorinsky model and the Smagorinsky dynamic model.

Turbulence models that are based on the Boussinesq hypothesis are known as *eddy viscosity models* and can be divided into four main categories:

- algebraic models (zero-equation models);
- one-equation models;
- two-equation models;
- second-order closure models.

Each class has a specific scope of applicability relative to the type of flow considered.

Algebraic models use algebraic equations to calculate μ_t, thus avoiding the need to solve differential equations. One-equation models involve solving a single differential transport equation for turbulent viscosity. Two-equation models involve solving two differential transport equations to calculate turbulent viscosity. The second-order closure models are the most computationally demanding as they solve six different transport equations, one for each component of the turbulent flow. The most widely used class is the two-equation models as they represent the best compromise between computational cost and accuracy of results.

It should be noted, finally, that turbulence models based on the Boussinesq hypothesis tend to provide inaccurate results in cases where the underlying assumptions may be unverified. Specifically, this occurs in the case of shock jets, as well as flows in ducts characterised by strong curvature or sudden changes in section.[2]

OpenFOAM® has the ability to perform calculations using one of the available turbulence models, among which there are:

- **LRR**: Launder, Reece and Rodi Reynolds-stress;
- **RNGkEpsilon**: Renormalisation group k-epsilon;
- **SpalartAllmaras**: Spalart-Allmaras one-equation;
- **kEpsilon**: standard k-epsilon;

[2] T.J. Craft et al., *Impinging jet studies for turbulence model assessment—II. An examination of the performance of four turbulence models*, International Journal of Heat and Mass Transfer, Elsevier, 1993.

- **LaunderSharmaKE**: k-epsilon modified by Launder and Sharma to also model the zones (near the walls) with a low Reynolds number;
- **SSG**: Speziale, Sarkar and Gatski based on the calculation of the Reynolds stress tensor;
- **kOmega**: standard k-omega;
- **kOmegaSST**: k-omega-SST (Shear Stress Transport);
- **kOmegaSSTLM**: 4-equation model of Langtry-Menter based on the k-omega-SST model;
- **kOmegaSSTSAS**: "Scale-Adaptive-Simulation" model based on k-omega-SST;
- **laminar**: laminar;
- **realizableKE**: Realizable k-epsilon;
- **qZeta**: Gibson and Dafa'Alla's two-equation q-Zeta model.

8.1.1 Standard $k - \epsilon$ Model

In this model, the turbulent viscosity and thermal diffusivity are expressed as

$$\mu_t = \rho C_\mu \frac{k^2}{\epsilon},$$
$$k_t = \frac{c_p \mu_t}{Pr_t}.$$

From the expression of the turbulent viscosity μ_t, it is noted that it is necessary to use a transport differential equation for each of the two terms: the turbulent kinetic energy k and the dissipation of turbulent kinetic energy per unit of mass and time ϵ, owing to viscous stresses. In other words, ϵ is the rate of transformation of turbulent kinetic energy into thermal energy per unit of mass and time due to molecular viscosity. To simplify the notation, we have eliminated the bar that indicates average quantities. The two equations used in this model are as follows:

- the transport equation of turbulent kinetic energy, k

$$\frac{\partial \rho k}{\partial t} + \nabla \cdot (\rho \mathbf{v} k) = \nabla \cdot \left(\mu_{eff,k} \nabla k\right) + P_k - \rho \epsilon;$$

- the empirical equation for the transport of the rate of dissipation of turbulent kinetic energy per unit of mass, ϵ

$$\frac{\partial \rho \epsilon}{\partial t} + \nabla \cdot (\rho \mathbf{v} \epsilon) = \nabla \cdot \left(\mu_{eff,\epsilon} \nabla \epsilon\right) + C_{\epsilon 1} \frac{\epsilon}{k} P_k - C_{\epsilon 2} \rho \frac{\epsilon^2}{k}$$

The following values are assigned empirically: $C_{\epsilon 1} = 1.44$, $C_{\epsilon 2} = 1.92$, $C_\mu = 0.09$, $\sigma_k = 1$, $\sigma_\epsilon = 1.3$, the turbulent Prandtl number $Pr_t = 0.9$. The compact form of the P_k term, representing the production of turbulent kinetic energy, is given by

$$P_k = \tau^R \cdot \nabla \mathbf{v}.$$

The terms $\mu_{eff,k}$ and $\mu_{eff,\epsilon}$ represent the effective viscosity which is the result of the sum of the fluid molecular viscosity and the turbulent viscosity of the flow. These terms are given by

$$\mu_{eff,k} = \mu + \frac{\mu_t}{\sigma_k}, \qquad \mu_{eff,\epsilon} = \mu + \frac{\mu_t}{\sigma_\epsilon}.$$

Since this model was initially designed to describe external flows in the absence of adverse pressure gradients (pressure gradients opposite to the flow velocity direction), with high Reynolds numbers and fully developed turbulence, it will provide the best results for such flows.

It is now possible to define the *Reynolds number of the turbulence* as:

$$Re_\tau = \frac{k^2}{\epsilon \nu}$$

in which the turbulent kinetic energy $k \sim u^2$ (where the symbol $\sim$ denotes "same order of magnitude") represents the velocity scale u of the turbulent vortices, while the dissipation term ϵ characterises the scale l of the turbulent vortices, given by $\epsilon \sim \frac{u^3}{l}$.

The standard $k - \epsilon$ model belongs to the class of turbulence models known as *high Reynolds number turbulence models*. In cases where turbulent phenomena occur near a contour where the velocity is set to zero (such as a wall), it is necessary to modify the standard $k - \epsilon$ model appropriately to ensure it provides accurate results in the vicinity of the wall, where the Reynolds number of the turbulence is lower. In general, turbulence models that are able to correctly describe the wall behaviour are called *low Reynolds number turbulence models*. This designation also applies to low Reynolds number versions of generic turbulence models, such as the one proposed by Launder and Sharma.[3] In the specific case of the $k - \epsilon$ model, damping functions are introduced. Damping functions modify the values of the constants $C_{\epsilon 1}$, $C_{\epsilon 2}$, and C_μ depending on the distance from the wall.

8.1.2 $k - \omega$ Model

In place of dissipation of turbulent kinetic energy ϵ, this model introduces the ω rate of energy conversion from turbulent kinetic energy to internal energy per unit volume and time. The use of this parameter allows for a better description of flows where there is an adverse pressure gradient resulting in boundary layer separation.

[3] B.E. Launder and B.I. Sharma, *Application of the energy-dissipation model of turbulence to the calculation of flow near a spinning disc*, Letters in Heat and Mass Transfer, Vol. 1, pp. 131–138, 1974.

The term ω is defined as

$$\omega = \frac{\epsilon}{C_\mu k}. \tag{8.5}$$

As in the case of the $k - \epsilon$ model, two equations are solved here, one for k:

$$\frac{\partial \rho k}{\partial t} + \nabla \cdot (\rho \mathbf{v} k) = \nabla \cdot \left(\mu_{eff,k} \nabla k\right) + P_k - \beta^* \rho k \omega \tag{8.6}$$

and one for ω:

$$\frac{\partial \rho \omega}{\partial t} + \nabla \cdot (\rho \mathbf{v} \omega) = \nabla \cdot \left(\mu_{eff,\omega} \nabla \omega\right) + C_{\alpha 1} \frac{\omega}{k} P_k - C_{\beta 1} \rho \omega^2 \tag{8.7}$$

in which

$$\mu_{eff,k} = \mu + \frac{\mu_t}{\sigma_{k1}} \qquad \mu_{eff,\omega} = \mu + \frac{\mu_t}{\sigma_{\omega 1}}.$$

The following values are empirically assigned: $C_{\alpha 1} = 5/9$, $C_{\beta 1} = 0.075$, $\beta^* = 0.09$, $\sigma_{k1} = 2$, $\sigma_{\omega 1} = 2$, the turbulent Prandtl number $Pr_t = 0.9$. The turbulent viscosity and thermal diffusivity are defined as

$$\mu_t = \rho \frac{k}{\omega},$$
$$k_t = \frac{\mu_t}{Pr_t}.$$

The advantages of using ω instead of ϵ are all related to its equation that:

- is more easily integrable;
- is able to correctly describe the turbulent phenomenon even near a wall;
- provides satisfactory results even in the presence of adverse pressure gradients.

The weakness of this turbulence model is the extreme sensitivity of the results to the undisturbed flow values set as boundary conditions for ω. This problem does not affect the $k - \epsilon$ model.

8.2 $k - \omega$ SST (Shear Stress Transport) Model

This model derives from the one known as *Baseline* $k - \omega$ ($k - \omega$ BSL) which attempts to combine the positive aspects of both $k - \epsilon$ and $k - \omega$ models. Specifically, from the $k - \omega$ model, this model tries to exploit

- the stability in handling flow areas close to the walls due to the low Reynolds number formulation;

- the ability to correctly describe flows characterised by the presence of adverse pressure gradients.

From the $k-\epsilon$ model, this model tries to exploit:

- the ability to correctly process turbulent flows in areas far from the walls;
- the insensitivity to the value set for the undisturbed flow.

In the $k-\omega$ BSL model, starting from the $k-\omega$ model, the Eq. 8.7 for ω is initially modified as

$$\frac{\partial \rho\omega}{\partial t} + \nabla\cdot(\rho\mathbf{v}\omega) = \nabla\cdot\left(\mu_{eff,\omega}\nabla\omega\right) + C_{\alpha 2}\frac{\omega}{k}P_k - C_{\beta 2}\rho\omega^2 + 2\sigma_{\omega 2}\frac{\rho}{\omega}\nabla k\cdot\nabla\omega \tag{8.8}$$

also modifying the value of the constants as follows: $C_{\alpha 2} = 0.4404$, $C_{\beta 2} = 0.0828$, $\sigma_{k2} = 1$, $\sigma_{\omega 2} = 0.856$, $Pr_t = 0.9$. Subsequently, a weighting function F_1 is introduced to obtain a weighted average $\tilde{\Phi}$ of the value of the constants according to the

$$\tilde{\Phi} = F_1\Phi_1 + (1-F_1)\,\Phi_2$$

where Φ_1 is the value of the constant assumed in the original $k-\omega$ model and Φ_2 is the value of the constant assumed in Eq. 8.8. The weighting function F_1 depends on the distance $d_\perp$ of the considered point from the nearest wall and it is defined as

$$F_1 = \tanh\left(\gamma_1^4\right)$$

with

$$\gamma_1 = Min\left[Max\left(\frac{\sqrt{k}}{\beta^*\omega\,(d_\perp)}, \frac{500\nu}{(d_\perp)^2\,\omega}\right), \frac{4\rho\sigma_{\omega 2}k}{CD_{k\omega}\,(d_\perp)^2}\right] \quad and$$

$$CD_{k\omega} = Max\left(2\rho\sigma_{\omega 2}\frac{1}{\omega}\nabla k\cdot\nabla\omega,\, 10^{-10}\right)$$

and so

$$\mu_t = \rho\frac{k}{\omega}, \qquad k_t = \frac{\mu_t}{Pr_t}, \qquad \mu_{eff,k} = \mu + \frac{\mu_t}{\tilde{\sigma}_k}, \qquad \mu_{eff,\omega} = \mu + \frac{\mu_t}{\tilde{\sigma}_\omega},$$

where $\nu = \mu/\rho$ denotes the kinematic viscosity of the fluid. So far, the $k-\omega$ BSL model has been described from which the $k-\omega$ SST model is derived. The first difference between these two models concerns the definition of the turbulent viscosity that is modified in the $k-\omega$ SST model according to

$$\mu_t = \frac{\rho a_1 k}{Max\left(a_1\omega, \sqrt{2}S_t F_2\right)}$$

in which $a_1 = 0.31$, $S_t = \sqrt{\mathbf{S}_t \cdot \mathbf{S}_t}$ is the modulus of the stress defined as $\mathbf{S}_t = \frac{1}{2}\left(\nabla\mathbf{v} + \nabla\mathbf{v}^T\right)$ and F_2 is a weighting function defined as

$$F_2 = \tanh\left(\gamma_2^2\right) \quad \textit{and} \quad \gamma_2 = Max\left(2\frac{\sqrt{k}}{\beta^*\omega\left(d_\perp\right)}, \frac{500\nu}{\left(d_\perp\right)^2\omega}\right).$$

The equations for k and for ω are the same as for the BSL model as well as identical is the use of the function F_1 and its definition. The second difference between the BSL and the SST model is in the production of turbulent kinetic energy term P_k in Eq. 8.6 of k which is replaced by

$$\tilde{P}_k = Min\left(P_k, c_1\epsilon\right)$$

in which ϵ is obtained from Eq. 8.5. The constants used in function F_1 to obtain the $k - \omega$ SST model constants are

$$C_{\alpha 1} = 0.5532,\ C_{\beta 1} = 0.0750,\ \sigma_{k1} = 2,\ \sigma_{\omega 1} = 2$$
$$C_{\alpha 2} = 0.4404,\ C_{\beta 2} = 0.0828,\ \sigma_{k2} = 1,\ \sigma_{\omega 2} = 1.186$$

The constants of the model are $\beta^* = 0.09$, $c_1 = 10$, $Pr_t = 0.9$. Finally, the expressions of k and ω use the effective turbulent viscosities calculated as

$$k_t = \frac{\mu_t}{Pr_t}, \qquad \mu_{eff,k} = \mu + \frac{\mu_t}{\tilde{\sigma}_k}, \qquad \mu_{eff,\omega} = \mu + \frac{\mu_t}{\tilde{\sigma}_\omega}.$$

8.3 The Boundary Layer

The momentum boundary layer is that portion of fluid in contact and close to a solid surface. In this area, in fact, the transition occurs between the undisturbed outer flow and the much slower one in contact with the solid wall. By convention, the *thickness of the boundary layer* (see Fig. 8.6) is defined as the portion of fluid whose velocity differs by 1% from the velocity of the undisturbed fluid (asymptotic velocity). The boundary layer originates where the solid surface begins. As shown in Fig. 8.6, the thickness of the boundary layer is divided into four zones:

- the viscous or laminar sublayer where there are no turbulent fluctuations;
- transitional sublayer;
- the fully turbulent region, also known as the logarithmic law region;
- the outer layer.

Using a specific law, the velocity profile can be described for each of these regions as the distance from the wall increases. Figure 8.7 shows the graph of the profile of the average value of the velocity component u parallel to the wall as a function

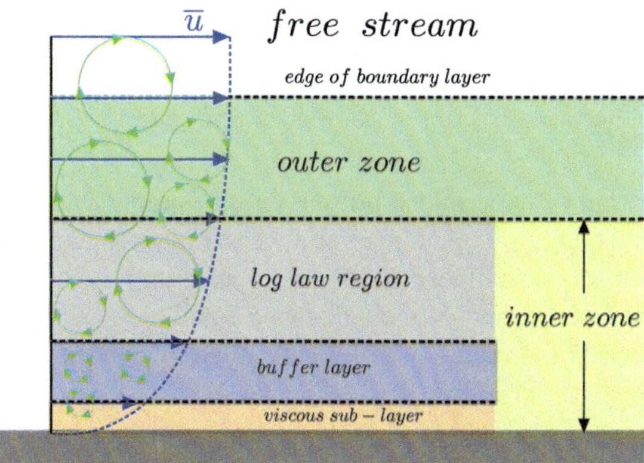

Fig. 8.6 Momentum boundary layer

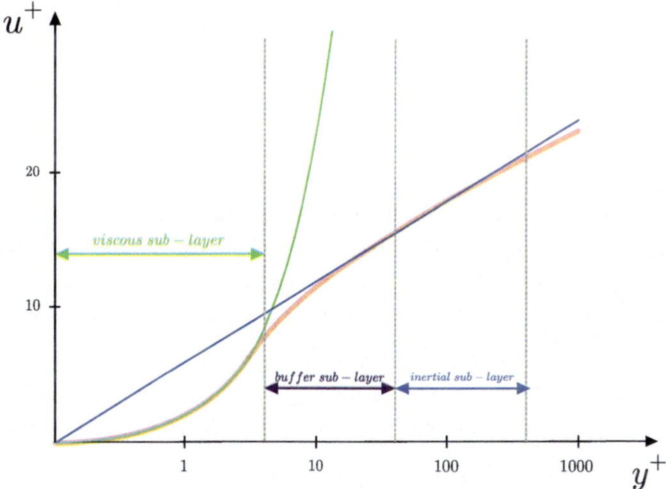

Fig. 8.7 velocity as a function of the distance from the wall in dimensionless units

of the distance $d_\perp$ from the wall itself. In this graph, the distance from the wall and the velocity are represented in dimensionless form. The distance from the wall, shown on a logarithmic scale in Fig. 8.7, is indicated by the symbol y^+ and it is adimensionalised as:

$$y^+ = \frac{d_\perp u_\tau}{\nu} \tag{8.9}$$

where u_τ is referred to as *shear or friction velocity* defined as

$$u_\tau = \sqrt{\frac{|\tau_w|}{\rho}} \tag{8.10}$$

where τ_w is the magnitude of wall shear stress. Denoted by u^+, the adimensional velocity is defined as:

$$u^+ = \frac{u}{u_\tau}.$$

In Fig. 8.7 the result of experimental observations is represented by the curve with the thicker line, which, in the laminar sub-layer, is in good agreement with the law $u^+ = y^+$ and appears curvilinear due to the representation on a logarithmic scale. The laminar sub-layer extends up to a distance equal to $y^+ = 5$. In the logarithmic region, characterised by values of y^+ greater than 30, and in the outer layer, the logarithmic law

$$u^+ = \frac{1}{\kappa}\log(y^+) + B = \frac{1}{\kappa}\log(Ey^+) \tag{8.11}$$

approximates the data observed experimentally in an acceptable manner. The constant $\kappa \approx 0.4$ is known as the *von Karman constant*, while $B \approx 5.5$ ($E \approx 0.98$) decreases as the roughness of the wall increases, according to empirical laws. Values of y^+ between 5 and 30 bound the transitional sub-layer, in which both of these laws approximate the data observed experimentally with lesser accuracy.

A first approach to correctly describe the velocity profile within the boundary layer involves the use of:

- a large number of cells to allow the correct resolution of strong gradients;
- a turbulence model capable of correctly describing the laminar sub-layer as well as the logarithmic region (the aforementioned low Reynolds number turbulence models such as the $k - \omega$ or the $k - \omega$ SST).

In this first approach, the laminar sub-layer is certainly correctly resolved. The centre of the first cell near the wall must have $y^+ \simeq 1$, and there is a real risk of obtaining computational grids with a number of cells that would result in prohibitive computational costs.

A second approach for calculating the velocity profile within the boundary layer involves:

- the use of so-called wall functions, i.e., the modelling of the velocity profile within the boundary layer rather than its actual resolution;
- the possibility of using a high Reynolds number turbulence model such as the $k - \epsilon$ model.

In this second approach, the centre of the first cell near the wall must have $y^+ > 30$. It is assumed that the velocity profile from the first cell centre near the wall is the one shown in Fig. 8.7. Based on this assumption, the gradient of velocity normal to the wall is calculated and consequently the shear stress. This type of approach allows the use of computational grids with a reduced number of cells. However, assuming the profile in Fig. 8.7 is not necessarily correct, for example, in the case of flow separation. Particular attention should be paid to the analysis of y^+ values. The presence of regions within the computational domain characterised by values between 5 and 30 should be avoided due to the low reliability of wall functions resulting from the direct resolution of the laminar sub-layer avoidance.

8.4 Wall Functions

In cases where the approach involving the modelling of the velocity profile within the boundary layer, rather than its actual resolution, is chosen, particular attention should be paid to determining the wall velocity gradient. This requires examining its definition.

$$\tau_w \equiv \mu \left(\frac{\partial u}{\partial y} \right)_{y=0},$$

It is clear that the wall shear stress τ_w is a function of the wall velocity gradient, as well as the distance y from the wall, as shown in Fig. 8.8. In this regard, it is crucial to determine the value y_p^+ of the centre of the first cell near the wall to identify the zone of the boundary layer in which it is located. With reference to Fig. 8.8 and formula 8.9, a possible approach involves considering

$$y_p^+ = \frac{u_\tau}{\nu} d_\perp = \frac{C_\mu^{1/4} \sqrt{k_p}}{\nu} y_p$$

In this context, k_p represents the turbulent kinetic energy calculated at the cell centre p, based on the corresponding velocity value. The threshold values for this quantity typically range between 11 and 12: lower values indicate that the cell centre p lies within the laminar sub-layer, while higher values suggest that it is located in the transitional sub-layer or the logarithmic region. Figure 8.8 illustrates this.

- The centre p of the cell, whose extension is represented by the shaded area;
- The centre of the face on which the no-slip boundary condition is applied;
- The value of the velocity component u parallel to the wall at the cell centre, obtained from the application of the equation of conservation of momentum;
- The real velocity profile, represented by the continuous thin line;
- The velocity gradient at the centre of the wall face ($y = 0$) as the slope of the segment with a dotted line, considering a simple linear approximation of the velocity values at the cell centre and at the wall face centre. Assuming a stationary wall, the velocity at the wall face centre is zero;
- The velocity gradient calculated at the centre of the wall face ($y = 0$) as the slope of the tangent—dashed segment—to the velocity profile at the centre of the wall face;
- The difference between the value of the gradient calculated with the linear approximation and the value of the gradient obtained as a tangent to the velocity profile, represented by a thick continuous line.

From Fig. 8.8 it is therefore evident that

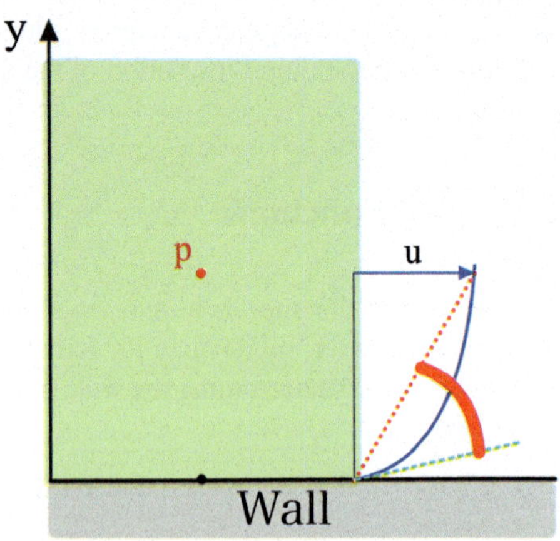

Fig. 8.8 Wall velocity gradient

$$\tau_w \equiv \mu \left(\frac{\partial u}{\partial y} \right)_{y=0} \neq \mu \frac{\Delta u}{\Delta y} = \mu \frac{u_P}{y_P}.$$

Choosing the linear approximation to express of the gradient of wall velocity as done for all of the interior faces of the computational domain, two approaches are possible:

- The addition of a specific source term in the equation of conservation of momentum;
- Considering a different value of viscosity μ_e (effective viscosity) to correct the error made in the calculation of the wall velocity gradient.

8.5 Distance from the Wall to the Centre of the First Cell Near the Wall

During the grid construction phase, it is necessary to determine the value to use for the distance from the wall to the centre of the first cell near the wall, so that the desired y^+ value is obtained once the simulation is completed. Positioning the centre of the first cell near the wall in such a way that the corresponding y^+ value matches the desired one is an iterative process and involves the following steps:

1. Initial estimate of the distance from the wall to the centre of the first cells;
2. Execution of the simulation and measurement of the obtained y^+ value. If the obtained values are acceptable, the process stops; if the obtained values are not acceptable, proceed to the next step;
3. Modification of the grid to move the centre of the first cells based on the values obtained in the previous step;
4. Return to step 2.

In this process, the initial estimate of the distance from the wall to the centre of the first cell can be obtained by calculating the free-flow Reynolds number for the actual case.

$$Re = \frac{\rho u_\infty L}{\mu}$$

where u_∞ is the free-stream velocity. Then, calculate the friction coefficient C_f for a flat plate of infinite dimensions.

$$C_f = 0.58 Re^{-0.2}.$$

Knowing the value of the friction coefficient, it is possible to calculate the wall shear stress as follows:

$$\tau_w = \frac{1}{2} C_f \rho u_\infty^2$$

This allows the deduction of the slip velocity.

$$u_\tau = \sqrt{\frac{\tau_w}{\rho}}.$$

Finally, from the definition of y^+,

$$y^+ = \frac{y u_\tau}{\nu}$$

it is

$$y = \frac{\mu y^+}{\rho u_\tau}$$

which is the distance from the wall to the centre of the first cell.

8.6 Wall Functions in OpenFOAM®

Based[4] on the type of calculation performed, the wall functions in OpenFOAM® can be classified into:

- those for which the value of the quantity (for example, turbulent kinetic energy) (Table 8.1) is calculated on the wall face of the considered cell;
- those for which the value of the quantity (for example, the dissipative term of the turbulence model used) is calculated at the centre of the cell.

---

[4] The article on the basis of which the boundary conditions that make use of wall functions were implemented in OpenFOAM® is Georgi Kalitzin, Gorazd Medic, Gianluca Iaccarino, and Paul Durbin. *Near-wall behavior of RANS turbulence models and implications for wall functions*, "Journal of Computational Physics", Vol. 204, pp. 265–291, 2005.

Table 8.1 Summary of wall boundary conditions in OpenFOAM®

ν_t	`nutUWallFunction`	ν_t value based on the value of the velocity at the centre of the wall cell
	`nutkWallFunction`	ν_t value based on the value of the turbulent kinetic energy at the centre of the wall cell
	`nutLowReWallFunction`	The value of ν_t is set to zero: to be used in the case where the centre of the first wall cell is inside the laminar sub-layer (low-Re)
	`nutUSpaldingWallFunction`	Spalding's law is used to continuously describe the value of ν_t as y^+ varies
k	`kqRWallFunction`	Based on the zero-gradient condition, this boundary condition provides the value of k, q and R assuming the centre of the wall cell to be within the logarithmic sub-layer (high-Re)
	`kLowReWallFunction`	Provides the value of k for the case of the centre of wall cell within the inertial sub-layer as well as it is in the viscous sub-layer
ϵ	`epsilonWallFunction`	Provides the value of ϵ under the assumption that the centre of the wall cell is in the logarithmic sub-layer (high-Re)
	`epsilonLowReWallFunction`	Provides the value of ϵ for the case of the centre of the wall cell is in the inertial sub-layer as well as it is in the viscous sub-layer using an approximate value of y^+
ω	`omegaWallFunction`	Provides the value of ω for the case of the centre of the wall cell is in the inertial sub-layer as well as it is in the viscous sub-layer using an approximate value of y^+

For quantities calculated with the first approach, a conservation equation is also solved in the wall cell, as in all other cells of the computational domain. For quantities calculated with the second approach (i.e., by providing the wall function value at the cell centre), the conservation equation is not solved. Among the quantities whose value can be calculated through wall functions are turbulent kinetic energy (kqR-WallFunctions), the dissipative term of the $k - \epsilon$ model (epsilonWallFunctions), the dissipative term of the $k - \omega$ model (omegaWallFunctions), and turbulent viscosity (nutWallFunction). The term "LowRe" in the name of the boundary conditions refers to the ability to correctly handle the case where the centre of the first wall cell is within the laminar sub-layer of the boundary layer. Due to the decrease in velocity in this sub-layer, the turbulent Reynolds number is necessarily reduced (see Sect. 8.1.1).

8.6.1 *kqRWallFunctions*

The calculation of turbulent kinetic energy at the centre of the cells having at least one face on the wall is performed by solving a conservation equation using the value provided by the wall function at the centre of the wall face. The wall functions for the calculation of turbulent kinetic energy in OpenFOAM® are:

- kqRWallFunction;
- kLowReWallFunction.

8.6.1.1 kqRWallFunction

With this boundary condition, it is assumed that, in the zone between the centre of the wall cell and the centre of the wall face of the same cell, the turbulent kinetic energy is constant and equal to the value calculated at the cell centre. In other words, it is assumed that the cell centre is in the inertial sub-layer of the boundary layer (see Eq. 20 of the aforementioned article by Kalitzin et al.). This is a Neumann boundary condition (zero gradient) for turbulent kinetic energy. This boundary condition is also used for:

- determining the value q of the square root of turbulent kinetic energy for turbulence models such as qZeta;
- determining the value of the Reynolds stress tensor R for turbulence models such as LRR.

In the case where the qZeta or LRR turbulence models are not used, the zero-gradient condition can be applied, as for y^+ values between 30 and 50, the turbulent kinetic energy varies slightly and can therefore be considered constant.

8.6.1.2 kLowReWallFunction

This boundary condition should be used in cases where it is possible that the centre of the first wall cell is within the viscous sub-layer of the boundary layer. In this case, the threshold value of y^+ that separates the viscous sub-layer from the turbulent one is first calculated. The buffer zone is then partly modelled with the linear law (for y^+ values lower than the limit value) and partly with the logarithmic law (for y^+ values higher than the limit value). Assuming a von Karman constant of 0.41 and a roughness coefficient of 0.9, the threshold value of y^+ is 11. In this boundary condition, the shear velocity is initially calculated for each wall cell, assuming that the cell centre is in the inertial sub-layer, and therefore using the formula:

$$u_\tau = C_\mu^{1/4}\sqrt{k}$$

where k is the turbulent kinetic energy at the centre of the considered wall cell, while $C_\mu = 0.09$. Knowing the value of the shear velocity, it is possible to calculate the value of y^+ for the cell centre, which is $y_p^+ = \frac{u_\tau}{\nu} d_\perp$ by definition. The assumption made about the positioning of the cell centre within the inertial sub-layer for the calculation of u_τ induces a negligible error in the case where the same cell centre is instead in the viscous sub-layer. The dimensionless turbulent kinetic energy on the wall face must now be considered. It is defined as $k^+ = k/u_\tau^2$. In the case where the value of y^+ is higher than the limit value, the following logarithmic law is used:

$$k^+ = \frac{C_k}{\kappa} \log(y^+) + B_k.$$

$C_k = -0.416$ and $B_k = 8.366$ are two empirical constants. In the case where the value of y^+ is lower than the threshold value, the following law is used

$$k^+ = \frac{2400}{C_{\epsilon 2}^2} C_f$$

to calculate the dimensionless turbulent kinetic energy on the wall face, $C_{\epsilon 2} = 1.9$ is an empirical constant. The following law is used to calculate the term C_f:

$$C_f = \frac{1}{(y^+ + C)^2} + \frac{2y^+}{C^3} - \frac{1}{C^2}$$

with $C = 11$ as an empirical constant. Knowing the values of k^+ and u_τ, the value of the turbulent kinetic energy on the wall face is obtained through the:

$$k = k^+ u_\tau^2$$

which allows for solving the transport equation for the turbulent kinetic energy within the wall cell using the same approach applied to all other cells of the computational domain.

8.6.2 *epsilonWallFunctions*

For the dissipative term of the $k - \epsilon$ model in the first wall cell, no transport equation is solved, and its value at the cell centre is calculated using the value provided by the wall function. The wall functions for the calculation of the dissipative term of the $k - \epsilon$ model in OpenFOAM® are:

- epsilonWallFunction;
- epsilonLowReWallFunction.

8.6.2.1 epsilonWallFunction

In this case, it is necessary to consider the possibility that the generic wall cell has more than one face on which the wall boundary condition is set. The value of ϵ at the cell centre is calculated based on the distance from the centre of the wall faces:

$$\epsilon = \frac{1}{W}\sum_{f=1}^{W}\left(\frac{C_\mu^{3/4}k^{3/2}}{\kappa y_f}\right)$$

where W is the number of faces of the cell on which the wall boundary condition has been set, k is the turbulent kinetic energy calculated at the centre of the cell, y_f is the distance from the cell centre to the considered face, κ is the von Karman constant, and $C_\mu = 0.09$.

8.6.2.2 epsilonLowReWallFunction

Similarly to what is done for the kLowReWallFunction boundary condition, for the calculation of ϵ at the centre of the wall cell, this boundary condition distinguishes the case of y^+ being less than the threshold value from that of y^+ being greater than the threshold value. In the case of y^+ greater than the limit value, the calculation is performed using the same formula as in the case of the boundary condition epsilon-WallFunction. In the case of y^+ being less than the limit value, the calculation is performed according to the formula below:

$$\epsilon = \frac{1}{W}\sum_{f=1}^{W}\left(\frac{2k\nu_f}{y_f^2}\right)$$

which is derived from the expression of ϵ^+ in the laminar sub-layer.

$$\epsilon^+ = 2\frac{k^+}{(y^+)^2}.$$

8.6.3 *omegaWallFunctions*

This boundary condition is the one to use when using the $k-\omega$ turbulence model to obtain the value of the dissipative term ω at the centre of the wall cells. Similarly to what has already been seen for the term ϵ, to compute the value of ω for the wall cells, the conservation equation is not solved. For this boundary condition, the value of ω at the wall face is first calculated using the following formula, relative to the case in which the cell centre is in the viscous sub-layer:

$$\omega_{vis} = \frac{6\nu}{\beta_1 y^2}$$

with $\beta_1 = 0.075$. Then, the value of ω at the wall face is calculated using the formula for the case where the cell centre is in the inertial sub-layer:

$$\omega_{log} = \frac{\sqrt{k}}{C_\mu^{1/4} \kappa y}$$

where y is the distance of the cell centre from the wall face and k is the turbulent kinetic energy calculated at the cell centre. Finally, the value at the cell centre is calculated as a combination of the two previously calculated values:

$$\omega = \sqrt{\omega_{vis}^2 + \omega_{log}^2}.$$

Similarly to what has already been seen for the case of the epsilonWallFunction boundary condition, in the case where the considered cell has more than one wall face, it will be:

$$\omega = \frac{1}{W} \sum_{f=1}^{W} \omega_f.$$

8.6.4 *nutWallFunctions*

The value of the wall shear stress is necessary for the solution of the momentum conservation equation for wall cells. By definition, the wall shear stress is:

$$\tau_w \equiv \mu \left(\frac{\partial u}{\partial y} \right)_{y=0}$$

having indicated with μ the molecular viscosity, with u the component of the velocity parallel to the wall, and with y the distance from the wall, as shown in Fig. 8.8. As already seen in Sect. 8.4, it is clear that:

$$\tau_w \equiv \mu \left(\frac{\partial u}{\partial y} \right)_{y=0} \neq \mu \frac{\Delta u}{\Delta y} = \mu \frac{u_p}{y_p}.$$

To compute the gradient of velocity at the face centre of any interior cell, the linear approximation is used. With the aim of using the same approach for wall cells, it is possible to

- use a large number of cells to ensure the validity of the linear approximation;

- consider a different value of viscosity μ_e (effective viscosity) to correct the error made in the calculation of the velocity gradient at the wall face using the value of the velocity at the centre of the wall cell:

$$\tau_w \equiv \mu \left(\frac{\partial u}{\partial y} \right)_{y=0} = \mu_{eff} \frac{u_P}{y_P}. \tag{8.12}$$

In order to obtain an expression for effective viscosity, we consider the definition of the dimensionless turbulent kinetic energy k^+:

$$k^+ = \frac{k}{u_\tau^2}$$

and its expression in the inertial sub-layer:

$$k^+ = \frac{1}{\sqrt{C_\mu}}$$

from which we obtain

$$u_\tau = C_\mu^{1/4} k^{1/2}.$$

By definition, the shear velocity is

$$u_\tau = \sqrt{\frac{\tau_w}{\rho}}$$

from which

$$\tau_w = \rho u_\tau^2.$$

Considering the definition of the dimensionless velocity $u^+ = u/u_\tau$, we get $u_\tau = u/u^+$, and therefore, recalling the expression of u^+ in the logarithmic sub-layer:

$$\tau_w = \rho u_\tau^2 = \rho u_\tau u_\tau = \rho u_\tau \frac{u}{u^+} = \rho u_\tau \frac{u_P}{\frac{1}{\kappa} \log(E y^+)} \tag{8.13}$$

in which u_P is the component parallel to the wall of the velocity at the centre of the wall cell.

Finally, considering Eqs. 8.12 and 8.13, it is

$$\mu_{eff} \frac{u_P}{y_P} = \frac{\rho u_\tau u_P}{\frac{1}{\kappa} \log(E y^+)}$$

that is, in terms of kinematic viscosity,

$$\nu_{eff}\frac{u_P}{y_P} = \frac{u_\tau u_P}{\frac{1}{\kappa}\log(Ey^+)}$$

from which, recalling the definition $y^+ = yu_\tau/\nu$

$$\nu_{eff} = \nu + \nu_t = \frac{y^+\nu}{\frac{1}{\kappa}\log(Ey^+)}$$

and so the expression of the turbulent kinematic viscosity is

$$\nu_t = \nu\left(\frac{\kappa y^+}{\log(Ey^+)} - 1\right) \tag{8.14}$$

which is used as a corrective term for the molecular kinematic viscosity to obtain the correct value of the wall shear stress. Depending on how the value of y^+ is calculated, in OpenFOAM®, different boundary conditions are available for ν_t.

This boundary condition sets the value of the turbulent kinematic viscosity to zero, and it is the one to use in cases where the grid is fine enough at the wall to correctly resolve the boundary layer.

8.6.4.1 nutkWallFunction

This boundary condition involves using the value of the turbulent kinetic energy at the centre of the wall cell to determine the value of y^+, which is used for calculating the turbulent kinematic viscosity. It is assumed that the centre of the wall cell lies in the logarithmic sub-layer for the calculation of y^+. By definition, $y^+ = \frac{yu_\tau}{\nu}$ and $k^+ = \frac{k}{u_\tau^2}$. In the inertial sub-layer, it will be:

$$k^+ = \frac{1}{\sqrt{C_\mu}}$$

from which, it is

$$u_\tau = C_\mu^{1/4}k^{1/2}$$

and therefore

$$y^+ = \frac{y}{\nu}C_\mu^{1/4}k^{1/2}.$$

As done for other boundary conditions, even for this one, the value calculated for y^+ is compared with the y^+ limit value:

- if y^+ is greater than the threshold value, the cell centre is considered to be in the logarithmic sub-layer and the turbulent kinematic viscosity is calculated using Eq. 8.14;

- if y^+ is less than the limit value, the cell centre is considered to be in the viscous sub-layer and $\nu_t = 0$ is assumed.

8.6.4.2 nutUWallFunction

For this boundary condition, the calculation of the turbulent kinematic viscosity is identical to that of the previous boundary condition. The difference lies in the calculation of the value of y^+, which this time is based on the value of the velocity at the centre of the wall cell. Recalling the definition of dimensionless velocity $u^+ = u/u_\tau$, it is assumed here, too, that the centre of the wall cell lies in the logarithmic sub-layer to calculate the value of y^+:

$$u^+ = \frac{u}{u_\tau} = \frac{1}{\kappa}\log(Ey^+)$$

from which

$$\frac{u}{yu_\tau/\nu}(y/\nu) = \frac{u}{y^+}(y/\nu) = \frac{1}{\kappa}\log(Ey^+)$$

and therefore

$$y^+\log(Ey^+) - \frac{\kappa y u}{\nu} = 0$$

which can be solved using the Newton-Raphson method to determine y^+ as:

$$y^+_{n+1} = y^+_n - \frac{f(y^+)}{f'(y^+)} = y^+_n - \frac{y^+_n\log(Ey^+_n) - \dfrac{\kappa y u}{\nu}}{1+\log(Ey^+_n)} = \frac{y^+_n + \dfrac{\kappa y u}{\nu}}{1+\log(Ey^+_n)}$$

Since the calculation of y^+ is based on the velocity value, this boundary condition can be used with both high and low Reynolds number turbulence models.

8.6.4.3 nutUSpaldingWallFunction

This boundary condition utilises the law

$$y^+ = u^+ + \frac{1}{E}\left[e^{\kappa u^+} - 1 - \kappa u^+ - \frac{1}{2}\left(\kappa u^+\right)^2 - \frac{1}{6}\left(\kappa u^+\right)^3\right] \tag{8.15}$$

to approximate the curve that links u^+ and y^+ in all three (laminar, buffer, and logarithmic) sub-layers of the boundary layer. Since the calculation of y^+ is based on the velocity value, this boundary condition can be used with both high and low Reynolds number turbulence models. Considering the definition of y^+ and u^+, it is

$$y^+ = \frac{y_P u_\tau}{\nu} \qquad u^+ = \frac{u_P}{u_\tau}.$$

Substituting into Eq. 8.15, we obtain an implicit equation in u_τ which, as with the previous boundary condition, can be solved using the Newton-Raphson method to obtain u_τ. Then, considering that by definition it is

$$u_\tau = \sqrt{\frac{\tau_w}{\rho}} \qquad \tau_w = \mu_{eff}\frac{u_P}{y_P},$$

we obtain

$$\nu_{eff} = \frac{u_\tau^2}{u_P/y_P}.$$

Keeping in mind that $\nu_{eff} = \nu_t + \nu$, the turbulent kinematic viscosity can be calculated as

$$\nu_t = \frac{u_\tau^2}{u_P/y_P} - \nu.$$

8.7 Implementation Aspects in OpenFOAM®

Regarding the imposition of wall boundary conditions and specifically the use of wall functions, attention must be given to the calculation of the turbulent kinematic viscosity ν_t. Indicated by the name `patch`, any surface to which a boundary condition is imposed, the file `0/nut` will contain the word `calculated` for all patches, except for those defined as walls. This is because the turbulent kinematic viscosity can be calculated as $\nu_t = k/\omega$ everywhere, but on the surface of a wall, where the velocity is imposed to be zero, the turbulent kinetic energy k will also be zero, and consequently ν_t will be zero. In the case where the centre wall cell is characterised by $y^+ > 30$, this would lead to a significant error in the calculation of the wall shear stress, defined as

$$\tau_w = \nu_{eff}\frac{\partial \overline{u}}{\partial y}.$$

When the wall is parallel to the x-axis, the error would derive not only from imposing $\nu_t = 0$, but also from an incorrect calculation of the velocity gradient normal to the wall. To overcome this problem, the implementation of the wall function in OpenFOAM® modifies the value of ν_t in order to correct the value of the velocity gradient normal to the wall, making it consistent with what is predicted by the curves shown in Fig. 8.7. The wall function normally used in OpenFOAM® is the one indicated by the word `nutkWallFunction`, which sets a velocity profile corresponding to the logarithmic law for values of y^+ of the wall cell centre greater

Table 8.2 Example of wall boundary conditions in the case of a boundary layer resolved by the grid (LowRe)

ν_t	`nutLowReWallFunction` or `calculated`
k	`fixedValue=0`
ϵ	`fixedValue=0` or `calculated`
ω	`fixedValue=0` or `calculated`

Table 8.3 Example of wall boundary conditions in the case of boundary layer not resolved by the grid (HighRe)

ν_t	`nutUWallFunction`
k	`kqRWallFunction` or `zeroGradient`
ϵ	`epsilonWallFunction`
ω	`omegaWallFunction`

than 30 and imposes $\nu_t = 0$ for values of y^+ less than 11. A different wall function, indicated by the word `nutUSpaldingWallFunction`, assigns ν_t values different from zero up to $y^+ = 0$. In the case of a grid fine enough to resolve the boundary layer (with the first cell at the wall having $y^+ \leq 1$), it is possible to set the value in the file `0/nut` to `calculated` also for the walls. Consistently, the value of k will be set to zero and `zeroGradient` will be applied for ω or ϵ on the walls (Tables 8.2 and 8.3).

8.8 Initial Values of Turbulent Quantities

Here, the $k - \omega$ turbulence model will be used to solve an incompressible flow, whose governing equations therefore do not include the term representing density. One consequence of this choice is that, instead of the dynamic viscosity μ, the kinematic viscosity ν will be considered, making the turbulent kinematic viscosity $\nu_t = k/\omega$. A key challenge in setting up a turbulent simulation is defining the initial and boundary values of the turbulent quantities. The turbulent length scale is a parameter that more intuitively represents the size of the turbulent vortices than other measures. It is defined as

$$L = C_\mu^{3/4} \frac{k^{3/2}}{\epsilon} \tag{8.16}$$

with $C_\mu = 0.09$.

The value of L is typically estimated empirically. Specifically, it is common practice to set L as a percentage of the characteristic dimensions of the problem. For a fully developed turbulent flow inside a duct, $L = 0.07 D_h$ is used, where D_h is the hydraulic diameter. The hydraulic diameter is defined as $D_h = 4A/P$, where

A is the cross-sectional area of the duct through which the flow passes, and P is the perimeter enclosing A. In the case of flows bounded by a single wall, the length of the turbulent scales is defined as $L = 0.4D_b$, where D_b is the boundary layer thickness at the section where the boundary condition is applied.

The estimation of turbulent kinetic energy k follows a similar approach to that used for L. The turbulent kinetic energy can be expressed in terms of turbulent intensity I:

$$k = \frac{3}{2}(\mathbf{v}I)^2$$

where the turbulent intensity is defined as

$$I = \frac{\sqrt{\frac{1}{3}\left(u'^2 + v'^2 + w'^2\right)}}{\sqrt{\overline{u}^2 + \overline{v}^2 + \overline{w}^2}}$$

which essentially quantifies velocity fluctuations relative to the average velocity. For a fully developed turbulent flow inside a duct, $1\% < I < 10\%$ is typically assumed, or an empirical formula can be used.

$$I = 0.16Re^{-1/8}$$

where $Re = \dfrac{|\mathbf{v}|\, D_h}{\nu}$. For external flows, $0.05\% < I < 1\%$ is typically assumed.

Given L and k, the following quantity can be determined

$$\omega = C_\mu^{-1/4}\frac{\sqrt{k}}{L}.$$

Using Eq. 8.16 it is possible to calculate

$$\epsilon = C_\mu^{3/4}\frac{k^{3/2}}{L}.$$

Once the values of k and ω have been calculated, it is always advisable to calculate the corresponding value of the turbulent kinematic viscosity $\nu_t = k/\omega$ to prevent obtaining excessively high values typical of very viscous fluids, such as honey, associated with velocity magnitudes on the order of 10 m/s, which would necessitate verifying the calculations and assumptions. Typical values of ν_t range between 10^{-6} and 10^{-2}.

8.9 Large Eddy Simulation (LES)

Unlike the RANS approach, where all turbulent vortices are modelled, the *Large Eddy Simulation* (LES) approach distinguishes large turbulent vortices (large eddies) from smaller ones, resolving the former without modelling and modelling the latter.

The fundamental principle of the LES approach is that large vortices exhibit characteristics strongly dependent on the type of flow, whereas smaller vortices have properties common to all flows, making their models universally valid. The scale of the vortices to be resolved is determined by applying a spatial statistical filter to the Navier–Stokes equations, yielding the *filtered Navier–Stokes equations*. In general, the filtering length scale, denoted by Δ, can be computed using various approaches, including those available in OpenFOAM®:

- cubeRootVol;
- maxDeltaxyz;
- maxDeltaxyzCubeRoot;
- smooth;
- vanDriest;
- Prandtl;
- IDDESDelta.

When the filter is solely determined by the cell size and its resolution capabilities, it is referred to as the *implicit filtering technique*. In this approach, the minimum size of the turbulent vortices is dictated by the cell size. Figure 8.9 illustrates, for a given discretised computational domain, the maximum size of resolvable turbulent vortices (largest circle), the minimum resolvable size (intermediate circle), and the largest vortices that remain unresolved due to the grid resolution (smallest circle). Modelled vortices are smaller than the grid cells. *Sub-Grid Scale (SGS) models* refer to models that account for the contribution of unresolved vortices in describing the flow. The SGS models available in OpenFOAM® include:

- Smagorinsky;
- kEqn;
- dynamicLagrangian;
- dynamicKEqn;
- WALE (Wall-Adapting Local Eddy-viscosity);
- DeardorffDiffStress.

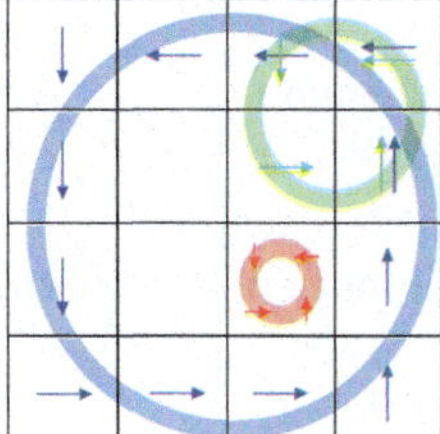

Fig. 8.9 Dimensions of turbulent vortices for a given computational discretised domain: maximum, minimum, and maximum unresolved (modelled) size

In electronic applications, filters are devices designed to modify a signal by eliminating unwanted components. In the LES context, a 'low-pass' filter is used, which leaves unchanged only the harmonics with frequencies lower than the so-called 'cut-off frequency'. The initial signal to which the filter is applied represents the temporal development of quantities, such as pressure or velocity.

A fundamental characteristic of a filter is the function used to represent it. In three-dimensional LES, the spatial filter function $G(\mathbf{x}, \mathbf{y})$ is defined as a function of the six spatial coordinates of two points, **x** and **y**. Applying a filter to a generic quantity $\phi(\mathbf{x}, t)$ yields the filtered version of the same quantity:

$$\overline{\phi}(\mathbf{x}, t) = \int G(\mathbf{x}, \mathbf{y})\phi(\mathbf{y}, t)d^3\mathbf{y}.$$

The quantity $\phi(\mathbf{x}, t)$ can therefore be decomposed into its filtered part, $\overline{\phi}(\mathbf{x}, t)$, and the remaining part, $\phi'(\mathbf{x}, t)$, called the '*residue*', expressed as:

$$\phi'(\mathbf{x}, t) = \phi(\mathbf{x}, t) - \overline{\phi}(\mathbf{x}, t). \qquad (8.17)$$

The filtered part is the *resolved part* (i.e., the turbulent vortices of larger dimensions), while the residue is the *modelled part* (i.e., the turbulent vortices of smaller dimensions). It should be noted that:

- the bar over the filtered quantity indicates the calculation of the spatial average through a three-dimensional integration, unlike the RANS case where the temporal average is calculated;
- filtering is a linear operation.

A commonly used example of a filter is the rectangular filter (box filter), defined as follows:

$$G(\mathbf{x}, \mathbf{y}) = \begin{cases} \dfrac{1}{\Delta^3} \quad for \quad |\mathbf{x} - \mathbf{y}| \leqslant \dfrac{\Delta}{2} \\ 0 \quad for \quad |\mathbf{x} - \mathbf{y}| > \dfrac{\Delta}{2} \end{cases}$$

where the symbol Δ represents the filter width, indicative of the minimum amplitude of turbulent vortices that should not be eliminated by the filter: in other words, Δ represents the dimension of the smallest resolved turbulent vortices that are not modelled. The expression commonly used to calculate Δ is:

$$\Delta = \sqrt[3]{\Delta x \Delta y \Delta z}$$

where Δx, Δy, Δz represent the dimensions of a generic cell in the three-dimensional grid.

For simplicity, the unfiltered Navier–Stokes equations are presented here in the absence of momentum sources for a fluid with constant viscosity.

The continuity equation is as follows:

$$\frac{\partial \rho}{\partial t} + \nabla \cdot (\rho \mathbf{u}) = 0.$$

The first component of the momentum conservation equation is as follows:

$$\frac{\partial \rho u}{\partial t} + \nabla \cdot (\rho \mathbf{u} u) = \mu \nabla \cdot (\nabla u) - \frac{\partial p}{\partial x}.$$

The second component of the momentum conservation equation is as follows:

$$\frac{\partial \rho v}{\partial t} + \nabla \cdot (\rho \mathbf{u} v) = \mu \nabla \cdot (\nabla v) - \frac{\partial p}{\partial y}.$$

The third component of the momentum conservation equation is as follows:

$$\frac{\partial \rho w}{\partial t} + \nabla \cdot (\rho \mathbf{u} w) = \mu \nabla \cdot (\nabla w) - \frac{\partial p}{\partial z}.$$

Considering the filter function $G(\mathbf{x}, \mathbf{y}) = G(\mathbf{x} - \mathbf{y})$ and exploiting the linearity of the filtering operation, we obtain the *filtered Navier–Stokes equations* for the incompressible case.

The filtered continuity equation is as follows:

$$\nabla \cdot \overline{\mathbf{u}} = 0$$

which coincides with the continuity equation used in LES.

The first component of the filtered momentum conservation equation is as follows:

$$\frac{\partial \rho \overline{u}}{\partial t} + \nabla \cdot (\rho \overline{\mathbf{u} u}) = \mu \nabla \cdot (\nabla \overline{u}) - \frac{\partial \overline{p}}{\partial x}. \tag{8.18}$$

The second component of the filtered momentum conservation equation is as follows:

$$\frac{\partial \rho \overline{v}}{\partial t} + \nabla \cdot (\rho \overline{\mathbf{u} v}) = \mu \nabla \cdot (\nabla \overline{v}) - \frac{\partial \overline{p}}{\partial y}. \tag{8.19}$$

The third component of the filtered momentum conservation equation as follows:

$$\frac{\partial \rho \overline{w}}{\partial t} + \nabla \cdot (\rho \overline{\mathbf{u} w}) = \mu \nabla \cdot (\nabla \overline{w}) - \frac{\partial \overline{p}}{\partial z}. \tag{8.20}$$

The ability to solve these four equations to obtain the filtered motion field ($\overline{u}$, $\overline{v}$, $\overline{w}$, $\overline{p}$) depends on the ability to calculate the terms $\nabla \cdot (\rho \overline{\mathbf{u} \phi})$ (with $\phi = u, v, w$), which can, however, be rewritten as:

$$\nabla\cdot(\rho\overline{\mathbf{u}\phi}) = \nabla\cdot(\rho\overline{\mathbf{u}}\overline{\phi}) + \left[\nabla\cdot(\rho\overline{\mathbf{u}\phi}) - \nabla\cdot(\rho\overline{\mathbf{u}}\overline{\phi})\right]. \tag{8.21}$$

Using Eq. 8.21, it is possible to rewrite Eqs. 8.18, 8.19, 8.20 as:

$$\frac{\partial\rho\overline{u}}{\partial t} + \nabla\cdot(\rho\overline{\mathbf{u}}\,\overline{u}) = \mu\nabla\cdot(\nabla\overline{u}) - \frac{\partial\overline{p}}{\partial x} - [\nabla\cdot(\rho\overline{\mathbf{u}u}) - \nabla\cdot(\rho\overline{\mathbf{u}}\,\overline{u})] \tag{8.22}$$

$$\frac{\partial\rho\overline{v}}{\partial t} + \nabla\cdot(\rho\overline{\mathbf{u}}\,\overline{v}) = \mu\nabla\cdot(\nabla\overline{v}) - \frac{\partial\overline{p}}{\partial y} - [\nabla\cdot(\rho\overline{\mathbf{u}v}) - \nabla\cdot(\rho\overline{\mathbf{u}}\,\overline{v})] \tag{8.23}$$

$$\frac{\partial\rho\overline{w}}{\partial t} + \nabla\cdot(\rho\overline{\mathbf{u}}\,\overline{w}) = \mu\nabla\cdot(\nabla\overline{w}) - \frac{\partial\overline{p}}{\partial z} - [\nabla\cdot(\rho\overline{\mathbf{u}w}) - \nabla\cdot(\rho\overline{\mathbf{u}}\,\overline{w})] \tag{8.24}$$

which are the conservation of momentum equations solved when performing a LES.

- the terms $\dfrac{\partial\rho\overline{\phi}}{\partial t}$ represent the time variation of the three components of the filtered momentum;
- the terms $\nabla\cdot(\rho\overline{\mathbf{u}}\,\overline{\phi})$ and $\mu\nabla\cdot(\nabla\overline{\phi})$ represent the convective and diffusive fluxes for the three components of the filtered momentum;
- the terms $-\frac{\partial\overline{p}}{\partial x}$, $-\frac{\partial\overline{p}}{\partial y}$, and $-\frac{\partial\overline{p}}{\partial z}$ represent the three components of the gradient of the filtered pressure.

In the RANS approach, the tensor of Reynolds stresses appears due to the temporal averaging of values in the Navier–Stokes equations. Similarly, given that $\overline{\mathbf{u}\phi} \neq \overline{\mathbf{u}}\,\overline{\phi}$, particular attention must be paid when dealing with the terms $\nabla\cdot(\rho\overline{\mathbf{u}\phi}) - \nabla\cdot(\rho\overline{\mathbf{u}}\overline{\phi})$ resulting from the filtered quantities (spatial averages). It is possible to write the i-th component of the tensor of stresses deriving from the filtered quantities as

$$\left[\nabla\cdot(\rho\overline{\mathbf{u}\phi}) - \nabla\cdot(\rho\overline{\mathbf{u}}\overline{\phi})\right] = \nabla\cdot\left(\rho\overline{\mathbf{u}\phi} - \rho\overline{\mathbf{u}}\overline{\phi}\right) = \nabla\cdot(\rho\overline{\mathbf{u}u_i} - \rho\overline{\mathbf{u}}\,\overline{u_i}) =$$
$$\nabla\cdot(\rho\overline{u_i\mathbf{u}} - \rho\overline{u_i}\,\overline{\mathbf{u}}) =$$
$$\frac{\partial(\rho\overline{u_i u} - \rho\overline{u_i}\,\overline{u})}{\partial x} + \frac{\partial(\rho\overline{u_i v} - \rho\overline{u_i}\,\overline{v})}{\partial y} + \frac{\partial(\rho\overline{u_i w} - \rho\overline{u_i}\,\overline{w})}{\partial z} = \frac{\partial\tau_{ij}}{\partial x_j}$$

where

$$\tau_{ij} = \rho\overline{u_i\mathbf{u}} - \rho\overline{u_i}\,\overline{\mathbf{u}} = \rho\overline{u_i u_j} - \rho\overline{u_i}\,\overline{u_j}.$$

The term τ_{ij} is called *modelled stress* (*subgrid-scale stress* or *subfilter-scale stress*) and represents the transport of momentum due to the interactions among the turbulent modelled vortices that are not resolved because they are too small compared to the minimum dimensions imposed by the filter value. Now, considering Eq. 8.17, it can be written

$$\phi(\mathbf{x},t) = \overline{\phi}(\mathbf{x},t) + \phi'(\mathbf{x},t)$$

from which

$$\rho\overline{u_i u_j} = \rho\overline{(\overline{u_i} + u'_i)(\overline{u_j} + u'_j)} = \rho\overline{\overline{u_i}\,\overline{u_j}} + \rho\overline{\overline{u_i} u'_j} + \rho\overline{u'_i \overline{u_j}} + \rho\overline{u'_i u'_j} =$$
$$\rho\overline{u_i}\,\overline{u_j} + \left(\rho\overline{\overline{u_i}\,\overline{u_j}} - \rho\overline{u_i}\,\overline{u_j}\right) + \rho\overline{\overline{u_i} u'_j} + \rho\overline{u'_i \overline{u_j}} + \rho\overline{u'_i u'_j}.$$

Thus, the total modelled stress is as follows:

$$\tau_{ij} = \rho\overline{u_i u_j} - \rho\overline{u_i}\,\overline{u_j} = \left(\rho\overline{\overline{u_i}\,\overline{u_j}} - \rho\overline{u_i}\,\overline{u_j}\right) + \rho\overline{\overline{u_i} u'_j} + \rho\overline{u'_i \overline{u_j}} + \rho\overline{u'_i u'_j}.$$

The last expression can be further modified by grouping some of its terms:

$$\tau_{ij} = L_{ij} + C_{ij} + R_{ij}$$

- $L_{ij} = \rho\overline{\overline{u_i}\,\overline{u_j}} - \rho\overline{u_i}\,\overline{u_j}$ represents the *Leonard stresses*, so-called after the name of the researcher who first discovered the method for the approximate calculation of this term starting from the filtered quantities. This term contains only filtered (resolved) quantities and therefore does not require modelling;
- $C_{ij} = \rho\overline{\overline{u_i} u'_j} + \rho\overline{u'_i \overline{u_j}}$ represents the stresses due to the interaction between the modelled and resolved turbulent vortices (*cross-stress*);
- $R_{ij} = \rho\overline{u'_i u'_j}$ represents the stresses due to the interaction among modelled turbulent vortices. This term, known as unresolved Reynolds stresses (*SGS Reynolds stress*), must be modelled using *sub-grid/sub-filter scale turbulence models.*

8.9.1 Smagorinsky-Lilly Modelling

Due to their small size, unresolved turbulent vortices can be assumed to follow the Smagorinsky hypothesis of isotropy. As with the Boussinesq hypothesis in the RANS approach (see Sect. 8.1), the local value of the unresolved Reynolds stresses is assumed to be proportional to the local value of the deformation velocity tensor of the filtered (i.e., resolved) part of the flow. Therefore, defining the deformation velocity tensor (see Sect. 1.2.1) of the filtered part of the flow as

$$\overline{S}_{ij} = \frac{1}{2}\left(\frac{\partial \overline{u}_i}{\partial x_j} + \frac{\partial \overline{u}_j}{\partial x_i}\right)$$

the original Smagorinsky model predicts:

$$R_{ij} = -2\mu_{SGS}\overline{S}_{ij} + \frac{1}{3}R_{ii}\delta_{ij} = -\mu_{SGS}\left(\frac{\partial \overline{u}_i}{\partial x_j} + \frac{\partial \overline{u}_j}{\partial x_i}\right) + \frac{1}{3}R_{ii}\delta_{ij} \qquad (8.25)$$

in which the proportionality coefficient μ_{SGS} represents the *dynamic SGS viscosity*; the term $\frac{1}{3}R_{ii}\delta_{ij}$ plays the same role as the term $-\frac{2}{3}\rho k\delta_{ij}$ in Eq. 8.3. In other words, the term $\frac{1}{3}R_{ii}\delta_{ij}$ allows the sum of the normal stresses—that is, the sum of the elements of the main diagonal of the tensor of the unresolved Reynolds stresses—to be equal to the turbulent kinetic energy of the unresolved turbulent vortices. In common practice, despite the differing nature of its elements, the modelled stress tensor τ_{ij} is calculated using Eq. 8.25:

$$\tau_{ij} = -2\mu_{SGS}\overline{S}_{ij} + \frac{1}{3}\tau_{ii}\delta_{ij} = -\mu_{SGS}\left(\frac{\partial \overline{u}_i}{\partial x_j} + \frac{\partial \overline{u}_j}{\partial x_i}\right) + \frac{1}{3}\tau_{ii}\delta_{ij}. \tag{8.26}$$

To determine the value of the dynamic SGS viscosity μ_{SGS}, we use the Prandtl mixing length model. The SGS kinematic viscosity is defined as $\nu_{SGS} = \mu_{SGS}/\rho$, with dimensions of length squared divided by time (m^2/s). Therefore, the SGS kinematic viscosity can be expressed, considering only physical dimensions, as the product of length l_0 and velocity v_0. For length, we could consider a characteristic length of the turbulent unresolved vortices, which, under the assumption of implicit LES, will be a fraction (C_{SGS}) of the dimension Δ of the actual cell: $l_0 = C_{SGS}\Delta$. Based on dimensional analysis, we can define velocity as the product of length and spatial derivative. Thus, defining the modulus of the deformation velocity tensor as $|\overline{S}| = \sqrt{2\overline{S}_{ij}\overline{S}_{ij}}$, the length will always be $l_0 = C_{SGS}\Delta$, while the spatial derivative is $|\overline{S}|$. In conclusion,

$$\mu_{SGS} = \rho l_0 l_0 |\overline{S}| = \rho (C_{SGS}\Delta)^2 \sqrt{2\overline{S}_{ij}\overline{S}_{ij}}.$$

Theoretical studies by Lilly on the decay rate of turbulent isotropic vortices in the inertial range of the turbulent energy spectrum have led to the consideration of reliable values between 0.17 and 0.21 for the dimensionless constant C_{SGS}. Subsequent studies have attributed different values to C_{SGS} for different types of flow, suggesting that the behaviour of the 'small' turbulent vortices may not be as 'universal' as initially hypothesised. Therefore, there are cases where more sophisticated techniques are required to model unresolved turbulent vortices.

8.9.2 Evaluation of LES Calculations

Considering that grids with very small cells are much more expensive in terms of computational costs compared to grids with larger cells at parity of total dimensions of the computational domain, it is desirable to determine the minimum cell size required to obtain acceptable LES results. In this context, it is necessary to familiarise oneself with the concept of energy cascade and related concepts. Given a certain turbulent vortex, assumed to be circular with diameter d, it is characterised by a wave number $\kappa = 2\pi/d$. From this definition, smaller turbulent vortices will have a higher wave number or spatial frequency. It should be noted that, although represented by the same

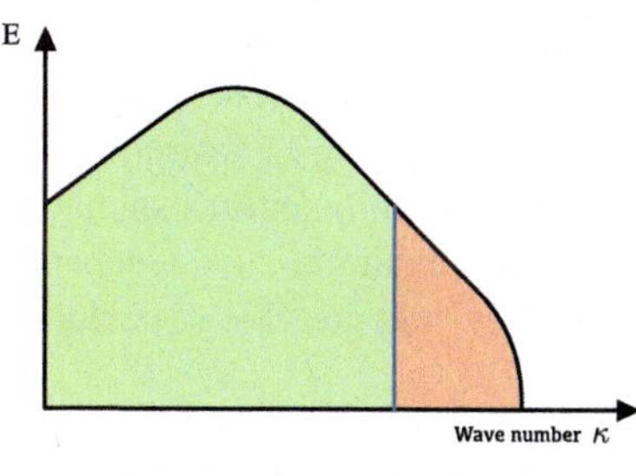

Fig. 8.10 Energy density spectrum: on the left, the resolved part; on the right, the modelled part in a LES

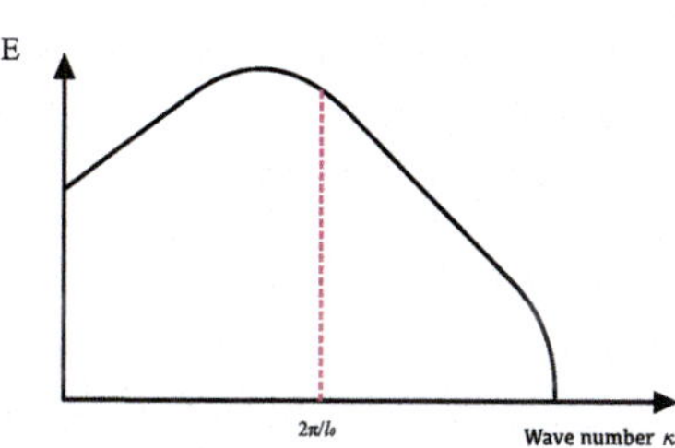

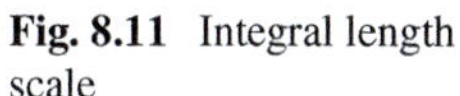

Fig. 8.11 Integral length scale

symbol, the wave number and turbulent kinetic energy should not be confused. In Fig. 8.1, the energy density spectrum is shown, which reports the density of turbulent kinetic energy E as a function of the wave number. From the analysis of the energy density spectrum, it is noted that turbulent vortices of larger dimensions—and lower wave number—are associated with a higher energy density. The area under the curve of the energy density spectrum represents the total turbulent kinetic energy of the flow at the considered point. This total turbulent kinetic energy is calculated using the RANS approach, where all turbulent vortices are modelled and indistinguishable.

In LES, vortices larger than the minimum cell size are resolved, while the remaining vortices are modelled. Considering the energy density spectrum diagram in Fig. 8.10, part of the total turbulent kinetic energy of the flow is resolved, and part is modelled. A LES is considered acceptable if at least eighty percent of the total turbulent kinetic energy is resolved. The energy density spectrum is a local characteristic of the flow. As a consequence, each cell of a computational grid will be characterised by a specific energy density spectrum. The *integral length scale* l_0 is defined in order to obtain a grid that resolves no less than eighty percent of the total turbulent kinetic energy of the flow at every cell in the computational domain. The integral length scale represents the dimension l_0 of a turbulent vortex whose energy is equal to the average value of the total turbulent kinetic energy present at the point in the computational domain. From a mathematical point of view it will be (Fig. 8.11)

$$l_0 = \frac{\int_0^\infty \frac{1}{k} E(k) dk}{\int_0^\infty E(k) dk}.$$

To better understand what has been written so far regarding the integral length scale and the variation of the energy density spectrum as a function of position within the computational domain, it is useful to observe Fig. 8.12, in which different zones are associated with different energy spectra and different integral length scales. Typically, before performing LES simulations, a RANS calculation is performed to obtain the distributions of quantities such as turbulent kinetic energy k and dissipation, which can be, for example, ϵ or ω, depending on the turbulence model used. Knowing these quantities, it is possible to calculate the value of the integral length scale as

$$l_0 = \frac{k^{3/2}}{\epsilon} \qquad \text{or} \qquad l_0 = \frac{k^{1/2}}{C_\mu \omega}.$$

A suitable data visualisation tool can be used to visualise the integral length scale distribution over the entire computational domain. C_μ is the constant used in the case where the $k - \omega$ turbulence model is used in the RANS simulation. In Fig. 8.12, it shows that the minimum size necessary to resolve a turbulent vortex of size l_0 varies depending on the position in the computational domain. Observing Fig. 8.9, it is noted that at least four cells are required to resolve a turbulent vortex. In other words, to resolve at least eighty percent of the turbulent kinetic energy in a generic point of the computational domain characterised by an integral length scale l_0, a reasonable initial estimate is to consider a grid whose cells have a maximum dimension Δ not exceeding $l_0/5$. Smaller values of the filter Δ will lead to resolving higher percentages of the total turbulent kinetic energy.

Figure 8.13 illustrates the case of a uniform Cartesian grid. In order to identify which areas of the computational domain are characterised by an inadequate level of grid refinement, it is appropriate to define the quantity.

$$f = \frac{l_0}{\Delta} = \frac{k^{3/2}}{\epsilon \Delta}$$

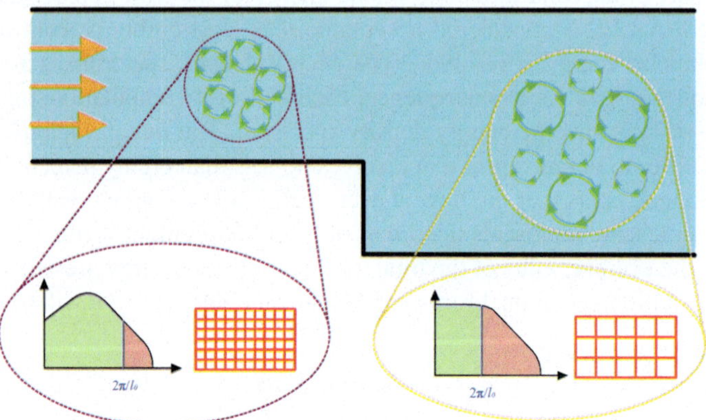

Fig. 8.12 Variation of the energy density spectrum and integral length scale as a function of position within the computational domain, and the corresponding cell size as a function of the integral length scale

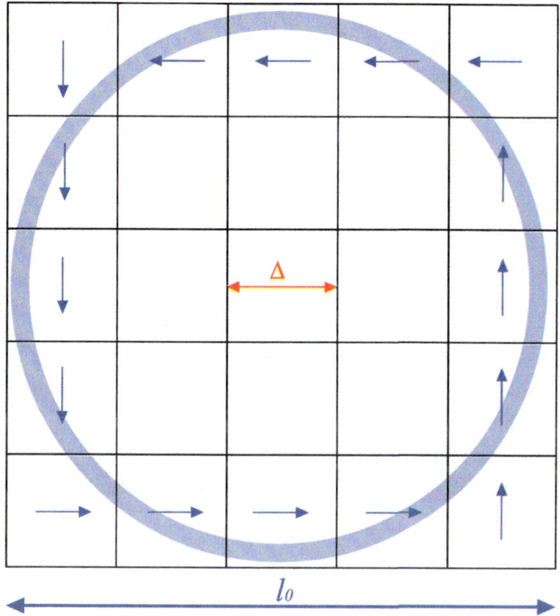

Fig. 8.13 Maximum cell size required to resolve at least eighty percent of the turbulent kinetic energy at a point in the computational domain characterised by an integral length scale l_0

in which the value of the filter Δ is calculated according to one of the methods described above. Once the quantity f is defined, it is possible to visualise its distribution in the computational domain by identifying the areas characterised by $f < 5$ as those in which it is necessary to reduce the size of the cells. Once the preparatory phase, in which a calculation grid is constructed based on the results obtained with previous RANS simulations, is completed, we now want to evaluate the quality of the grid based on the results obtained with the LES. The evaluation of aspects such as the integration time interval or the total number of time intervals performed will be discussed later. Once again, a possible evaluation criterion is based on the concept of turbulent kinetic energy, which, for an acceptable LES, must be resolved for not less than eighty percent of the total at every point in the computational domain (see Fig. 8.10). The process of evaluating the quality of the grid based on LES results involves a series of steps. The first consists of observing that, considering for example only the velocity, the results of the simulation are the instantaneous values at each point in the computational domain. Figure 8.14 shows the polygonal chain representing the modulus U of the velocity component along one of the three coordinate axes at a specific point in the computational domain as a function of time. The dashed curve instead provides the temporal average of these instantaneous values. In order to obtain the correct average velocity value in each cell, it is necessary to continue the simulation until reaching a statistically stationary value of the velocity itself. From this moment on, it will be possible to use the obtained values to calculate the correct average velocity value.

Knowing the average velocity field (see Fig. 8.16), it is possible to subtract it from the instantaneous velocity field (see Fig. 8.15) to obtain the field of instantaneous velocity fluctuations u' (see Fig. 8.17). As these fluctuations were obtained, it is clear that they are derived from the turbulent vortices resolved by the grid, rather than those modelled. Observing Fig. 8.2, it is evident that making a temporal average of u' results in a value of exactly zero, as u' is the fluctuation component. In formula: $\overline{u'} = 0$. Consequently, it follows that $\overline{u'} \cdot \overline{u'} = 0$, while $\overline{u'u'} \neq 0$.

What has been considered so far is only one of the three components of velocity. Therefore, it will be necessary to consider all three components, namely

$$u' = U - \overline{U}, \qquad v' = V - \overline{V}, \qquad w' = W - \overline{W}.$$

Considering that the kinetic energy per unit mass is generally defined as half the product of the velocity vector with itself, to calculate the kinetic energy of the fluctuations or the turbulent kinetic energy, it will be necessary to consider the sum of the products of the various components. The turbulent kinetic energy resolved by the grid will be

$$k_{res} = \frac{1}{2}\left(\overline{u'u'} + \overline{v'v'} + \overline{w'w'}\right)$$

or, equivalently

$$k_{res} = \frac{1}{2}\left(\overline{(u')^2} + \overline{(v')^2} + \overline{(w')^2}\right). \tag{8.27}$$

To better understand how to obtain this value if OpenFOAM® is used, it is necessary to consider that the nine possible products between the components of the instantaneous velocity fluctuation

$$u'u', \quad u'v', \quad u'w', \quad v'u', \quad v'v', \quad v'w', \quad w'u', \quad w'v,' \quad w'w'$$

Representing the instantaneous Reynolds stresses and their time average value, the Reynolds stress tensor per unit mass is organised in matrix form as follows:

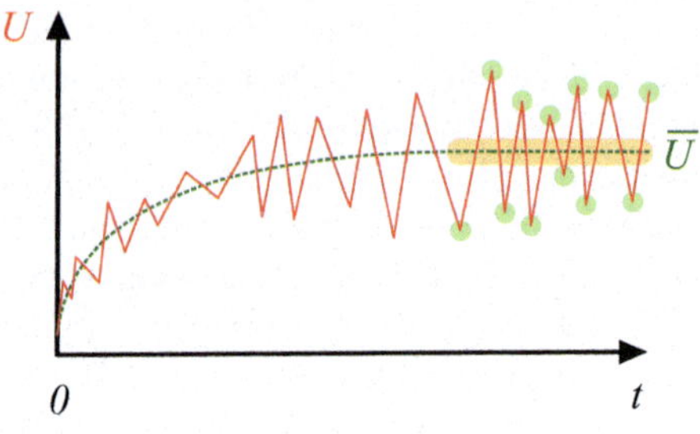

Fig. 8.14 Velocity value obtained from an LES at a point in the computational domain and its average value over time

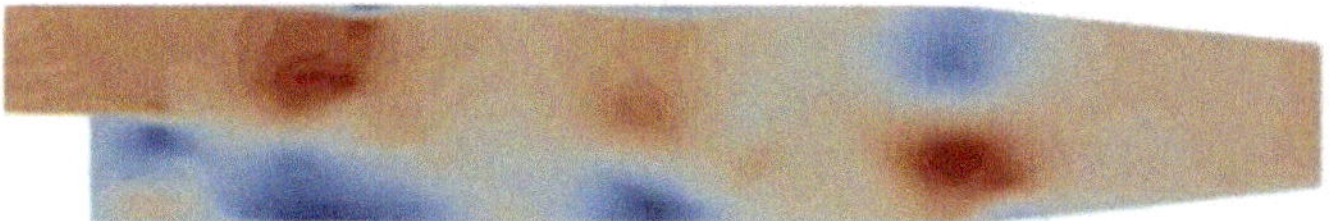

Fig. 8.15 Example of instantaneous velocity field

Fig. 8.16 Example of average velocity field

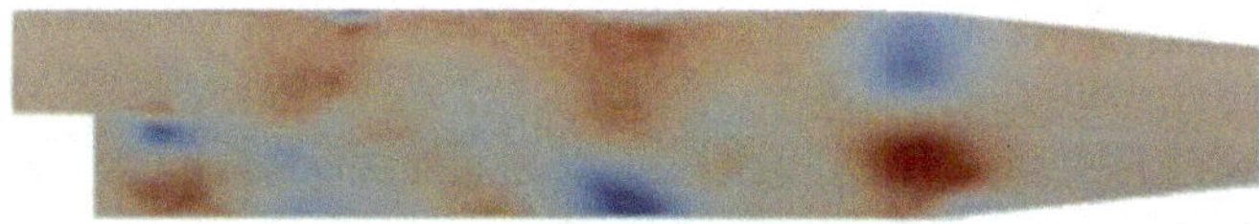

Fig. 8.17 Example of instantaneous velocity fluctuation field

$$\frac{R_{ij}}{\rho} = \begin{bmatrix} \overline{u'u'} & \overline{u'v'} & \overline{u'w'} \\ \overline{v'u'} & \overline{v'v'} & \overline{v'w'} \\ \overline{w'u'} & \overline{w'v'} & \overline{w'w'} \end{bmatrix}.$$

Considering that

$$\overline{u'v'} = \overline{v'u'} \qquad \overline{u'w'} = \overline{w'u'} \qquad \overline{v'w'} = \overline{w'v'},$$

we obtain the symmetric tensor

$$\frac{R_{ij}}{\rho} = \begin{bmatrix} \overline{u'u'} & \overline{u'v'} & \overline{u'w'} \\ & \overline{v'v'} & \overline{v'w'} \\ & & \overline{w'w'} \end{bmatrix}$$

The elements of this tensor have units of measure m^2/s^2.

Among the various results of an LES performed using OpenFOAM®, the values of this tensor will be contained in the file `uPrime2Mean`. Taking advantage of the symmetry of this tensor and to reduce the computer memory usage required, only six, and not nine, components will be stored. Once the values of the Reynolds stress tensor are obtained, it is possible to calculate the value of the turbulent kinetic energy resolved by the grid using Eq. 8.27. Data visualisation software typically allows creating new quantities from those initially available. Therefore, a new quantity can be defined in the Data visualisation software as `0.5*(UPrime2Mean_XX2̂ + UPrime2Mean_YY2̂ + UPrime2Mean_ZZ2̂)`. As an example, Fig. 8.18 shows the case in which the data analysis software is Paraview®. In the same figure, highlighted at the bottom, there is a button to choose from the six possible components of the tensorial field `UPrime2Mean`. Once the turbulent kinetic energy resolved by the grid is available, we aim to determine the value of the modelled turbulent kinetic energy k_{sgs} over the entire computational domain. In other words, we seek to determine the value of turbulent kinetic energy derived from turbulent vortices that have dimensions smaller than those minimum necessary for the grid to resolve them. In the case of a LES performed using OpenFOAM®, such values are written in the file `k`. It is then possible to analyse the distribution of two distinct fields—k_{res} and k_{sgs}—as shown in Figs. 8.19 and 8.20.

Noting k_{res} and k_{sgs} it is possible to define the new field

$$\frac{k_{res}}{k_{res} + k_{sgs}}$$

Representative of the percentage of resolved turbulent kinetic energy relative to the total turbulent kinetic energy, as shown in Fig. 8.21. Zones of the computational domain characterised by values lower than eighty percent will need to undergo grid refinement. It is useful to remind that reductions in the minimum cell size lead to an increase in the percentage of resolved turbulent kinetic energy compared to the total turbulent kinetic energy.

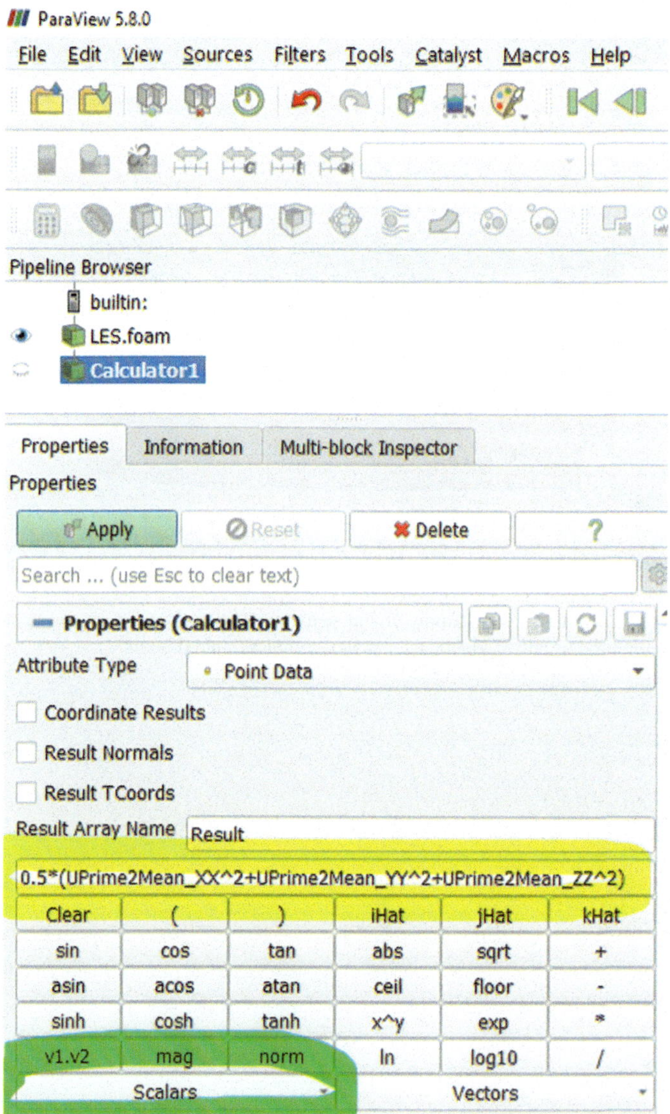

Fig. 8.18 Calculation of the resolved turbulent kinetic energy through paraview

Fig. 8.19 Example of k_{res} distribution in a computational domain

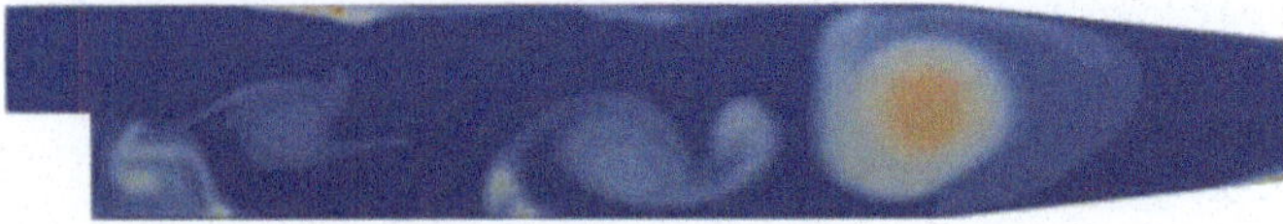

Fig. 8.20 Example of k_{sgs} distribution in a computational domain

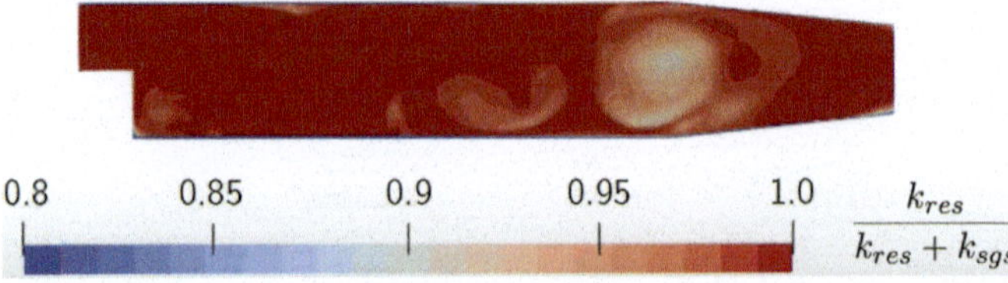

Fig. 8.21 Example of the distribution of the percentage of resolved turbulent kinetic energy compared to the total turbulent kinetic energy in a computational domain

Bibliography

Anderson, J. (2009). Governing equations of fluid dynamics. In J. F. Wendt (Ed.), *Computational fluid dynamics* (pp. 15–51). Berlin Heidelberg: Springer.

Blazek, J. (2005). *Computational fluid dynamics: Principles and applications*. Elsevier Science.

Ciofalo, M. (2021). *Thermofluid dynamics of turbulent flows: Fundamentals and modelling*. Springer International Publishing.

Davidson, P. (2015). *Turbulence: An introduction for scientists and engineers*. Oxford University Press.

Durran, D. (2010). *Numerical methods for fluid dynamics: With applications to geophysics*. New York: Springer.

Ferziger, J., Perić, M., & Street, R. (2019). *Computational methods for fluid dynamics*. Springer International Publishing.

Greenshields, C. (2023). *Openfoam v11 user guide*. The OpenFOAM Foundation. https://doc.cfd.direct/openfoam/user-guide-v11

Greenshields, C., & Weller, H. (2022). *Notes on computational fluid dynamics: General principles*. CFD Direct Ltd.

Holzmann, T. (2019). *Mathematics, numerics, derivations and openfoam(r)*. Holzmann CFD.

Jasak, H. (1996). *Error analysis and estimation for the finite volume method with applications to fluid flows*. Imperial College London (University of London).

Kee, R., Coltrin, M., Glarborg, P., & Zhu, H. (2018). *Chemically reacting flow: Theory, modeling, and simulation*. Wiley.

Maliska, C. (2023). *Fundamentals of computational fluid dynamics: The finite volume method*. Springer International Publishing.

Maric, T., Hpken, J., & Mooney, K. (2014). *The openfoam technology primer*. Stan Mott.

Mazumder, S. (2015). *Numerical methods for partial differential equations: Finite difference and finite volume methods*. Elsevier Science.

Moore, R. (2017). *Fluids, waves and optics*. CreateSpace Independent Publishing Platform.

Moukalled, F., Mangani, L., & Darwish, M. (2015). *The finite volume method in computational fluid dynamics: An advanced introduction with openfoam and matlab*. Springer International Publishing.

Mueller, J. (2020). *Essentials of computational fluid dynamics*. CRC Press.

Pulliam, T., & Zingg, D. (2014). *Fundamental algorithms in computational fluid dynamics*. Springer International Publishing.

Quartapelle, L., & Auteri, F. (2013a). *Fluidodinamica comprimibile*. CEA.

Quartapelle, L., & Auteri, F. (2013b). *Fluidodinamica incomprimibile*. CEA.

G. Caramia and E. Distaso, *A Practical Approach to Computational Fluid Dynamics Using OpenFOAM®*, https://doi.org/10.1007/978-3-031-88957-8

Quarteroni, A., Saleri, F., & Gervasio, P. (2017). *Calcolo scientifico: Esercizi e problemi risolti con matlab e octave*. Springer Milan.

Rodriguez, S. (2019). *Applied computational fluid dynamics and turbulence modeling: Practical tools, tips and techniques*. Springer International Publishing.

Saad, Y. (2003). *Iterative methods for sparse linear systems* (2nd ed.). Society for Industrial; Applied Mathematics. for Industrial; Applied Mathematics.

Salsa, S. (2016). *Equazioni a derivate parziali: Metodi, modelli e applicazioni*. Springer Milan.

Sasoh, A. (2020). *Compressible fluid dynamics and shock waves*. Springer Nature Singapore.

Toro, E. (2009). *Riemann solvers and numerical methods for fluid dynamics: A practical introduction*. Berlin Heidelberg: Springer.

Trottenberg, U., Oosterlee, C., & Schuller, A. (2001). *Multigrid methods: Basics, parallelism and adaptivity*. Elsevier Science.

Versteeg, H., & Malalasekera, W. (2007). *An introduction to computational fluid dynamics: The finite volume method*. Pearson Education Limited.

The manufacturer's authorised representative in the EU is Springer Nature Customer Service Centre GmbH, Europaplatz 3, 69115 Heidelberg, Germany. If you have any concerns regarding our products, please contact ProductSafety@springernature.com

Printed and bound by CPI Group (UK) Ltd, Croydon, CR0 4YY

03/07/2026

02155806-0002